AUTOMATIC
CONTROL
SYSTEMS
AND COMPONENTS

AUTOMATIC CONTROL SYSTEMS AND COMPONENTS

James R. Carstens, P.E.

ASSISTANT PROFESSOR
School of Technology
Michigan Technological University

PRENTICE HALL Englewood Cliffs, New Jersey 07632

Library of Congress Cataloging-in-Publication Data

Carstens, James R.
    Automatic control systems and components / James R. Carstens.
        p.   cm.
    ISBN 0-13-054297-0
    1. automatic control. 2. Process control. I. Title.
  TJ213.C293   1990
  629.8—dc20   89-37977
                                            CIP

Editorial/production supervision and
  interior design: **Diane M. Delaney**
Cover design: **Wanda Lubelska Design**
Manufacturing buyer: **Dave Dickey**

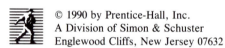 © 1990 by Prentice-Hall, Inc.
A Division of Simon & Schuster
Englewood Cliffs, New Jersey 07632

Printed in the United States of America

10 9 8 7 6 5 4 3 2 1

ISBN   0-13-054297-0

Prentice-Hall International (UK) Limited, *London*
Prentice-Hall of Australia Pty. Limited, *Sydney*
Prentice-Hall Canada Inc., *Toronto*
Prentice-Hall Hispanoamericana, S.A., *Mexico*
Prentice-Hall of India Private Limited, *New Delhi*
Prentice-Hall of Japan, Inc., *Tokyo*
Simon & Schuster Asia Pte. Ltd., *Singapore*
Editora Prentice-Hall do Brasil, Ltda., *Rio de Janeiro*

To my wife, Sandy, and to my two children, Sig and Lesley,
all of whom have given meaning to my life.

Contents

Preface

This book is intended to be a textbook for students in engineering and engineering technology. However, it is also intended to be a textbook for those engineers in industry who find themselves involved in a project dealing with automatic controls and would like to have a more-or-less self-taught refresher course. I can't count the number of times I found myself in exactly this position while in industry and unable to find a practical book on the subject so that I could become an instant expert and impress my peers. While I certainly don't pretend to call myself an expert on automatic controls, I hope that I have remedied this situation for a substantial number of inquiring minds by writing this book.

I have always felt that physics was the great key to unlocking engineering problems. This is one subject where theory is so important. I have my own theory that says, if a person is observant enough, and if he or she can relate a seemingly complicated technical explanation to an easily observed everyday living example, this would help in understanding that explanation and it would certainly help in making that person feel more at home with the explanation. And speaking of theories, the aforementioned is exactly the theory behind this book. A fairly good background in basic physics coupled with a keen eye for observing things will go a long way in understanding many of the concepts that I presented in this book.

There is a fair quantity of math involved in my discussions of control theory. I have attempted to keep the math on a moderate level by circumventing much of the differential and integral calculus normally used in dealing with this subject. I feel that if one can learn a subject using a low-powered approach, why destroy the interest in that subject by using a much higher powered approach? I do love calculus; it's fun and illuminating to use. However, when it comes to explaining everyday happenings, there are better approaches to take. Why use a bulldozer in preparing a flower bed when a spade shovel will get the job done? This book is intended to be a straightforward hands-on attack on automatic controls. I don't believe calculus is the tool to use for that particular undertaking, at least at this beginning stage. There are more important things to be understood at this point.

Chapter 1 may appear to be an intense review of electronics, and that is exactly what I intended it to be. I have attempted to cover as much electronic theory as possible that deals with the area of controls. If you feel weak in this area, this chapter will boost your electronics knowledge level to the point of understanding the material that follows. You will undoubtedly find some portions of this review not being directly applicable to the rest of the book. I believe, however, that you can't know too much about electronics when delving into control systems. Someday you will need and want this additional information.

In Chapter 3 I have introduced a math concept for those of you who feel especially comfortable with algebra. The concept is called Laplace transformation. Laplace transforms are necessary in order to avoid differential and integral calculus. Including this concept was certainly not an original idea of mine, since other automatic control text authors have done it before me. However, I like to think that I have applied these transformations to a greater extend than perhaps some of the others have done. Using transforms is like using logarithms to avoid multiplication and division. For those of you who are familiar with computers, you might want to think of Laplace transforms as macros. Many of the more often used calculus expressions have been transformed into these macros to reduce the number of software statements and keystrokes. Transforms allow you to maintain your sanity while working through some of the more difficult moments encountered in automatic control system theory. They represent a very neat concept that is readily understood by most technical people. And a concept understood is a concept that's a friend for life, I always say.

Control systems are fun and impressive to watch. There is a certain amount of beauty in watching a system perform a function and then make self-adjustments and then continue functioning with no outside supervision. With the advent of the microprocessor, the automatic control system is certainly approaching what is best described as the closest thing we have for artificial intelligence (or real ignorance, depending on its application). Chapter 12 deals specifically with this subject of computers and automatic control systems.

In Chapter 4 you will find considerable material on electrical components, namely, transducers. There is also an extensive discussion of the transfer function. That discussion, along with the discussions on electrical transducers, is a keystone to understanding the later material.

In the following chapters, I introduce many design problems along with their solutions. I tried presenting them in a manner which I felt closely resembles the thought paths taken by a typical design engineer. That is, the first idea for a solution may not be the final solution used, but instead forms the basis for another approach.

At the end of this book are appendices that contain some rather interesting, but simple, software programs. If you have a computer, try them. You'll have a lot of fun with them, proving out the concepts discussed in the text.

I want to give special thanks to those individuals who have made this entire writing venture possible. I thank professors Walter Anderson and Robert Stebler for their vocal encouragement and for making much-needed writing resources and secretarial help available. I also thank students Heather Baab and Todd Coulter for the dedication and eagerness they showed in generating the much-needed artwork this book required, using the School of Technology's CAD facilities. It's young people like these that make engineering technology and teaching so enjoyable and full of hope.

I sincerely thank one of my manuscript reviewers, P. Erik Liimatta, professor of engineering and technologies at Anne Arundel Community College in Arnold, Maryland, for his excellent comments and suggestions in reviewing this manuscript. I know it was a time-consuming task, and I am indebted to his patience which certainly contributed in making this a better book.

But most of all, I thank the members of my family. They showed tremendous patience while my many hours at the keyboard resulted in my being an absentee husband and father. Certainly, without their encouragement, patience, and understanding, none of this would have been possible.

And finally, it is my hope that you, the reader, will get as much enjoyment out of reading and understanding this material as I had in writing it.

J.R.C.
Houghton, Michigan

AUTOMATIC
CONTROL
SYSTEMS
AND COMPONENTS

1

Review of Electronic Fundamentals

1-1 INTRODUCTION

This chapter is meant to be a review of electronic circuit theory. It's assumed that you have a basic knowledge of atomic structure from a previous encounter with a physics course or electronics course. This chapter reviews those basic principles usually encountered in a typical automatic control circuit and presents some old concepts in a somewhat different light for easier understanding. Few derivatives are presented. These can be obtained, instead, by referring to any good basic electronics text. Only the major highlights of circuit fundamentals are presented in this chapter.

1-2 THE DC CIRCUIT

For all the debates over whether or not the use of water flow in a pipe to explain the flow of electrical current in a wire is scientifically proper or not, the anology does make the point. It is used here to demonstrate the many characteristics of electron flow behavior. However, to begin this discussion, don't think of a pipe carrying water, but instead, think of a pipe carrying marbles. Assume that this pipe is completely filled with them. Now, if you were to place one more marble in the pipe's end, certainly, a marble will pop out of the other end, as seen in Figure 1-1. Note that this would not be the same marble that was placed in the pipe's other end. Current flow can be described similarly. A current flow is comprised of electrons bumping or displacing adjacent electrons along a confined path such

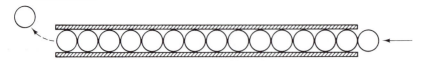

Figure 1-1   An electron flow analogy.

that a flow takes place inside this path. What causes this bumping or flow to take place is discussed shortly. For the present, however, we use this analogy to help explain current flow.

1-2.1 The Ampere

The electric charge found on an electron is the quantity that accounts for the attractive and repulsive forces that exist between adjacent electrons and protons. The electric charge carried by just one electron is 1.6×10^{-19} coulombs. It takes 6.242×10^{18} electrons multiplied by 1.6×10^{-19} coulombs to form one coulomb of electric charge. This means that if 6.242×10^{18} electrons flow past a particular point in a conductor in one second (that is, one coulomb each second), one ampere of charge-flow is flowing. This flow, in amperes, can be found by the following equation:

$$I = \frac{Q}{t} \tag{1-1}$$

where I = current (amperes, or simply "amps")
  Q = charge (coulombs)
  t = time (seconds)

EXAMPLE 1-1

Calculate the amount of current, in amperes, resulting from 39 coulombs flowing through a conductor past a given point in that conductor, in 16 seconds.

Solution:

Using Eq. (1-1), the amount of current, in amperes, would be:

$$I(\text{amps}) = \frac{39 \text{ coulombs}}{16 \text{ secs}}$$
$$= 2.438 \text{ amps}$$

1-2.2 The Volt

In order to force electrons to flow through a substance, a pressure difference is needed, just as a pressure differential is needed to force water through a conduit such as a pipe. This pressure can be created through an electrochemical process (batteries) or it can be created electronically (electrical power supplies). In both cases, a reservoir of electrons is created opposite a similar reservoir that has a deficiency of electrons. Since this situation represents an unnatural or unstable state (every chemically stable thing that occurs in nature is electrically balanced or neutral, so to speak), the overabundant numbers of electrons will strive to find a path back to the deficient electron area in an attempt to "balance things up" once again. The pressure, or potential difference as it is sometimes called, needed to separate these electrons represents an expended energy or work. If the unit of energy is the joule, and the unit of pressure is called the volt, then the following equation shows the relationship between the work expended, the coulombs, moved, and the pressure needed to move these electrons:

$$V = \frac{W}{Q} \tag{1-2}$$

where V = potential difference or pressure (volts)
$\quad\quad W$ = work expended (joules)
$\quad\quad Q$ = charge (coulombs)

EXAMPLE 1-2

Assume that it required 136 joules of work in order to keep a charge of 12 coulombs from recombining with the parent ionized atoms in a battery. Calculate the voltage developed during this process.

Solution:

Using Eq. (1-2), calculate the voltage:

$$V(\text{volts}) = \frac{136 \text{ joules}}{12 \text{ coulombs}}$$

$$= 11.333 \text{ volts}$$

1-2.3 The Voltage Source

The electronic symbol used for a direct current voltage source (that is, a voltage source that "pumps" or moves electrons through a circuit in one direction only) is illustrated in Figure 1-2. The longer side of the symbol is understood to be the positive (+) terminal, while the shorter side of the symbol is the negative (−) terminal side. Remember, the negative side contains an overabundance of electrons, whereas the positive side has a deficiency of electrons. The voltage supplied by this source is usually stated alongside the symbol. It is important to note that this source is usually considered *ideal*. That is, its voltage output never

Figure 1-2   Electronic symbol for a voltage source.

varies. We think of battery voltages as "running down" after prolonged usage, but not this voltage source. The reason for assuming a "perfect battery" for this particular source is merely to simplify circuit performance characteristics. Things can get fairly complicated if we work with a voltage source whose output varies while we are trying to make measurements to prove a point.

1-2.4 Current Direction

It is important to pause for a moment to discuss a very important aspect of DC circuit behavior. This concerns the direction of electron (or current) flow resulting from a pressure difference (i.e., voltage difference) in a circuit. Remembering that it is the electrons that do the traveling, so to speak, in a circuit, it's logical to assume that electron flow in any given circuit is from the negative side of the voltage source to its positive side. This is logical, because the negative side represents a packed accumulation of electrons waiting to break out when given the chance. Of course, in order for these electrons to travel from the (−) side to the (+) side, they have to go by way of any circuit devices that happen to be in their path.

So what is current flow? Since many years ago it was thought that current flow was from positive to negative (hence the expression, "shorting something to ground," where ground was, and still is today, thought to be a negative sump or reservoir that electrons readily flow into), this is now referred to as *conventional current flow*. This convention is used consistently throughout this text. So from now on, instead of referring to electron flow as flowing from (−) to (+), we refer to conventional current flow, or simply current, that flows from (+) to (−).

1-2.5 Determining Polarity

Once the direction of current flow has been established in a circuit, the polarity of the voltage drop across a device resulting from that current is readily obtained. Look at Figure 1-3. The circuit devices through which current is flowing are all resistors. We would like to know the polarities of all the voltage drops occurring across each resistor. In order to do this, simply remember the rule of thumb:

Current goes in positively and comes out negatively.

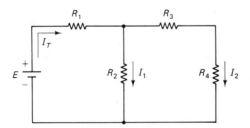

Figure 1-3   Current flow through a circuit of resistive devices.

Using this rule, the polarities developed across each resistor would look like Figure 1-4. This rule can also be used in reverse. If the polarities are given across the circuit components and you had to determine the current directions, merely determine the directions that would generate these polarities.

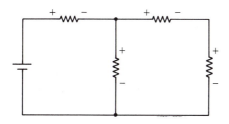

Figure 1-4   Resultant voltage polarities of current flowing in a circuit.

1-2.6 Ohm's Law

The relationship that exists between voltage and charge-flow, as the voltage (pressure) forces the charge down a conductor, depends on how much of an impediment that the conductor's material decides to create for the flow of current. Obviously, no substance is going to allow a flow of electrons, especially if it's a large flow, to move through it without creating some sort of internal resistance in the process. This resistance is in the form of friction, and friction generates heat. (This problem is discussed in detail in Section 1-13.) Let's now take a look at the relationship that exists between this resistance, which has been given the unit of the *ohm,* the volt, and electron flow or current. That relationship is expressed in the following equation, called *Ohm's law:*

$$I = \frac{E}{R} \tag{1-3}$$

where I = current (amps)
$\quad\quad E$ = voltage (volt)
$\quad\quad R$ = resistance to current or electron flow (ohm)

EXAMPLE 1-3

Find the current, in amps, flowing through a 270-ohm resistor that experiences a voltage drop of 17.4 volts.

Solution:

Using Eq. (1-3), the current would be equal to:

$$I(\text{amps}) = \frac{17.4 \text{ volts}}{270 \text{ ohms}}$$

$$= 0.064 \text{ amps}$$

1-2.7 *Resistivity*

The electrical resistance of any material is determined by four major factors. These are:

1. the type of material (i.e., its chemical make-up),
2. its temperature,
3. its length, and
4. its cross-sectional area.

All four of these factors are related to one another by the following expression:

$$R = \frac{kl}{A_{CM}} \tag{1-4}$$

where R = resistance (ohms)

k = resistivity factor or proportionality constant, which is temperature dependent (ohms-circular mils/ft, or ohms-circular mils/meter)

A_{CM} = cross-sectional area (circular mils, or CM. See following discussion.

Table 1-1 lists the resistivity constants, expressed in the English system of measurement, for several common materials.

In order to convert any given area into units of *circular mils* (CM), the following equation can be used (remember that a linear mil, or simply "mil," is a linear measurement in inches that has been transformed to thousandths of inches. This is done by merely multiplying the linear amount by 1,000. The area,

**TABLE 1-1 RESISTIVITY
CONSTANTS FOR SEVERAL
COMMON MATERIALS (VALUES
ARE FOR 68° F)[1]**

Material	Resistivity (ohms-CM/ft)
Silver	9.9
Copper	10.37
Gold	14.7
Aluminum	17.0
Tungsten	33.0
Nickel	47.0
Iron	74.0

[1] Donald G. Fink, *Electronics Engineer's Handbook*, (New York, N.Y.: McGraw-Hill, Inc., 1975), pp. 6–4, 6–5.

expressed in square mils, can then be found by multiplying the linear mil figures as you would if you were using normal measurement figures):

$$A_{CM} = (A_{sq.\ mil}) \left(\frac{4}{\pi}\right) \tag{1-5}$$

where A_{CM} = area expressed in circular mils (CM)
  $A_{sq.\ mil}$ = area expressed in linear square inches (in²) multiplied by 1,000,000 or 10^6

EXAMPLE 1-4

Find the resistance of a copper bar having a length of 14.5 feet and whose width and height are 3.28 inches and 2.13 inches respectively.

Solution:

Step 1. Calculate the bar's cross-sectional area using thousandths of inch units or mils to calculate sq. mils:

$$\text{Area} = (3.28\ \text{in} \times 1,000)(2.13\ \text{in} \times 1,000)$$
$$= 6.986 \times 10^6\ \text{mils}^2$$

(Notice that this is nothing more than taking the "normal" or linear area of 3.28 in. × 2.13 in. and multiplying the result by 1,000,000 or 10^6.)

Step 2. Convert the linear sq. mil area to CM area:
  Using Eq. (1-5):

$$A_{CM} = (6.986\ \text{in}^2) \left(\frac{4}{\pi} \times 10^6\right)$$
$$= 8.895 \times 10^6\ \text{CM}$$

Step 3. Determine the resistivity of copper:
  Referring to Table 1-1, the resistivity constant for copper at 68°F is 10.37 CM-ohms/ft.

Step 4. Calculate the resistance:
  Using Eq. (1-6):

$$R = \frac{(10.37\ \text{CM-ohms/ft})(14.5\ \text{ft})}{8.895 \times 10^6\ \text{CM}}$$
$$= 16.9 \times 10^{-6}\ \text{ohms}$$

  or   0.0000169 ohms

Once the concepts of current, voltage, resistance, and resistivity are understood, the next concept to review and understand is the DC circuit. Refer to Figure 1-5. This figure shows several devices that utilize varying amounts of voltage and current for their proper operation. One such device, the resistor, is now described.

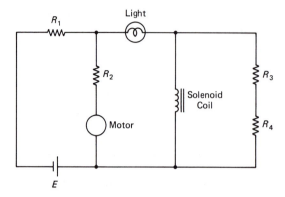

Figure 1-5 A typical DC circuit.

1-3 THE RESISTOR

The resistor is a device specifically designed to be inserted into a conductor to impede the flow of electrons in that conductor. The resistor can be constructed either from a length of conductor itself, or it can be manufactured from a wide variety of material mixtures whose electrical conductivities can be tightly controlled by varying the mixture proportions. A common mixture is comprised of carbon-like material that is highly conductive, with a clay or ceramic-like substance that is highly resistive to current flow. Varying the mixing proportions of these two substances will produce varying amounts of resistances. The resistive value associated with each resistor can be ascertained from a series of colored bars printed directly on the body of the resistor if the resistance value itself isn't already printed there. Usually there are four bands. The first three bands indicate the resistance, while the fourth band indicates the manufacturing tolerance. The first two bands are the significant digits, while the third band is the decade multiplier. Table 1-2 shows the color coding scheme that is widely used.

**TABLE 1-2   THE
RESISTOR COLOR CODE**

Black	0	Green	5
Brown	1	Blue	6
Red	2	Violet	7
Orange	3	Gray	8
Yellow	4	White	9
	Gold 0.1		
	Silver 0.01		

When the gold and silver bands are the third band on the resistor, they act as multipliers, as shown previously.

Tolerances: Silver ± 10%
Gold   ± 5%

When the gold and silver bands are the fourth band on the resistor, they act as tolerance indicators, as shown previously.

EXAMPLE 1-5

Find the resistance of a resistor having the following color bands: first band = red, second band = red, third band = orange, fourth band = gold.

Solution:

Referring to Table 1-2, the resistor would have a resistance of 22,000 ohms (first two digits are 2 and 2, and the decade multiplier is 3 or 10^3, which is 10,000). The tolerance of this resistor would be ±5%.

The physical size of a resistor has a considerable bearing on the amount of current and voltage it can safely handle before overheating and destroying itself. We discuss this problem further in Section 1-14. Some commonly used sizes or "wattages" are shown in Figure 1-6.

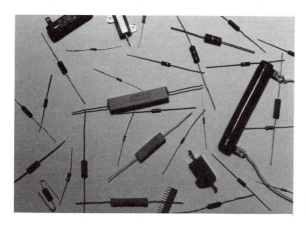

Figure 1-6   Resistor sizes used in electronic circuits.

1-4 SERIES AND PARALLEL RESISTANCES AND HOW THEY ARE CALCULATED

There are two major types of resistive circuits: the *series* circuit and the *parallel* circuit. There is also a third kind of circuit which is a hybrid of the first two, a combination of the series and parallel circuit.

1-4.1 The Series Resistance Circuit

Resistors placed in series have the following characteristics: The total current (I_T) flowing in the circuit shown in Figure 1-7 flows through all the resistors, R_1, R_2, R_3, and R_4. If you were to measure the current at point c, point d, or point e, you would find the measured current to be the same as I_T. This is also the same current flowing back up into the negative side of the voltage source, E_s. If E_s were

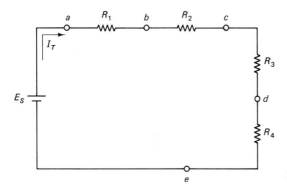

Figure 1-7   A series DC resistive circuit.

thought of as being a water pump, the water leaving the pump must be the same water entering the pump, assuming no losses (leaks) along the circuit of course. Furthermore, as I_T flows through each resistor, a voltage drop occurs across each resistor (Figure 1-8).

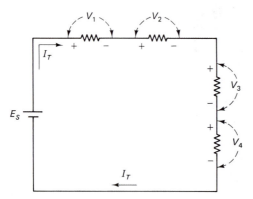

Figure 1-8   Voltage drops developed in a series circuit.

Each voltage drop amount is directly dependent on the value of each R across which the drop occurs. And furthermore, the sum of $V_1 + V_2 + V_3 + V_4$ must equal E_s. The preceding conditions are true only because the current entering and leaving each resistor is the same current amount. The condition just described defines the series resistance circuit. Summarizing:

$$E_s = V_1 + V_2 + V_3 + \cdots V_x \tag{1-6}$$

also

$$R_T = R_1 + R_2 + R_3 + \cdots R_x \tag{1-7}$$

where R_T = total resistance in a series circuit as measured between terminals a and b in Figure 1-7 (ohms)

R_1, R_2, R_3 etc. = individual series resistors (ohms)

EXAMPLE 1-6

Refer to Figure 1-9. Here we see five resistors all wired in series with each other. Determine the total circuit resistance (being aware that if this were an actual circuit, you would want to remember to remove the voltage source, E_s, while this is being done. This is so the internal DC resistance of the supply will not be included in the actual measuring of the circuit's total DC resistance.). The circuit's total resistance would then be measured with an ohmmeter placed between terminals a and b. However, we want to calculate the total resistance, not measure it. Also, in addition to this, we want to calculate the individual voltage drops across each resistance.

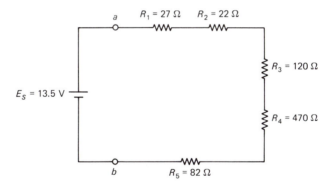

Figure 1-9   Circuit for Example 1-6.

Solution:

Step 1.   Verify that this is in fact a series circuit. Do this by noting that the total current (I_T) must flow through all the resistors in order to get back to the other side of the voltage source, E_s. This means that each resistor will experience the same circuit going into and coming out of itself.

Step 2.   Determine the circuit's resistance as measured between a and b assuming that E_s has been temporarily removed for this determination: Using Eq. (1-7), let R_1 = 27 ohms, R_2 = 22 ohms, R_3 = 120 ohms, R_4 = 470 ohms, and R_5 = 82 ohms. Therefore,

$$R_T = 27 \text{ ohms} + 22 \text{ ohms} + 120 \text{ ohms} + 470 \text{ ohms} + 82 \text{ ohms}$$
$$R_T = 721 \text{ ohms}$$

Step 3.   Determine the circuit's total current, I_T:

Using Ohm's law (Eq. 1-3) and the fact that the total supply voltage in the circuit is 13.5 volts, then,

$$I_T = \frac{13.5 \text{ volts}}{721 \text{ ohms}}$$
$$= 0.019 \text{ amps,}$$

or   19 milliamps

Step 4.   Determine now the individual voltage drops across each resistor.

Since each resistor has a particular resistive value with a known current flowing through it, we can again use Ohm's law (Eq. 1-3) to determine the individual voltage drops. To find the voltage drop across R_1: Since $I = E/R$, then $E = IR$.

Note: The designation for voltage in Ohm's law is E, and V in other expressions. The two letters are often used interchangeably to mean the same thing. E is often used for voltage sources, whereas V is used for voltage drops occurring across circuit resistances or impedances.

$$V_1 = (0.019 \text{ amps})(27 \text{ ohms})$$
$$= 0.513 \text{ volts}$$

To find the voltage drop across R_2:

$$V_2 = (.019 \text{ amps})(22 \text{ ohms})$$
$$= 0.418 \text{ volts}$$

To find the voltage drop across R_3:

$$V_3 = (.019 \text{ amps})(120 \text{ ohms})$$
$$= 2.280 \text{ volts}$$

To find the voltage drop across R_4:

$$V_4 = (.019 \text{ amps})(470 \text{ ohms})$$
$$= 8.930 \text{ volts}$$

And finally, to find the voltage drop across R_5:

$$V_5 = (.019 \text{ amps})(82 \text{ ohms})$$
$$= 1.558 \text{ volts}$$

According to Kirchhoff's voltage law, all five voltages must equal the supply voltage in this circuit. When totalling these voltages, we get 13.70 volts. We see that this is not 13.5 volts, our supply voltage. The error resulted in our rounding off of our five voltage calculations.

1-4.2 The Parallel Resistance Circuit

Refer to Figure 1-10. It would help to refer back to our water pipe analogy at this point in order to envision the concept of parallel resistance. E_s is now our water pump supplying pressure to force the water through the various circuit devices (in this case, resistors). Note that the total water flow (i.e., current I_T) must divide at point a, since there is a device (R_1) allowing some of the current to return to the intake side of our pump (the $(-)$ terminal of E_s) and a device, R_2, that is similarly situated allowing I_2 to flow back to the pump, as does I_{3-4}. Note that all three current divisions recombine back at point b to again form I_T before returning to the pump. The point is this: If devices are connected across a voltage source such that the $(+)$ and $(-)$ terminals are common to these devices, then these devices are said to be in parallel with each other. Another observation is: If the voltage source's current divides and then recombines going to and coming from each circuit device, then these devices are again said to be in parallel with each other.

Let's take another look at Figure 1-10. Are resistors R_2, R_3, and R_4 in parallel with each other? The answer is no. R_3 and R_4 have the same current flowing through them; consequently, R_3 and R_4 are in series with each other. However, it *is* proper to say that R_3 plus R_4 are in *parallel* with R_2 (and R_1 too, as far as that goes).

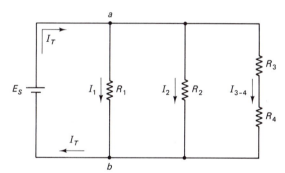

Figure 1-10   A typical parallel resistance circuit.

To calculate the equivalent resistance of resistors that are in parallel with each other, the following equation should be used (see Figure 1-11):

$$\frac{1}{R_T} = \frac{1}{R_a} + \frac{1}{R_b} + \frac{1}{R_c} + \cdots \frac{1}{R_x} \tag{1-8}$$

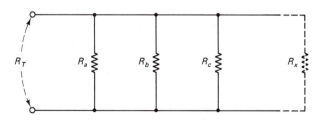

Figure 1-11   Circuit used for Eq. (1-8).

where R_T = the total equivalent resistance that could replace all the parallel resistances without affecting the circuit (ohms)

R_a, R_b, R_c, etc. = the individual parallel resistors (ohms)

EXAMPLE 1-7

Refer to Figure 1-12. Calculate the total equivalent resistance that would be measured across terminals a and b, and determine the voltage drops across each resistor.

Note: Remember to imagine E_s removed from the circuit during this resistance measuring so that its internal resistance doesn't interfere with the measuring. You would normally remove E_s if you had to make this resistance measurement in an actual circuit.

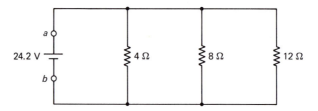

Figure 1-12   Circuit for Example 1-7.

Solution:

Letting the 4-ohm resistor = R_a, the 8-ohm resistor = R_b, and the 12-ohm resistor = R_c, we can use Eq. (1-8) to solve for R_T:

$$\frac{1}{R_T} = \frac{1}{4} + \frac{1}{8} + \frac{1}{12}$$
$$= 0.458$$
$$R_T = 2.183 \text{ ohms}$$

Another equation can can be used for calculating the total parallel resistance of *two* resistors, is the following:

$$R_T = \frac{R_a R_b}{R_a + R_b} \tag{1-9}$$

Equation (1-9) can be modified for larger numbers of parallel resistors; however, the equation's complexity increases to a point of unwieldiness. Consequently, for circuits involving more than two parallel resistors, use Eq. (1-8) instead.

As for the voltage drops across each resistor, since these resistors are all in parallel with each other, they will experience the same voltage drop. In other words, the voltage drop will be the source voltage in this case, which is 24.2 volts.

1-4.3 The Combined Series–Parallel Resistance Circuit

This type of circuit is shown in Figure 1-13. This is where a combination of series and parallel resistors are found, and perhaps more realistically describes a typically encountered circuit of resistors. If the circuit in Figure 1-13 is understood, then you shouldn't have any further problems with understanding series and parallel resistors and with being able to distinguish one circuit type from another.

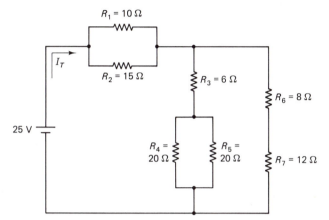

Figure 1-13   Circuit for Example 1-8.

EXAMPLE 1-8

Find the magnitude of current I_T in Figure 1-13.

Solution:

In order to find I_T, you must first find the total equivalent resistance, R_T. Then,

$$I_T = \frac{R_T}{E_s} \text{ (Ohm's law)} \tag{1-10}$$

But in order to find R_T so that Eq. (1-10) can be solved, we begin by combining series resistances and parallel resistances so that the circuit can be reduced to a simpler form.

$$\text{Let } R_1 \parallel R_2 = R_A$$

  Note: When you see the designation $\parallel$ used in conjunction with resistor discussions, the symbol means "in parallel with."

Then

$$R_A = \frac{R_1 R_2}{R_1 + R_2}$$

therefore,

$$R_A = \frac{(10)(15)}{10 + 15}$$

$$= 6 \text{ ohms}$$

Similarly, we let $R_4 \parallel R_5 = R_B$

Then

$$R_B = \frac{R_4 R_5}{R_4 + R_5}$$

therefore,

$$R_B = \frac{(20)(20)}{20 + 20}$$

$$= \frac{400}{40}$$

$$= 10 \text{ ohms}$$

Now let $R_6 + R_7 = R_C$ (these resistors are in series).

$$R_C = 8 \text{ ohms} + 12 \text{ ohms}$$
$$= 20 \text{ ohms}$$

The resulting simplified or reduced series–parallel circuit now looks like the circuit shown in Figure 1-14. Upon inspecting Figure 1-14 we see that we can do

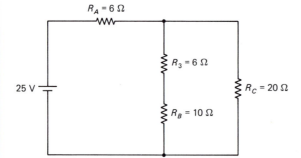

Figure 1-14

some more circuit reduction. We can combine R_3 with R_B since they are in series. Therefore,

let

$$R_3 + R_B = R_D$$

or

$$R_D = 6 \text{ ohms} + 10 \text{ ohms}$$
$$= 16 \text{ ohms}$$

which now results in a more simplified circuit shown in Figure 1-15. Since R_D is now $\parallel$ with R_C, we can combine those two to get

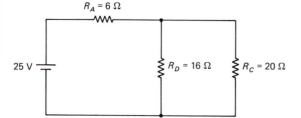

Figure 1-15

$$R_E = \frac{(16)(20)}{16 + 20}$$
$$= 8.889 \text{ ohms}$$

resulting in the further reduced circuit of Figure 1-16.

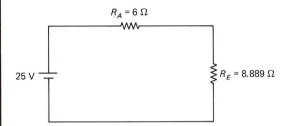

Figure 1-16

Since R_A and R_E are now in series with each other, we can combine those two resistors to give us a final equivalent resistance of:

$$R_{\text{equiv}} = 6 \text{ ohms} + 8.889 \text{ ohms}$$
$$= 14.889 \text{ ohms}$$

Essentially, we are saying that if all the resistors of Figure 1-13 were replaced with a 14.889-ohm resistor, the voltage source, E_S, would not know the difference since it would experience the same current, I_T, being drawn from it. However, we still haven't finished our problem here. We still have to find the magnitude of I_T. By going back and using Eq. (1-10) we can now calculate I_T as

$$I_T = \frac{14.889 \text{ ohms}}{25 \text{ volts}}$$
$$= 0.596 \text{ amps (or 596 mA)}$$

1-5 KIRCHHOFF'S LAWS

We now review two very important rules or laws developed by Gustav Kirchhoff back in the mid-nineteenth century. These laws have to do with the behavior of circuit voltages and currents.

1-5.1 Kirchhoff's Voltage Law

Earlier it was stated that the sums of all the voltage drops occurring across resistors in a circuit must be equal to the supply voltage at the voltage source. This is because of *Kirchhoff's voltage law*. His law states that *the algebraic sum of the potential rises and drops around a closed loop in a circuit is zero.* Let's take a look at Figure 1-17. We see three resistances, R_1, R_2, and R_3 all in series and wired to E_s. The voltage drop polarities and values have all been determined

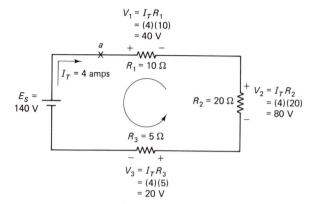

Figure 1-17   Using Kirchhoff's voltage law.

based on the direction and magnitude of I_T, the total current. Note too that we are working here with a closed loop of current flow. That is, the current, I_T, loops back to where it "began." Selecting a point, say point a on the circuit as a starting point, and picking a direction of summation, say counterclockwise as shown, begin summing all the rises and drops in voltages. We find, therefore, that:

$$-140v \text{ (a drop)} + 20v \text{ (a rise)} + 80v \text{ (a rise)} + 40v \text{ (a rise)} = 0$$

Picking any starting point and traveling in either direction, clockwise or counterclockwise, will always produce the same results according to Kirchhoff's voltage law.

1-5.2 Kirchhoff's Current Law

Kirchhoff has a comparable law to his voltage law that deals with current. Simply stated, it is this: *All current flowing into and out of a junction comprised of conductors must equal zero.* Or, stating it even more simply, *what goes into something must also come out of it.*

Refer to Figure 1-18. Note point a. This is a *junction* because of the joint formed by three wires or conductors. The current amounts flowing into and out of the junction have all been determined. Using Kirchhoff's current law and assuming that currents flowing into the junction are (+) and currents coming out are (−) (we could just as easily have picked the opposite signs and directions if we wanted to and we would not have affected the outcome of the answer); the following results would be obtained:

$$+10A \text{ (entering)} - 6A \text{ (leaving)} - 3A \text{ (leaving)} - 1A \text{ (leaving)} = 0$$

Now pick point b to see if Kirchhoff's current law works there also.

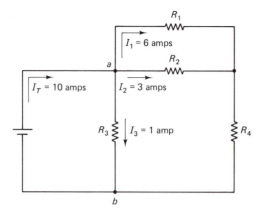

Figure 1-18   Using Kirchhoff's current law.

1.6 THE VOLTAGE DIVIDER RULE

When working with resistive networks in circuits, a handy way to calculate voltage drops across resistors is by using the *voltage divider rule*. Figure 1-19 shows a typical situation where this rule can be applied. The voltage divider rule states that:

$$V_{out} = V_{in} \frac{R_{out}}{R_{in}} \tag{1-11}$$

where V_{out} = the divider's output voltage (volts)
   V_{in} = the divider's input voltage (volts)
   R_{out} = the total resistance across which V_{out} is found (ohms)
   R_{in} = the total resistance across which V_{in} is found (ohms)

In the preceding example, $R_{out} = R_5 \parallel (R_3 + R_4)$ and $R_{in} = R_5 \parallel (R_3 + R_4) + R_1 + R_2$. However, R_{out} could be comprised of any type of resistor configuration, not just the one shown in Figure 1-19. The same can be said of R_{in} also.

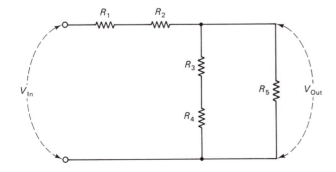

Figure 1-19   Using the voltage divider rule.

EXAMPLE 1-9

Find the output voltage, V_{out}, for the circuit shown in Figure 1-20.

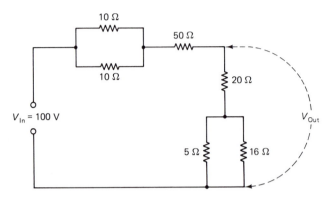

Figure 1-20   Circuit for Example 1-9.

Solution:

We see that V_{in} = 100 v. We must then find R_{in}. This value happens to be the circuit's total resistance across which the 100 volts is appearing. It is equal to:

$$(10 \parallel 10) + 20 + 50 + (5 \parallel 16) = 78.81 \text{ ohms}$$

To find V_{out}, we must determine R_{out}, which equals $(5 \parallel 16) + 20$, or 23.81 ohms. Finally, according to Eq. (1-11),

$$V_{out} = 100 \left[\frac{23.81}{78.81} \right]$$

$$= 30.21 \text{ volts}$$

1.7 POWER SUPPLY REGULATION

Earlier we stated that the voltage sources used in our example problems and circuit explanations were ideal sources. That is, they are unlike our car battery, where, when we start the engine with the headlights on, the battery's terminal voltage drops, thus dimming the lights. Our ideal voltage source would not lower its terminal voltage, but would instead remain constant. In the case of our car's battery, the starter's increased current demand placed on the battery causes the voltage to drop at its terminals. Batteries have an internal resistance that is inherent and unavoidable in their construction. A voltage drop develops across this internal resistance just as if it were an external resistance like other circuit resistors. Since the sum of all the voltage drops across the series resistances must equal the voltage source's voltage (this is Kirchhoff's voltage law just explained in Section 1-5.1), the greater the voltage drop across this internal resistance due to increased current flow, the less voltage there is available at the voltage source's terminals. Look at the illustrations in Figure 1-21. Notice that when a current of 8 amps is demanded from the battery, this creates a voltage drop of 4 volts across the battery's internal resistance, because

$$V_{R(\text{INT})} = (I_{\text{LOAD}})(R_{\text{INT}})$$
$$= (8)(0.5)$$
$$= 4.0 \text{ volts}$$

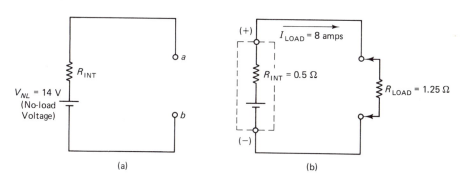

Figure 1-21   A typical battery circuit. **(a)** Before load is applied. **(b)** After load is applied.

And according to Kirchhoff's voltage law, this leaves only 10 volts to drop across the load resistance ($14V - 4V = 10V$). This is the same voltage that appears across the battery's terminals at *a* and *b*. The battery's regulation is not good enough to keep its output voltage at a constant value because of its internal resistance. Obviously, it would be nice if it were possible to manufacture batteries with little or no internal resistance.

Our ideal voltage source doesn't have a regulation problem. We purposely stated that the voltage stays constant regardless of current demand in order to simplify our problem solving. Once the problem is solved under these ideal conditions, we can then analyze the same problem using a not-so-ideal voltage source such as a battery or electronic power supply to see how the circuit actually behaves. But this complicates the problem considerably, and we'll avoid it for the present.

If we were to plot the voltage regulation characteristics of our car battery versus an ideal voltage source, we would get something that would look like Figure 1-22. Note that, in the case of the ideal voltage source, as the current demand increases the voltage output of the source remains constant. However, in the case of the car battery, as more current is demanded its output becomes lower.

The *percent of voltage regulation* is calculated using the following equation:

$$\% \text{ VR} = \frac{V_{NL} - V_{FL}}{V_{FL}} \times 100 \qquad (1\text{-}12)$$

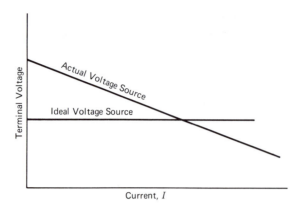

Figure 1-22   Load characteristics for an ideal and actual voltage source.

where % VR = percent of voltage regulation (%)

V_{NL} = the no-load voltage at the terminals (volts)

V_{FL} = the full-load voltage at the terminals (volts)

EXAMPLE 1-10

Using the information in Figure 1-21, calculate the percent voltage regulation for the battery shown.

Solution:

Using Eq. (1-12), the percent voltage regulation would be calculated as follows:

$$V_{NL} = 14v$$

$$V_{FL} = 10v$$

Therefore,
$$\% \ VR = \frac{14v - 10v}{10v} \times 100$$

$$= 40\%$$

1-8 THE AC CIRCUIT

Our next review subject has to do with alternating current or voltage. Up to this point, we have been dealing with current flow in one direction only. We now discuss the condition in which current makes direction reversals in the circuit where it is flowing.

1-8.1 Generating an AC Signal

There are only two major methods used for generating an alternating current or voltage. One is by using rotating mechanical machinery and the other is by using electronic oscillators. Figure 1-23 shows a mechanical device, a *generator,* or

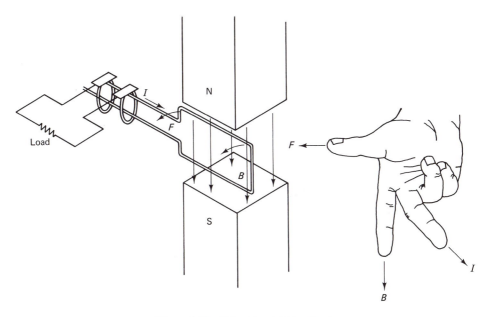

Figure 1-23 Using the right-hand rule.

alternator, in which a looped inductor is rotated through a magnetic field thereby generating an electric current within the loop. The rotation direction of the wire loop (F) and the magnetic field strength (B) determines the amount of current (I) generated. The right-hand rule is an aid in remembering the relationship that exists between the F, B, and I quantities. The revolutions per second made by the rotating loop determine the frequency of current reversals within the circuit attached to the generator. If the loop is rotated at 3,600 rpm (revolutions per minute), the number of current reversals is 3,600 times per minute, or 60 times per second. Each reversal is called a *cycle,* or *hertz.*

The second method of generating an AC signal is through the use of an oscillator. There are hundreds of oscillator designs in use, and we won't get into that particular discussion, but again, any basic electronics design book is a good source for these designs. In the oscillator, the frequency of oscillation is usually determined by the value of a resistor, a capacitor, an inductor, or by an oscillating piezoelectric crystal, or a combination of each. Each of these components is reviewed later. For now, we want to review the properties and nomenclature associated with AC waveforms.

1-8.2 Waveform Definitions and Conventions

The following is a list of definitions that apply to any alternating or repeatable waveform. The *sinusoidal* waveform is used as a model for this discussion, but any other periodic or repeatable waveform would do as well. (See Figure 1-24.)

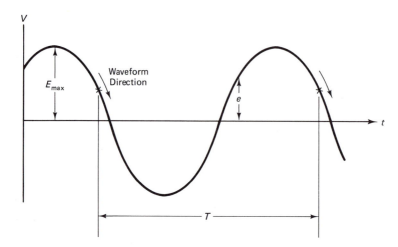

Figure 1-24   Important features of a cyclic waveform.

DEFINITIONS

WAVEFORM   The trace or path made by a measurable quantity such as voltage or current, plotted as a function of angular position, or time, pressure, or temperature, etc.

INSTANTANEOUS VOLTAGE, CURRENT OR POWER   The magnitude of the quantity (i.e., the voltage, current, or power) measured at an instant in time. This is synonymous with taking a snapshot of the waveform's height at a particular instant in time in order to record an otherwise fleeting value. Usually, small letter symbols are used to represent these quantities, such as e, i, p for instantaneous voltage, current, and power, respectively.

PERIODIC WAVEFORM   A waveform that is cyclic; i.e., it repeats itself after a period of time.

AMPLITUDE   The maximum or peak value of a waveform. We use the designation of E_{max} or I_{max} to signify this quantity in this book.

PERIOD   The length of time between identical adjacent points of a repeating waveform. The waveform must be traveling in the same direction for these points to be identical. The capital letter T is used for this quantity.

CYCLE   The portion of a waveform contained in *one period* of that waveform. The unit of the Hz is used here.

FREQUENCY   The number of cycles that occur in one second. Frequency and period are related by the equation:

$$f = \frac{1}{T} \qquad (1\text{-}13)$$

Frequency has units of Hz, kHz, MHz, etc.

Because the electrical generator was generally recognized as the first of the two methods used for frequency generation, it was common to plot waveforms of current and voltage versus an angular function such as degrees or radians. The angular function could then be related to the angular rotation per unit time of the generator's rotating armature or loop. This is still done to this day, not only for mechanical generators but also for electronic oscillators.

The following mathematical relationships are often used in AC circuit analyses:

$$2\pi \text{ radians} = 360° \text{ (See Figure 1-25)} \qquad (1\text{-}14)$$

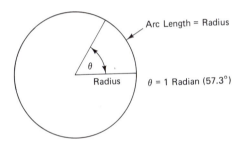

Figure 1-25   The radian.

$$1 \text{ radian} = 57.3° \text{ (approximately)} \qquad (1\text{-}15)$$

To convert degrees to radians: $\text{radians} = \dfrac{\pi \times \text{degrees}}{180}$ $\qquad (1\text{-}16)$

Or, for a close approximation: $\text{radians} = \dfrac{\text{degrees}}{57.3}$ $\qquad (1\text{-}17)$

To convert radians to degrees: $\text{degrees} = \dfrac{180° \times \text{radians}}{\pi}$ $\qquad (1\text{-}18)$

Or, for a close approximation: $\text{degrees} = \text{radians} \times 57.3$ $\qquad (1\text{-}19)$

$$\text{Angular velocity, } \omega, \text{ (rad/sec)} = \frac{\text{distance (rad)}}{\text{time (secs)}} \qquad (1\text{-}20)$$

Also, $\text{angular velocity, } \omega = \dfrac{2\pi}{T}$ $\qquad (1\text{-}21)$

But since $T = \dfrac{1}{f}$ (a rearrangement of Eq. 1-13) where f = frequency in *Hz*,

then $$\omega \ (\text{rad/sec}) = 2\pi f \tag{1-22}$$

The basic general mathematical format or "template" for a sinusoidal waveform is:

$$y = A_{max} sin(\omega t) \tag{1-23}$$

where y = the instantaneous amplitude on a waveform at ωt
  ω = angular velocity (rad/sec)
  t = instantaneous time (sec)
  A_{max} = maximum amplitude of waveform (expressed in volts, current, power, etc.)

EXAMPLE 1-11

Refer to Figure 1-24. Assume that the maximum amplitude of the displayed waveform is 8.3 volts. The frequency is 35 kHz. Write the mathematical function that describes the waveform having the given characteristics.

Solution:

Using Eq. (1-23) and Eq. (1-22), we write

$$y = A_{max} sin(2\pi ft)$$

Since A_{max} = 8.3v and f = 35 kHz,

then $$y = 8.3 sin(219, 912t)$$

or

$$y = 8.3 sin(2.2 \times 10^5 t)$$

1-8.3 The Phase Angle

Quite often it is necessary to know how much a particular signal is "out of phase" with another signal at a given time. As an example, consider the signal again in Figure 1-24. Notice that it doesn't appear to start at the zero axis of the v-t coordinates, but instead appears to be shifted to the left somewhat. The amount of shift is usually measured in radians or in degrees. If the shift is expressed in radians, then multiples of π are often used. (Remember, π radians = 180°; i.e., 3.1416 rad. $\times$ 57.3° = 180°, approximately.) By convention, a sinusoidal waveform shifted to the left is said to be *leading* the other waveform by so many degrees or by so many radians. A sinusoidal waveform shifted to the right is said to be *lagging* by so many degrees or by so many radians. Mathematically, Eq. (1-23) can be modified to denote these phase shifts in the following ways:

For a leading waveform: $$y = A_{max} sin(\omega t + \theta) \tag{1-24}$$

and for a lagging waveform $y = A_{max}sin(\omega t - \theta)$ (1-25)

where θ = phase angle shift (rad)

EXAMPLE 1-12

Refer to Figure 1-24 once again. Assume that the curve shown has been shifted 45 degrees (or $\pi/4$ radians) from the origin of the shown coordinates. Write the mathematical expression describing this curve. Assume a maximum amplitude of 4.3 amp and a frequency of 60 Hz.

Solution:

Since the curve in Figure 1-24 has been shifted to the left to indicate a leading waveform, we use Eq. (1-24). Also, we were told that A_{max} = 4.3 amps and $\theta = \pi/4$ radians.

Then	$y = 4.3 sin(\omega t + \pi/4)$	(from Eq. 1-24)
and since	$f = 60$ Hz	
and	$w = 2\pi f$	(from Eq. 1-22)
then	$w = 2\pi 60$	
	$= 377$ rad	
Therefore,	$y = 4.3 sin(377t + \pi/4)$	

Note that the trigonometric function ''cos'' is frequently used instead of ''sin.'' Since a cosine function waveform is 90° out of phase with a sine function waveform (Figure 1-26), we could rewrite all $A_{max}sin(\omega t)$ functions by changing the sine to a cosine and subtracting 90° or $\pi/2$ radians. In other words,

$$A_{max}sin\omega t = A_{max}cos(\omega t - \pi/2)$$ (1-26)

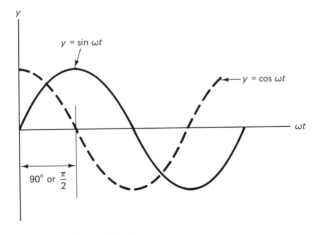

Figure 1-26   The sin *x* and cos *x* curves.

EXAMPLE 1-13

Rewrite the solution found in Example 1-11 so that the cosine function is used instead of the sine function.

Solution:

Using Eq. (1-26), we rewrite the expression from Example 1-11, which was $y = 8.3 sin(2.2 \times 10^5)$ to now read:

$$y = 8.3 cos(2.2 \times 10^5 t - \pi/2)$$

Now refer to Figure 1-26. Notice that shifting the cosine function to the right will place that curve back in phase with the original sine function.

1-9 THE CAPACITOR

In the next few sections we discuss the characteristics of the *capacitor,* how it functions, and how to calculate circuit total capacitance.

1-9.1 Capacitor Characteristics

The capacitor is a temporary electronic storage device that is synonymous with a water storage tank for a municipality or building. The capacitor stores electric charge. The unit of capacity is the *farad,* but because of it being such a large unit of measure, we instead speak in terms of *microfarads* (abbrev. μF, or mF) and micro-microfarads (abbrev. $\mu\mu$f, or mmf, or more recently, *picofarads,* pf). The electronic symbol for a capacitor is shown in Figure 1-27. The earlier symbol is still seen in some older schematics, while the later version is the one now being used. The lower curved line of the later version's symbol is usually connected to *earth or chassis ground* when grounding becomes necessary. This curved side is also the negative (−) side of the capacitor. Some capacitors are polarized and require this polarized connection for their proper operation. Variable capacitors are used primarily in electronic communications or in the controlling of timing circuits or oscillator circuits in general.

The relationship between capacity, electric charge, and voltage, is the following:

$$C = \frac{Q}{V} \tag{1-27}$$

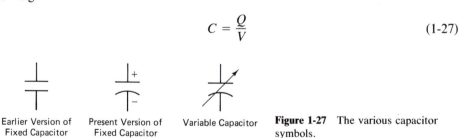

Earlier Version of   Present Version of   Variable Capacitor   **Figure 1-27** The various capacitor
Fixed Capacitor   Fixed Capacitor   symbols.

where C = capacity, farads (F)

    Q = electric charge, coulombs (c)

    V = voltage, volts (v)

Capacitance is also calculated using the following relationship:

$$C = \frac{8.85 \times 10^{-12} \, e_r A}{d} \tag{1-28}$$

where C = capacitance (F)

    A = plate area in square meters (m^2)

    d = distance separating the opposite plates (meters)

    e_r = relative permittivity, sometimes called the dielectric constant (no units)

The dielectric constants of various substances used in the manufacturing of capacitors are listed in Table 1-3.

TABLE 1-3 THE DIELECTRIC CONSTANTS OF SEVERAL SUBSTANCES

Substance	Dielectric constant (e_r)
Air	1.0006
Barium-strontium titanite	7500.0
Porcelain	6.0
Transformer oil	4.0
Bakelite	7.0
Rubber	3.0
Paper, parafinned	2.5
Teflon	2.0
Glass	7.5
Mica	5.0

EXAMPLE 1-14

Find the capacitance of two flat aluminum plates each having an area of 100 cm^2 and whose faces are parallel and directly opposite each other and separated by a sheet of glass 0.15 cm thick.

Solution:

From Table 1-3, we see that the e_r for glass is 7.5. Since Eq. (1-28) requires that the area of the plates be expressed in meters rather than in centimeters, we must first convert the given area to meters. Since there are 10^4 cm^2 in a square meter, then there are $100/10^4$ m^2, or 0.01 m^2 of capacitor plate. Also, we must convert the glass thickness, in cm, to meters. Therefore, 0.15 cm is the same as 0.0015 m. Therefore, using Eq. (1-28),

$$C = \frac{(8.85 \times 10^{-12})(7.5)(0.01\text{m})}{0.0015\text{m}}$$

$$= 44.25 \times 10^{-11} \text{ farads}$$

or

$$= 442.5 \text{ pF}$$

An important characteristic of a capacitor is that it *blocks* a flow of DC current. This is because of its construction. The capacitor is comprised of two conductors separated by a nonconducting barrier or dielectric. A DC current can't pass through such a construction. However, a capacitor *will* allow a portion of an AC current to pass through it. We will have more to say on this subject later in Section 1-13.

1-9.2 Capacitors in Parallel

Capacitors that are installed in parallel are identified as such just as resistors in parallel with each other. However, that is where the similarity ends. To calculate an equivalent capacitance for parallel capacitors, all the individual capacitors are *added* together. That is,

$$C_T = C_A + C_B + C_C + \cdots C_x \tag{1-29}$$

where C_T = the total capacitance (farads, μf, pf, etc.)
C_A, C_B, C_C, etc. = the individual $\parallel$ capacitors (farads, μf, pf, etc.)

EXAMPLE 1-15

Refer to Figure 1-28. Calculate the equivalent capacitance for capacitors C_8, C_9, C_{10}, and C_{11} that are all in parallel with one another.

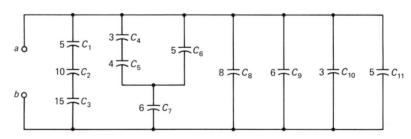

Note: All Capacitances in pf.

Figure 1-28 Circuit for Example 1-15.

Solution:

Using Eq. (1-29) we calculate:

$$C_T = C_8 + C_9 + C_{10} + C_{11}$$

$$C_T = 8 \text{ pf} + 6 \text{ pf} + 3 \text{ pf} + 5 \text{ pf}$$
$$= 22 \text{ pf}$$

1-9.3 Capacitors in Series

Capacitors that are installed in the series configuration are identified as such, again, just as resistors in series. But again, the comparison must stop here. To calculate an equivalent capacitance for series capacitors, you must treat them as resistors in parallel by using the following equation:

$$\frac{1}{C_T} = \frac{1}{C_A} + \frac{1}{C_B} + \frac{1}{C_C} + \cdots \frac{1}{C_x} \tag{1-30}$$

where C_T = the total equivalent capacitance (farads, μF, pF, etc.)
C_A, C_B, C_C, etc. = the individual series capacitors (farads, μF, pF, etc.)

EXAMPLE 1-16

Calculate the equivalent capacitance for the series capacitors C_1, C_2, and C_3 in Figure 1-28.

Solution:

Using Eq. (1-30):

$$\frac{1}{C_T} = \frac{1}{C_1} + \frac{1}{C_2} + \frac{1}{C_3}$$

$$= \frac{1}{5} + \frac{1}{10} + \frac{1}{15}$$

$$\frac{1}{C_T} = \frac{6 + 3 + 2}{30}$$

$$= \frac{30}{11}$$

$$= 2.727 \text{ pF}$$

1-9.4 Series–Parallel Capacitance

The treatment of combination series–parallel capacitor circuits is similar to that of series–parallel resistors. (Refer to Section 1-4.3 to review this process.) Using the results from Examples 1-15 and 1-16, and combining them with the remaining capacitors of Figure 1-28, you should be able to calculate the total equivalent circuit capacitance measured between terminals *a* and *b*. This value will be 27.895 pF. Verify this result before going on to new material.

1-10 THE INDUCTOR

The next component that we study and review is the *inductor*. As you will soon learn, there are similar behavior characteristics as compared to the resistor and capacitor.

1-10.1 Inductor Characteristics

The inductor is another temporary electronic storage device like the capacitor, except that it is usually thought of as having a shorter term storage capability, relatively speaking. The storage takes place in the electromagnetic field that surrounds the inductor when a current flows through it. An inductor is comprised of one or more insulated turns of conductive material such as copper wire wrapped around a center core having a known *permeability*. The permeability of a material is an indication of how easy or difficult it is for magnetic flux lines to become established in the material. As a result, there is a similarity between permeability and conductivity in electrical circuits.

The symbols for an inductor are shown in Figure 1-29. The variable inductor is used primarily in electronic communications.

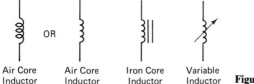

| Air Core Inductor | Air Core Inductor | Iron Core Inductor | Variable Inductor |

Figure 1-29 Inductor symbols.

The unit of inductance is the *henry*. Typical variations of the henry are *millihenries* and *microhenries*.

The behavior of the inductor when encountering AC- and DC-type signals is exactly opposite of that of the capacitor. The inductor, being comprised of a continuous inductor wound around a central core, will readily allow a DC current to flow through it. On the other hand, it will make an attempt to resist any change of direction of current as is typical in an AC current flow. As far as the AC current is concerned, it experiences an *impedance* to its flow just as if someone had placed a resistor in its path. As we review later, the amount of impedance experienced by the AC signal is dependent on the signal's frequency in addition to the amount of inductance possessed by the inductor.

The calculating of the *self-inductance* or simply *inductance* of a coil can be done using the following equation:

$$L = \frac{N^2 \mu A}{1} \qquad (1\text{-}31)$$

where L = the inductance in henries (H)
N = number of turns of conductor
μ = permeability of core material in Webers/Amp turns-meter (Wb/At-m)
A = area of core in meters² (m²)
l = length of core in meters (m)

EXAMPLE 1-17

Calculate the inductance of the coil in Figure 1-30. The permeability of air is 12.57 × 10^{-7} Wb/At-m.

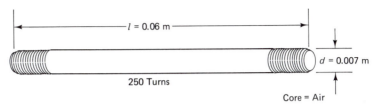

$l = 0.06$ m

250 Turns

$d = 0.007$ m

Core = Air

Figure 1-30   Inductor for Example 1-17.

Solution:

Applying Eq. (1-31) and using the information given in Figure 1-30, we get:

$$L = \frac{(250^2)(12.57 \times 10^{-7})(A)}{0.06}$$

Note: The units have been dropped in the preceding equation to simplify its solution. We continue to do this from now on.

We now have to calculate A, the area of the inductor's core:

$$A = \frac{\pi d^2}{4}$$

Therefore,

$$A = \frac{\pi (0.007^2)}{4}$$

$$= 3.849 \times 10^{-5} \text{ m}^2$$

and finally,

$$L = \frac{(250^2)(12.57 \times 10^{-7})(3.849 \times 10^{-5})}{0.06}$$

$$= 504 \times 10^{-7} \text{ H}$$

$$= 50.4 \ \mu\text{H}$$

1-10.2 Inductors in Series

Inductors in series are easily recognized as behaving just like resistors as far as becoming mathematically combined. That is, their total equivalent inductance is calculated just like totalling series resistances:

$$L_T = L_A + L_B + L_C + \cdots L_x \qquad (1\text{-}32)$$

where L_T = the total equivalent inductance in henries, millihenries, or microhenries (H, mH, μH)

No example is given here to demonstrate solving a series inductive circuit, since the solution method is virtually identical to the method outlined in Section 1-4.1 for series resistances.

1-10.3 Inductors in Parallel

Inductors in parallel with one another are recognized as being so, again, just like resistors. Parallel inductors can be replaced with an equivalent inductor whose inductive value can be calculated using the same form of equation as used with parallel resistors. That is,

$$\frac{1}{L_T} = \frac{1}{L_A} + \frac{1}{L_B} + \frac{1}{L_C} + \cdots \frac{1}{L_x} \tag{1-33}$$

where L_T = the total equivalent inductance in henries, millihenries, or microhenries (H, mH, μH)

L_A, L_B, L_C, etc. = individual parallel inductors in henries, millihenries, or microhenries (H, mH, μH)

Again, no example is given here to demonstrate the solution of a parallel inductive circuit, since the solution is virtually identical to the one used in Section 1-4.2 for parallel resistors.

1-10.4 Series–Parallel Inductors

The analysis of a combination series–parallel inductor circuit is identical to that of a series–parallel resistive circuit. Consequently, no example is given here to demonstrate a typical solution. Instead, refer back to Section 1-4.3 to see how this is done, if this method is unclear to you.

1-11 THE TRANSFORMER

The next component we review is the *transformer*. This device is used for changing AC voltage and current amounts in a circuit. The next three sections discuss the transformer's characteristics and its ratings.

1-11.1 Transformer Characteristics

Transformers are comprised of two inductors constructed close enough to each other so that the flux field surrounding the one encompasses the windings of the second inductor. In this manner, a voltage that has been developed across the *primary* winding of a transformer can become inductively coupled to the *secondary* winding (Figure 1-31). In order for the flux field to develop, however, we need an alternating current or a changing current for inductive coupling to take place. A direct current of constant magnitude will not create a changing flux field. For this reason, transformers cannot transform DC voltages.

Note the symbol used in Figure 1-31 to denote an AC signal source. This is the symbol that we use from now on to indicate a voltage source whose output is a

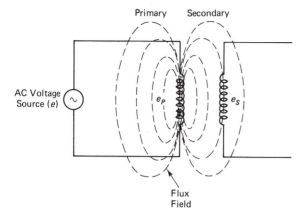

Primary   Secondary

AC Voltage
Source (*e*) e_p e_s

Flux
Field

Figure 1-31 Magnetic coupling of a
transformer's coils.

varying AC signal, either in the form of a voltage output or a current output. You
will always be told what form of output is being generated whenever this symbol is
seen. You will see either an *e* or an *i* followed by its magnitude or value.

Transformers come in many sizes and shapes, as do all the other components that we have reviewed so far. However, transformers can be broken down
roughly into three groups. These are:

1. the *iron-core transformer,* used primarily for the purpose of transforming
   significant quantities of current and voltage at audio or "line" frequencies;
2. the *air-core transformer,* used mainly for communication applications; and
3. The *variable core transformer,* which may be used for either power or communications applications.

1-11.2 The Turns Ratio

The following relationship holds true for a transformer's primary and secondary
winding turns ratio and the voltages being induced:

$$\frac{E_p}{E_s} = \frac{N_p}{N_s} \tag{1-34}$$

where E_p = the primary winding voltage (V)
   E_s = the secondary winding voltage (V)
   N_p = the number of turns of wire on the primary
   N_s = the number of turns of wire on the secondary

EXAMPLE 1-18

A transformer has 35 turns of wire on its primary winding. The winding is supplied
with 115 VAC. How many turns are necessary on the secondary to develop 40
VAC?

Solution:

Using Eq. (1-34), and letting $N_p = 35$ turns, $E_p = 115$ VAC, and $E_s = 40$ VAC, then

$$\frac{115}{40} = \frac{35}{N_p}$$

Solving for N_p:

$$N_p = \frac{(40)(35)}{115}$$
$$= 12.2 \text{ turns}$$

The following is the relationship existing between a transformer's turns ratio and the induced current:

$$\frac{I_p}{I_s} = \frac{N_s}{N_p} \qquad (1\text{-}35)$$

where I_p = the primary winding current (A)
I_s = the secondary winding current (A)
N_p = the number of turns of wire on the primary
N_s = the number of turns of wire on the secondary

EXAMPLE 1-19

If the *turns ratio* (the number of turns on the primary divided by the number of turns on the secondary) is 3 to 1, and the primary current is 12 amps, find the output current.

Solution:

Using Eq. (1-35) and noting that $I_p = 12$ amps and $N_p/N_s = 3$, then

$$\frac{12}{I_s} = \frac{1}{3}$$

Solving for N_s:

$$I_s = 12 \times 3$$
$$= 36 \text{ amps}$$

1-11.3 Power Transformer Ratings

It is very important to remember that a transformer doesn't "step-up" or "step-down" power as it does to voltage and current. The power remains the same regardless of what happens to the current or voltage.

In the rating of a power transformer, it is often the practice to mention the term *VA* rating or *volt–amp* rating. This is the voltage and current values of either the primary or secondary windings multiplied together. In other words, the primary VA rating must equal the secondary VA rating. However, the VA rating is *not* the transformer's true power rating. We review this fact in Section 1-14.

1-12 TRANSIENT BEHAVIOR

Both the capacitor and inductor, because of their signal storage capabilities, exhibit a rather unique charging or discharging characteristic (or filling or emptying characteristic, if you prefer) which is momentary, called *transient behavior*. To explain this, it's better to start with the capacitor, since its behavior is a little easier to visualize than the inductor's transient behavior.

1-12.1 The R–C Circuit

Figure 1-32 represents, in simplified form, a type of circuit in which a capacitor, C, is charged through a resistor, R, whose purpose is to slow down the charging rate of the capacitor. If the resistor were missing and the capacitor were allowed to charge directly from a voltage source, the capacitor would become instantly charged. Assume that the switch, S, were thrown to position A. The charging current, i, would have to flow through R to reach C. The higher the resistive value of R, the slower the charging rate of C. Or, to word it another way, the slower would be the voltage build-up, v_c, of the capacitor. The time that it takes for v_c to build up to the source voltage, V_s, is

$$t = 5RC \tag{1-36}$$

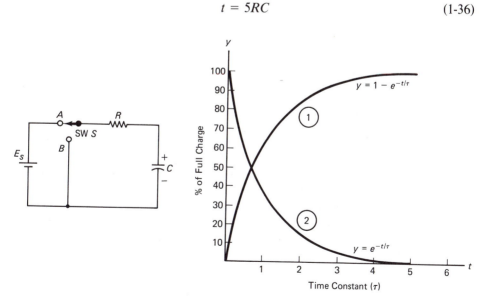

Figure 1-32 The RC time constant.

where t = time to fully charge capacitor (sec)
  R = value of resistor (ohms)
  C = value of capacitor (farads)

Note: Many textbooks on control theory and electronics state that $t = 5RC$, as Eq. 1-36 shows. However, in theory, it takes at least six time constants to fully charge a capacitor. In practice, though, it is commonly agreed that five time constants is sufficiently accurate for most applications.

The *time constant* for the preceding circuit (sometimes referred to as an *R–C circuit*), is simply:

$$\tau = RC \tag{1-37}$$

where τ = *one* time constant (sec)

At the end of one time constant, the capacitor has reached 63.2% of its fully charged value. The charging curve takes on the appearance of curve 1 in Figure 1-32. The equation for this curve is:

$$y = 1 - e^{-t/\tau} \tag{1-38}$$

where $\tau = R \times C$ (sec)

If switch S in Figure 1-32 were thrown to position B, capacitor C would then discharge at a rate already fixed by the same RC time constant. The discharge voltage curve would look like curve 2 in Figure 1-32. The equation for this particular curve is:

$$y = e^{-t/\tau} \tag{1-39}$$

where $\tau = R \times C$ (sec)

After one time constant, the capacitor will have lost 63.2% of its full voltage and will be within 36.8% of being completely discharged.

1-12.2 The R–L Circuit

The discussion for the *resistance–inductance* circuit, or simply, *R–L circuit,* is similar to that of the *R–C* circuit. Referring to Figure 1-32 once again, merely change the capacitor to an inductor and place switch S back to position A. We now monitor the voltage, V_L, across the inductor, instead of V_C as we did with the capacitor. When switch S is thrown to position A in Figure 1-32, the voltage, V_L, immediately becomes equal to the supply voltage, E_s. The time it takes for the flux field initially surrounding the inductor when the switch was closed to completely collapse to zero, is

$$t = \frac{5L}{R} \tag{1-40}$$

where t = time to fully collapse flux field, or, time for V_L to drop from a value of
E_s to zero (sec)
L = inductance (henries)
R = resistance (ohms)

The time constant for this L–R circuit is simply:

$$\tau = \frac{L}{R} \qquad (1\text{-}41)$$

where τ = one time constant (sec)

As in the R–C circuit, at the end of one time constant the inductor's voltage has become reduced to 63.2% of its original full voltage, E_s. The curve representing this deteriorating voltage across the inductor resembles curve 2 in Figure 1-32. It's important to note at this point that while V_L was approaching zero, the voltage across the resistor, R, was building up to a value equal to E_s. Kirchhoff's voltage law demands that this happen. We have a situation where, for all practical purposes, the inductor is now acting as a short and only resistor R is present for creating a voltage drop within our circuit.

Now let's see what happens when switch S is thrown to position B. Since $V_R = E_s$, and this continues to be true for an instant after switch S is in position 2 due to the current flowing through the inductor and the inductor current's inability to change instantly, the voltage across the inductor now becomes equal to the voltage across R, except now it is opposite in polarity. However, the flux field surrounding the inductor now begins to collapse. This reduction now produces a decrease in V_L, but reversed in polarity, across the inductor. The characteristic of this decrease is similar again to curve 2 in Figure 1-32, except for the reversed polarity just referred to. In other words, curve 2 must be "flipped over" so that it is shown decreasing from a negative value up to zero as time goes on.

1-13 REACTANCE AND IMPEDANCE

Reactance and *impedance* are the two major factors that complicate AC circuit analysis as compared to the relatively simple DC circuit analysis. Various components used in AC circuitry show a tendency to "react" to AC signals in varying degrees depending on that component's capacitive or inductive nature, and on the frequency of the signal itself. This reaction is in the form of a blockage or pseudoresistance. As a matter of fact, this pseudoresistance has been assigned the unit of the ohm to emphasize the similarity. Unfortunately, this commonality of units has also led to a lot of confusion for people just beginning in electronics. This is where the similarity ends when comparing this so-called impedance of current to the *resistance* of a resistor. To begin with, an ideal resistor does not react to an AC signal other than presenting the same resistance to it as to a DC signal. Let's now take a critical look at this apparent lack of reaction of a resistor to an AC signal.

1-13.1 The Resistor's Reactance to AC Waveforms

When an alternating current waveform is sent through a resistor, an alternating voltage waveform develops across that resistor. Let's assume that we have placed an oscilloscope across the resistor to read this voltage drop, and another oscilloscope to read current, placed in series with the resistor. (In order to do this, you want to be sure to use the proper current shunts in conjunction with the scope's probes to create the voltage drop needed, since an oscilloscope is a voltage, not a current sensing device.) Refer to Figure 1-33. The resultant waveforms are drawn alongside the oscilloscope symbols depicting what we would observe on that scope's screen. As a matter of fact, we would see nothing unusual. But we should point out that the two observed waveforms are *in phase* with each other. In other words, as the current increases from zero to its maximum value, so does the voltage waveform. This is an important point to remember when we discuss the capacitor and inductor's behavior.

1-13.2 The Capacitor's Reactance to an AC Waveform

Now let's substitute a capacitor for the resistor in our experiment by referring to Figure 1-34. Notice the waveforms produced by the oscilloscopes. The voltage and current waveforms are 90° out of phase with each other. While it may not be evident in comparing the drawn waveforms, the voltage actually *lags* the current waveform by 90°. The reason is surprisingly simple. When the current flow into the capacitor is at a maximum, the voltage developed across the capacitor is at a minimum. It's only when the current is at minimum or zero does the capacitor

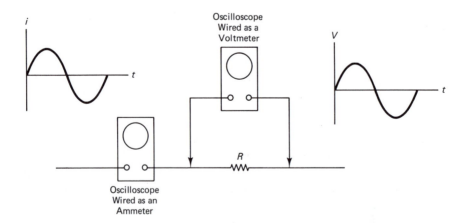

Figure 1-33 The current–voltage relationship for a resistor.

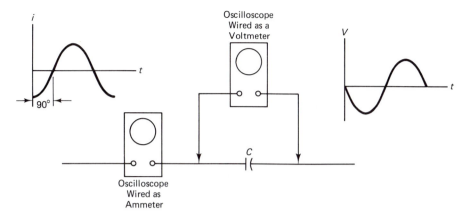

Figure 1-34   The current–voltage relationship for a capacitor.

have its fully developed voltage. This is because when the capacitor becomes fully charged, a kind of resistance or back pressure is created such that the current stops flowing. This is *capacitive reactance*. The capacitor's size determines how soon the capacitor fills, and the waveform's frequency determines when the filling rate is going to create this resistive reactance. So obviously, capacitive reactance is dependent on at least two items that we have mentioned here: (a) The size of capacitor used, and (b) the waveform's frequency. The actual relationship is expressed in the following equation:

$$X_C = \frac{-j}{2\pi f C} \tag{1-42}$$

where X_C = capacitive reactance (ohms)
$\quad f$ = frequency (Hz)
$\quad C$ = capacitance (farads)
$\quad -j$ = a *j*-operator, which means that there is a 90° phase angle difference between current and voltage. The minus sign indicates that capacitive reactance is drawn 90° downward in a phasor or impedance diagram. (See Section 1-13.5.)

Whenever a reactance figure is stated for a component (in this case a capacitor) you must also be careful to state the phase angle. For a capacitor it is −90°. For instance, a capacitor may have a voltage drop across it having the value of 5.34v ∠−90°.

Quite often, Eq. (1-42) is written in the following form:

Since $\qquad\qquad \omega = 2\pi f \qquad$ (from Eq. 1-21)

then $\qquad\qquad X_C = \frac{1}{\omega C} \angle -90° \tag{1-43}$

EXAMPLE 1-19

Assume that an AC current of 250 mA at a frequency of 400 Hz is passing through a 0.1 mf capacitor. Calculate the reactance produced by this capacitor.

Solution:

Using Eq. (1-42):

$$X_C = \frac{-j}{2\pi(400)(0.1 \times 10^{-6})}$$

Note that the 0.1 mf capacitor has been converted to farads according to the requirements of Eq. (1-42).

$$X_C = -j3978.9 \text{ ohms}$$

or,   3978.9 ohms $\angle -90°$

Remember that the reactance of 3978.9 ohms is *not* measurable with an ordinary DC ohmmeter. This produced quantity, instead, is a reactance to an AC current's frequency for a given capacitance. If you were to measure the DC resistance of this capacitor, or any capacitor, you would read an infinite resistance on an ohmmeter (or close to it, since you might read a very high resistance due to the capacitor's dielectric material inside). A capacitor acts as an infinitely high resistance to DC regardless of any current flow.

1-13.3 The Inductor's Reactance to an AC Waveform

Let's now substitute an inductor in place of the resistor in Figure 1-33. This is shown in Figure 1-35. Again we see that a 90° phase shift has taken place between the current and voltage waveforms. And again this is due to the "filling" of the storage flux envelope surrounding the inductor. However, we have a radically

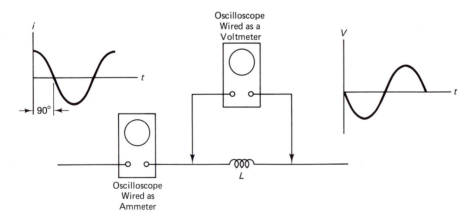

Figure 1-35   The current–voltage relationship for an inductor.

different sequence of events going on here. In general, when a flux field is in the process of expanding around the inductor, the lines of flux cut across the inductor's windings causing an induced voltage of opposite polarity to develop. This creates back pressure that tends to reduce or diminish any further current or voltage build-up. When the flux field is at maximum build-up, that is, no more windings are being cut by moving flux lines, current begins to flow again through the inductor since the back pressure has now been relieved. The voltage measured across the inductor is now zero (remember that this is a pure inductor with no DC resistance associated with the coil windings). Now when the field collapses, the flux lines once again cut across the turns of inductor wire but now generate a forward pressure that helps to pull or push the current through it. Furthermore, this forward pressure or voltage represents a reversal in polarity across the inductor, that is, a reversal as compared to the voltage's polarity during the flux field build-up. If all of this sounds complicated, just remember that we now have a situation that shows the current *lagging* the voltage by 90°. The amount of *inductive reactance* is dependent on the amount of inductance and the waveform's frequency. This relationship is expressed by:

$$X_L = +j2\pi fL \tag{1-44}$$

where X_L = the inductor reactance (ohms)
  f = frequency of current being supplied (Hz)
  L = the inductance (henries)
  $+j$ = a j-operator, which means that there is a 90° phase angle difference between current and voltage. The (+) sign indicates that the inductive reactance is drawn 90° upward in a phasor or impedance diagram.

EXAMPLE 1-20

Calculate the reactance created by a 20-mH inductor subjected to a current whose frequency is 1.2 MHz.

Solution:

Using Eq. (1-44):
$$X_L = +j2\pi(1.2 \times 10^6)(20 \times 10^{-3})$$

Remember: All units must be in *henries* and *Hertz*.

$$X_L = +j150.8 \times 10^3 \text{ ohms}$$

or,
$$150.8 \times 10^3 \text{ ohms } \angle +90°$$

Often, Eq. (1-44) is seen written in the following form:

Since
$$\omega = 2\pi f \qquad \text{(from Eq. 1-21)}$$

then,
$$X_L = +j\omega L$$

or,
$$\omega L \angle +90° \tag{1-45}$$

1-13.4 What Is Impedance?

Impedance is the term used to describe the resultant of combining inductive and capacitive reactances with resistances. Look at Figure 1-36. Here we have a combination of inductance, capacitance, and resistance all being supplied with a 10 VAC source at 490 Hz. The problem is this: How do you find the current *i*, and how would you keep track of the phase shifts created by the combined effects of the capacitor and inductor? We can't neglect the effect of the resistor. Even though we know that a resistor doesn't cause a phase shift between the current passing through it and the voltage dropping across it, it *will* definitely affect the overall phase relationship between the output voltage and current of a circuit containing both resistors, capacitors, and/or inductors. To solve this problem, we first must devise a way of combining all three components into one component, so to speak, much like we did in our discussion of DC circuits where we combined all the resistors together and then, using Ohm's law, found the total current. In the case of an AC circuit, however, where inductors, capacitors, and resistors are all involved, we combine them all into what is called the circuit's total impedance. And this impedance, which will have the unit of the ohm just as do reactance and resistance, will have associated with it a phase angle just as we found in the cases of the two reactances just studied. Assuming that we have found this total equivalent impedance at some phase angle, can we still use Ohm's law to solve for the current, *i*? The answer is yes. Ohm's law will look like the following expression for AC circuits:

$$I = \frac{E}{Z \angle \theta} \tag{1-46}$$

where I = the current (amps)
E = the voltage (volts)
$Z \angle \theta$ = the circuit's impedance, in *polar form* (ohms)

Before we can proceed with our discussion, we must do some more reviewing of terminology concerning *j*-operators, polar forms, and the difference between *i* and *I* and *e* and *E*.

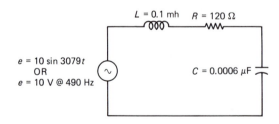

Figure 1-36   An RLC circuit.

1-13.5 Understanding the J-Operator

Reactance values and impedance values are considered *vectors* in electronics. A capacitive reactance of, say $-j172$, has both magnitude and direction, just as an airplane flying at 225 mph NW has both magnitude and direction. Figure 1-37 shows the coordinate system for the *j*-operator. Actually, most *j*-operators have associated with them a *real* number along with a *j*-operator number, or what some mathematicians call an *imaginary* number. A typical complete *j*-operator plus real number would look something like $110 - j172$. This entire number is often referred to as a *complex* number. A plot of $-j172$ and $110 - j172$ are shown in Figure 1-37, which also shows their vector quantities. Notice the angle, θ. This is the phase angle for $110 - j172$. The phase angle for simply $-j172$ would be $-90°$. To calculate the phase angle, θ, for any given complex number,

$$\theta = \text{ArcTan}\,\frac{\pm jB}{A} \qquad (1\text{-}47)$$

where $\pm A$ and $\pm jB$ are the components of any complex number.

To find the magnitude of any complex number (in our case, the magnitude of the reactance of impedance),

$$Z = \sqrt{A^2 + B^2} \qquad (1\text{-}48)$$

where, again, $\pm A$ and $\pm jB$ are the components of any complex number.

By now you may have guessed or recalled that the A term in our complex number is a resistive value that always gets plotted on the R axis of our *j*-operator coordinate system. The B term is the reactive or impedance value that always

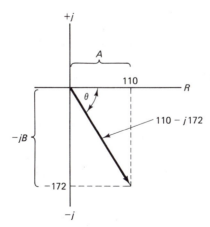

Figure 1-37 How complex and polar forms are related.

gets plotted on the *j* axis. Let's now calculate both the magnitude and direction of our example $110 - j172$. Using Eq. (1-47), we find the phase angle to be:

$$\theta = \text{ArcTan } \frac{-172}{110}$$
$$= -57.4°$$

And by using Eq. (1-48), we can calculate the magnitude of the impedance:

$$Z = \sqrt{110^2 + (-172^2)}$$
$$= 204.2 \text{ ohms}$$

The usual convention used for writing our completed expression is: 204.2 $\angle -57.4°$. This answer is called a *polar* form and is read "204.2 ohms at a phase angle of minus 57.4 degrees." If, at this point, you feel a little uncertain about using *j*-operators, or *phasors,* as they are sometimes called, refer to Appendix A at the end of this book to review the conversion process. The complex number and *j*-operator occur again in later chapters.

1-13.6 *Instantaneous Versus RMS Voltages and Currents*

Before we get back into our impedance discussion, there is one additional matter of convention we must discuss and review. This concerns RMS and instantaneous voltages and currents and how to distinguish one from the other.

Up to this point in our review of electronic circuitry, we have assumed that whenever we were discussing alternating voltages or currents, we were discussing the waveforms in their true size and shape as one would see them on, say, an oscilloscope screen. If you were to hand-plot a waveform described by the expression, $y = 14sin(377t + \pi/4)$, you would graph a sinusoidal waveform whose frequency was 60 Hz ($f = \omega/2\pi$) and whose maximum and minimum amplitude was exactly ±14 volts with a phase shift of 45°, leading. You would, in this case, be plotting the *instantaneous* values of that particular waveform. Consequently, whenever a waveform is presented in the form of $y = A_m sin(\omega t \pm \theta)$ or $y = A_m cos(\omega t \pm \theta)$, you will know that you will be dealing with instantaneous values. As a matter of fact, notations using the small script *e* or *i* for voltage and current respectively, are also an indication that instantaneous values are being used.

All of the preceding discussion is for the purpose of distinguishing instantaneous voltage and current references from another kind of voltage and current designation, called *RMS* voltage and current. RMS voltage and current (RMS standing for the mathematical term *root mean square*) are used in discussions dealing with power consumption or heat dissipation problems. This is where it is necessary to determine an equivalent voltage or current amplitude that would generate the same heat in a purely resistive network of resistors as would be generated by a DC voltage or current source. As it turns out, for a sinusoidal waveform, simply taking 0.707 times the maximum instantaneous value will give you that equivalent heating value. This is called the RMS value, or sometimes the *effective* value. In the case of our preceding example where we looked at the graphed instantaneous waveform of $y = 14sin(377t + \pi/4)$, the RMS value would be expressed as 0.707×14 or 9.898 volts $\angle +45°$. Since we are speaking of RMS values here, we used the polar form of the sinusoidal expression. That is the convention we use in this book. So, whenever you see the polar form being used to describe voltages or currents, you will know that we are speaking about RMS or effective values.

The best way to understand all of these concepts is to consider another example.

EXAMPLE 1-21

Solve for the current, I, in Figure 1-36.

Solution:

We solve this problem in a step-by-step manner.

Step 1.   Find X_C: Using Eq. (1-42),

$$X_C = \frac{-j}{2\pi(490 \times 10^3)(0.0006 \times 10^{-6})}$$

$$= \frac{-j}{1.8472 \times 10^{-3}}$$

$$= 0.541 \times 10^{-3}$$

$$= -j541 \text{ ohms}$$

Step 2.   Find X_L: Using Eq. (1-44),

$$X_L = 2\pi(490 \times 10^3)(0.10 \times 10^{-3})$$

$$= j307.87 \text{ ohms}$$

Step 3.   Combine all the j-terms and the real terms:

$$Z_T = j307.87 - j541 + 120$$

$$= 120 - j233.13$$

Step 4.   Find Z_T by expressing in polar form: Using Eq. (1-48) and Eq. (1-47),

$$Z_T = \sqrt{120^2 + (-233.3^2)}$$

$$= 262.2 \text{ ohms}$$

and

$$\theta = ArcTan \frac{120}{-233.33}$$

$$= \angle -62.8°$$

Therefore, the impedance will be $262.2 \angle -62.8°$

Step 5.   Determine current, I: Using Eq. (1-46),

$$I = \frac{10}{262.2 \angle -62.8°}$$

$$= 0.038 \angle +62.8°$$

Notice that if we wanted to express our answer in an instantaneous value instead, we would want to use the sinusoidal form, which would look like this:

$$i = 0.038 \times 1.414 sin(3.079 \times 10^6 t + 62.8°)$$

or, $$i = 0.054 sin(3.079 \times 10^6 t + 62.8°)$$

where multiplying the RMS current by 1.414 converted it back to an instantaneous value.

This resulting waveform is telling us that because of the impedance created by the inductor, capacitor, and resistor in our circuit, the current is now being caused to *lead* the voltage by 62.8°.

Impedances in circuit networks may be combined just as if they were resistors in series or parallel using the same series and parallel equations as used for resistors. The only additional difficulty is the mathematical manipulating of the attached phase angles with each impedance value. Again, it is suggested that you review the information in Appendix A if you feel unsure of yourself in handling these kinds of problems.

1-14 POWER

What makes a discussion on power so confusing is that there are so many different types of power. The two power types that we are mostly concerned with are the following.

DEFINITIONS

APPARENT POWER: The "power" resulting from merely multiplying voltage times amps together as you would in a DC circuit. That is,

$$P_a = VI \tag{1-49}$$

where P_a = apparent power volt–amps)
   V = RMS value of the voltage (volts)
   I = RMS value of the current (amps)

AVERAGE POWER, EFFECTIVE POWER, OR REAL POWER: This is the power resulting from taking into account the phase angle existing between the voltage and current waveforms.

$$P = P_a cos\theta \tag{1-50}$$

where P = average, effective, or real power (watts)
   P_a = apparent power
   θ = phase angle existing between the voltage and current waveforms (degrees)

The term $\cos\theta$ in Eq. (1-50) is called the *power factor*. In general, electrical equipment is rated in volt–amperes instead of watts, since it is not known by the manufacturer what sort of electrical load or impedance is going to be attached to the equipment by the customer. This would have a direct bearing on the phase angle alteration that is most likely to happen when the equipment is installed. In addition, it is quite possible for a customer to operate his or her equipment within a seemingly safe power limit according to an attached wattmeter. However, the current limitations placed on the equipment could still be greatly exceeded only because the power factor was unusually low for some reason (perhaps due to a highly inductive load). As a result, most manufacturers will rate their electrical equipment in terms of volt–amperes (VA) instead of real power consumption, leaving the real power consumption determinations up to the customer.

1-15 RESONANT FREQUENCY

The concept of resonant frequency plays a very important role in electronics and in automatic controls in particular. A resonant frequency is one which is generated by an electrical circuit (or mechanical circuit) that acquires a portion of its operating energy through a more or less self-sustaining feedback loop or circuit. Perhaps one of the best examples of such a system is shown in Figure 1-38. This is a simplified diagram of an old radio circuit devised by an experimenter named Armstrong back in 1922, called a *regenerative receiver,* or more appropriately, the Armstrong oscillator. The theory of operation was quite simple and ingenious. By merely "dumping" or feeding some of the amplified radio signal (amplified by the rf or radio frequency amplifier) back to the antenna, the amplified signal could become reamplified and reamplified again, until very weak signals could be detected and heard. However, if too much reamplification took place, the entire

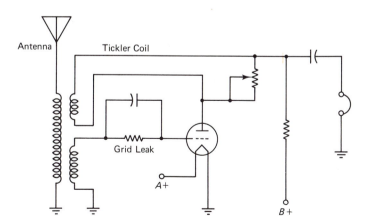

Figure 1-38   The Armstrong regenerative radio receiver.

circuit would oscillate or resonate at a particular frequency determined by the radio's tuning circuits. This was undesirable from the standpoint of radio reception, but nevertheless an interesting side development for Armstrong. He later used this principle for radio transmission experiments. We return to this example later when we discuss automatic controls.

Any circuit that contains an inductance and a capacitance is usually a prime candidate for resonating at a particular frequency. The principle is rather simple. If a certain sized capacitor is allowed to discharge into an inductor, and the inductor's field is allowed to build up as a result and then to collapse so as to reintroduce a charge-flow back into the capacitor, and then the capacitor is allowed to discharge back into the inductor, and so on, we have a resonant frequency condition. The frequency of this back-and-forth oscillating is determined by the amount of capacitance and inductance in the circuit. The actual resonant frequency can be calculated by the following equation:

$$f_r = \frac{1}{2\pi\sqrt{LC}} \tag{1-51}$$

where f_r = the resonant frequency (Hz)
  L = the inductance (henries)
  C = the capacitance (farads)

EXAMPLE 1-22

Refer to Figure 1-39. A signal generator wired as shown has its output frequency varied over a range of frequencies. Determine the frequency that produces the highest output voltage across the resistor as measured by the peak in the waveform seen on the oscilloscope's screen. In other words, find the circuit's resonant frequency.

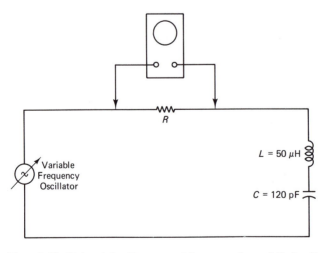

Figure 1-39   Determining the resonant frequency for an LC circuit.

Solution:

Using Eq. (1-51),

$$f_r = \frac{1}{2\pi \sqrt{(50 \times 10^{-6})(120 \times 10^{-12})}}$$

$$= \frac{1}{48.67 \times 10^{-8}}$$

$$= 2.055 \times 10^6 \text{ Hz}$$

or, $= 2.055$ MHz

A frequency of 2.055 MHz from the signal generator will produce the highest amplitude waveform on the oscilloscope's screen.

Obviously, the back-and-forth oscillating of this electron charge cannot continue forever by itself. Otherwise, we would have the makings of a rather controversial device, a perpetual motion machine. These have yet to be proven workable. Internal circuit friction (resistance in this case), which is the usual downfall of all such schemes, eventually puts an end to the oscillating action. However, the oscillations can be continued for an indefinite period of time by simply injecting a fresh charge of electrons at just the right time from an outside source. This is exactly what an oscillator does.

1-16 OSCILLATORS

The oscillator has valuable circuit applications in electronics, especially in the area of timing. Oscillators become the circuit clocks that synchronize or coordinate the intricate functions of computer operations and of control circuits in general. These types of precision clocks are usually further aided in their accuracy through the utilization of *crystals*. These are precisely machined or etched pieces of quartz-like material that vibrate at a resonant frequency either with or without the aid of a resonant *L–C* circuit.

Another type of electronic oscillator that is frequently used takes advantage of the *R–C* or *R–L* time constant. By varying the resistance of this type of circuit, the circuit's time constant can be varied over a wide range of values to simulate a resonant frequency condition. Again, this can be used to produce the necessary timing frequencies in many control circuitry applications.

1-17 SOLID STATE DEVICES

When working with automatic control devices, one can't help but come in contact with some sort of electronic solid state device being used in the control circuitry. For the purposes of understanding automatic control theory, it isn't necessary to fully understand the internal workings of these devices, although admittedly, it helps. What we do here instead is deal with the devices pragmatically. Just knowing in general how they work and what they do is certainly more than sufficient for the purpose of obtaining a good grasp on how and why a system's design behaves the way it does.

Figure 1-40 shows the most often drawn symbols used in automatic control designs, along with their identification. The following is a quick summary of what each component does in a circuit.

DEFINITIONS

DIODE The diode is used for allowing current to flow in only one direction, the direction indicated by the symbol's arrow.

ZENER DIODE The zener diode is similar to the diode except that it is used to regulate or maintain a particular voltage level in a circuit.

NPN AND PNP TRANSISTOR The transistor is used primarily for amplifying *current* signals. The transistor type (npn or pnp) determines the direction of current flow within the emitter circuit as indicated by the emitter symbol.

FET TRANSISTOR This is a type of transistor that is a *voltage* signal amplifier as opposed to the current signal amplifier qualities of the npn or pnp transistor. The characteristics of a FET are similar to those of a vacuum tube. There are several types of FET transistors; the type shown in Figure 1-40 is a dual-gate MOSFET (the MOS in MOSFET standing for "metal oxide semicon-

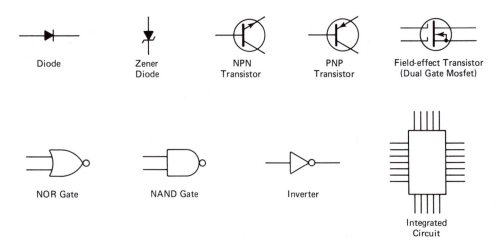

Figure 1-40 Typical solid state components encountered in automatic control circuits.

ductor''). This device is used frequently in applications where it is necessary to mix two signals to form a third signal. It is noted for its very small physical size.

NOR, NAND, INVERTER GATES These are logic devices designed to make a logical decision based on a number of input choices given to them.

INTEGRATED CIRCUIT (IC) The IC covers a tremendously wide field of devices too numerous to mention here. It is quite often difficult to determine how a circuit that contains an IC behaves unless you are well versed in circuit design or have used that particular device in another circuit. It is often necessary to refer to reference manuals supplied by the IC's manufacturer in order to find out how the device functions. ICs are complete electronic circuits comprised of various components all packaged into a small epoxied container designed to be soldered into or socketed into a larger circuit.

SUMMARY

Chapter 1 was intended to be a summary of basic electronic circuitry theory. Enough theory was presented to enable you to develop more confidence in interpreting the various automatic control system schemes that are presented later in the coming chapters. We discussed the definitions of DC and AC current and voltage, along with resistance, capacitance, inductance, reactance, and impedance. We also defined transients and their characteristics. We discussed transformer action, power, and resonant frequency. We also identified the many different solid state components and their behavior in an effort to help you understand the electronic circuitry that we discuss later.

REFERENCES

American Radio Relay League, *The ARRL Radio Amateur's Handbook,* Newington, Conn.: ARRL Pub., 1987.

BOYLESTAD, ROBERT L., *Introductory Circuit Analysis,* Columbus, Ohio: Charles E. Merrill Publishing Co., 1977.

JONES, THOMAS H., *Electronic Components Handbook,* Reston, Va.: Reston Publishing Co., 1979.

2

Control Theory Basics

The problem with learning many modern-day engineering concepts is that the concepts can be quite intimidating if you are not acquainted with the theory that goes along with them. Rather than be intimidated, a little knowledge about the subject goes a long way in clearing up mysteries and making you feel a little more at home with the subject. Automatic control theory can be just one of those threatening subjects if not approached properly. Therefore, with that bit of philosophy in mind, let's now try to relate the basics of automatic control theory with some everyday observations.

You may be surprised to find out that the ordinary stereo amplifier used in playing recordings is an excellent source of automatic control theory basics. It is necessary to consider the amplifier's electronics at this point, but we'll keep it relatively simple for the present so as not to lose sight of what we're trying to accomplish here.

2-1 FREQUENCY RESPONSE

Many of us, at one time or another, have had the unpleasant experience of listening to our favorite recording played on an inexpensive amplifier or player system and have remarked, "Wow! That really sounds awful!" What we were doing was really criticizing that system's inability to faithfully play back the audio frequencies whose true sounds we have memorized. We can thank our ears and brain for that. Our ears have an amazingly flat frequency response to a broad range of frequencies from about 16 Hz to about 20,000 Hz (Figure 2-1). The stereo manu-

54

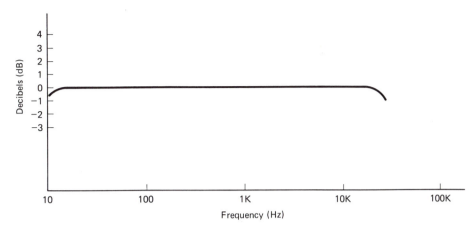

Figure 2-1   Frequency response curve for the normal human ear.

facturer tries to match this response characteristic as closely as possible, or as close as his or her design and manufacturing budget will allow. If the manufacturer doesn't do this, our ears can usually detect the difference in sound quality.

Let's look at Figure 2-1 a little more closely. Notice that in this figure we have plotted frequency versus *decibels*. Notice especially that the scale used in making this response curve is logarithmic. This is because the human ear is logarithmic in its response to sounds. Also, because of the extremely large span of frequencies that the ear responds to (20,000 Hz–20 Hz, or a span of 19,980 Hz), the logarithmic scale allows the plotting of this large span of data. Another interesting feature of the ear is the extremely flat frequency response throughout its entire hearing range. That is to say, there's very little detectable variation of loudness within this flat range.

Ideally, an audio amplifier will possess this same output characteristic. Any variation in the amplifier's output will surely be detected by the ear. Figure 2-2 shows a frequency response curve of an audio amplifier. In reality, a typical frequency curve would not be quite as smooth and symmetrical as the one shown here. Notice that the curve shows *amplification* versus frequency. (Ignore the second vertical scale alongside the graph for now. We talk about that one later.) Amplification is often referred to as *gain*. This is nothing more than a ratio of the amplifier's output voltage divided by its input voltage. The question is, how much gain variation can the amplifier get by with without the human ear being able to detect it? The ear can just detect a variation of two or three decibels, but what about gain? Equation (2-1) shows the relationship that exists between voltage gain and decibel output:

$$\text{Decibel output} = 20 \, log\left[\frac{V_{\text{out}}}{V_{\text{in}}}\right] \qquad (2\text{-}1)$$

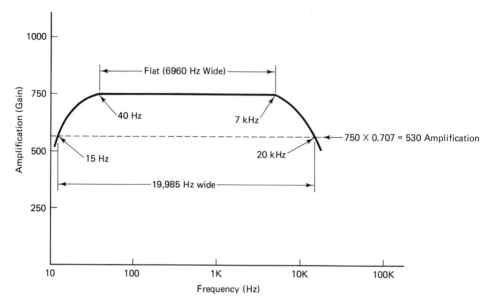

Figure 2-2   Frequency response curve of a typical audio amplifier.

Consequently, in order to define frequency response, we must set up some sort of criteria concerning the flatness of the response curve so that we know what portions of the curve to include in our frequency response statement. In other words, we have to know how much gain variation the ear can tolerate before it detects noticeable and annoying changes in gain. Up to this point, our discussion has concerned audio amplifiers. However, frequency response is a concern for other kinds of systems also. Mechanical systems have a frequency response characteristic just as pronounced as electronic systems. These systems are also capable of generating a frequency response curve and have similar tolerable response limits just as the audio amplifiers. However, instead of the ear being the judge as to what is tolerable and what is not, our general opinion based on observation and experience dealing with vibration and motion is used. As we'll soon learn in the discussions ahead, the criteria used in analyzing automatic control systems is virtually identical to that used for the amplifier.

2-2 GAIN

We have already defined the term *gain* in our discussion of frequency response. But to formalize the definition in the shape of an equation, we can say the following:

$$\text{Gain} = \frac{\text{output voltage}}{\text{input voltage}} \qquad (2\text{-}2)$$

Since it is generally assumed that the ear can just detect a 3-decibel (dB) change in voltage level or loudness, you have only to place 3 dB into Eq. (2-1) and solve for V_{out}/V_{in} to find that 3 dB represents a voltage ratio change of 1.414 or 0.707, depending on whether the change was an increase (i.e., +3 dB) or a decrease (−3 dB). More often than not, the voltage ratio change will often be a decrease occurring at the extreme ends of the response curve. As a result, these points are often referred to as the curve's *3-dB down points*. (Refer again to Figure 2-2.)

2-3 BANDWIDTH

The total span, or *bandwidth,* of frequencies in our audio amplifier includes the frequency range spanned by the 3-dB down points in the amplifier's specifications. In other words, this adds an additional frequency range to the flat portion of the curve. The ear will most likely be unable to detect this change in gain within the 3-dB down points, and as far as the manufacturer is concerned, it's an additional bonus for them in their advertising! Therefore, as a formal definition, *the bandwidth of a device is the span of frequencies that is included within the 3-dB down points of that device's frequency response curve.*

2-4 LINEARITY

There are a number of things that determine the bandwidth of a device and they all have to do with the circuit design and the types of components used in the device's design and construction. The device's *linearity* is determined by components such as diodes, transistors, integrated circuits, vacuum tubes, capacitors, inductors, and diodes. In general, the linearity of a device such as an amplifier refers to that device's ability to faithfully reproduce the input signal at its output. This is done in such a manner so that the only variation in the output signal as compared to the input signal is its magnitude.

2-5 PHASE

The phenomenon called *phase* is often a source of confusion. In order to better understand this term, spend a moment studying Figure 2-3. This figure represents a test setup that can be assembled by anyone wishing to explore the characteristics of an audio amplifier in order to better understand the terms defined and used up to this point. We see our audio amplifier with its input not only attached to a signal source, but it is also attached to a dual trace oscilloscope. This is a kind of oscilloscope that allows you to look at two waveforms from two different signal sources simultaneously. The amplifier's output is also wired to the scope such

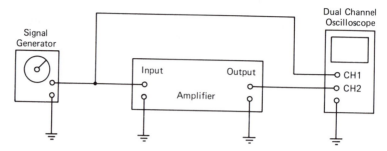

Figure 2-3 Hook-up for analyzing the performance characteristics of an audio amplifier.

that we can now view both the amplifier's input and output waveforms at the same time. The amplifier's input is wired to the output of an audio sine wave signal generator having an output frequency capability of 1 Hz to, say, 30 KHz. Furthermore, let's assume that the output audio signal from the amplifier has been attenuated such that its amplitude or voltage output is identical to the input signal's amplitude. We can make this adjustment to the output's gain level somewhere in the amplifier's middle frequency range where we know the amplifier's gain is flat. For instance, if we know ahead of time that our amplifier has a flat frequency response up to, say, 20 KHz, we would then adjust the output gain to match the amplifier's input at approximately 10 KHz. Let's now analyze our first results appearing on the oscilloscope's screen as depicted in Figure 2-4.

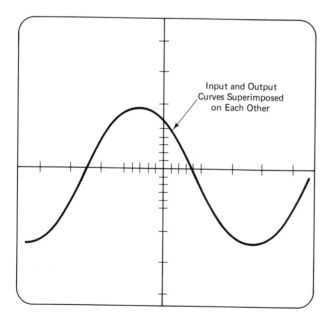

Input and Output
Curves Superimposed
on Each Other

Figure 2-4 Input and output waveforms that are in phase with each other.

As we compare the input and output signals at our middle frequency value where we adjusted the amplitudes to be purposely the same, we really don't see anything all that unusual. As a matter of fact, the signals are identical in every respect, indicating that the amplifier is faithfully reproducing the signal being supplied to it. Moreover, the positive and negative peaks of both curves are occurring at precisely the same spots; that is, the two curves *are in phase with each other*.

Let's now change the frequency output of our signal generator. Let's reduce the frequency to some very low value such as 10 Hz and again compare the two resultant curves on our oscilloscope. Figure 2-5 shows us the outcome. Immediately, we notice two things that have changed. For one, the output amplitude has not diminished when compared to the input; for another, there now appears to be a *shift in phase* between the two curves. More precisely, we note that the output curve is *leading* the input curve by some measurable amount.

Once again, let's change the signal generator's frequency to a now much higher frequency representing the upper end of our amplifier's frequency response curve, say, 18,000 Hz. Figure 2-6 represents the results obtained with our oscilloscope. We now see that the output waveform is *lagging* the input waveform and the output is again considerably less in amplitude as compared to the input waveform. We can now list at least two observable things that happened in our experiment while changing the frequency of our signal generator:

1. The output signal diminished considerably at both the low end and high end of the audio amplifier's frequency response range.

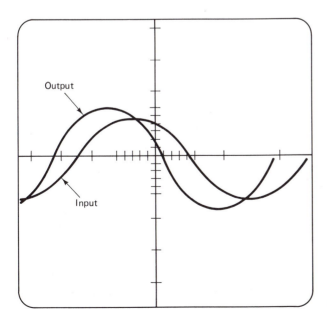

Figure 2-5 The output waveform leading the input waveform.

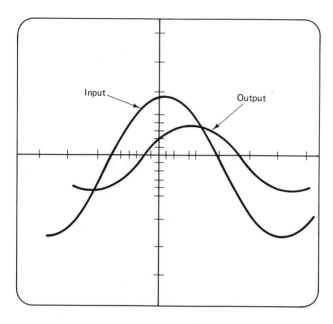

Figure 2-6 The output waveform lagging the input waveform.

2. The phase relationship between the two curves became either leading or lagging at the ends of the response curve.

The results of our test, assuming that we recorded output amplitudes and phase variations throughout the amplifier's entire frequency range, can be summarized in the two curves shown in Figure 2-7.

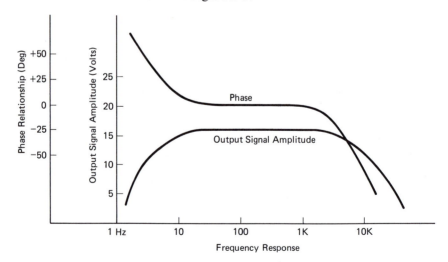

Figure 2-7 Frequency response curves for an audio amplifier not using negative feedback.

Up to this point, we have only observed a phase shift phenomenon but have not really explained what it is or what causes it. To begin with, phase shift is caused by components whose outputs are frequency sensitive, or to say it another way, are frequency dependent. Examples are capacitors, inductors, transistors, certain integrated circuits, etc. In other words, any electrical component that has a nonlinear range of operation will display this phase shift characteristic. Phase shift occurs because of the operating of these components in their nonlinear range. There are certainly other reasons that cause phase shift too. One common cause is the different lengths of propagation paths over which the two signals must travel. If one signal path is longer than the other, then the two signals will arrive at different times and will be out of phase with each other. In the case of our amplifier, however, this probably isn't the case. What is interesting about our phase shift problem, though, is that it can be substantially reduced if not entirely corrected. This can be done through the use of negative feedback circuits.

2-6 FEEDBACK

We are all familiar with feedback. If a microphone is placed too close to the speaker, the audio coming from the speaker is coupled back into the microphone to be amplified once again by the amplifier's system. This produces a louder speaker output which, in turn, is fed back once again through the microphone, and the procedure of amplification is repeated. The result is a very loud howling or screeching noise indicating that the audio system is out of control and needs adjustment. This kind of feedback is referred to as *positive feedback*. It is a type of feedback where the output signal is fed back *in phase* with the input signal, either on purpose or by accident, as the input signal enters the system or amplifier. Let's analyze this system a little more closely to see if we can describe what is going on in mathematical terms.

Figure 2-8 depicts our amplifier with no feedback. It is simply an amplifier with an input and an output having an amplification or gain of A. E_{in} is the input signal voltage and E_o is the output signal voltage. This system is called an *open-loop system*. In other words, no feedback is present. The gain of any amplifier is simply $A = E_o/E_{in}$. Now let's add a feedback circuit to the amplifier (Figure 2-9). Notice in Figure 2-9 that we have installed a feedback path by using a circuit that connects the amplifier's output back to its input through another circuit labeled *FC* (feedback control). The feedback control is nothing more than a variable resistor that controls the amount of output signai being dumped back into the

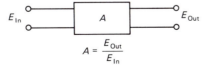

$$A = \frac{E_{Out}}{E_{In}}$$

Figure 2-8

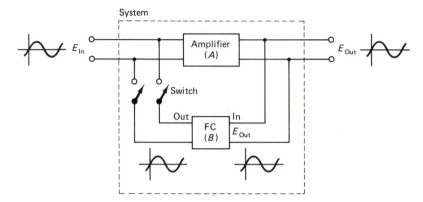

Figure 2-9   A positive feedback system setup.

amplifier's input circuit. Let's call the gain of this particular circuit *B*. What is especially important to notice here are the indicated waveforms that you would find if you placed an oscilloscope at the locations indicated. We have also added a switch at the amplifier's input. Notice that when the switch is closed, the output signal from the amplifier will be fed back through *FC in phase* with the original input signal coming from the signal generator, and both signals, added together, are fed back to the amplifier's input.

Consider what happens when the switch is opened. With the switch in its opened position, we have an open-loop amplifier. That is, no feedback is present within the system. This is similar to the situation in Figure 2-8. However, when the switch is closed, the amplifier is introduced to a feedback circuit having the following characteristics beginning at the input of *FC*: (a) The input voltage of circuit *FC* is the same as E_o; (b) The output voltage being supplied back to the amplifier's input is equal to E_o times *B* (the gain of the feedback control circuit), or simply BE_o; and (c)

$$E = E_{in} + BE_o \tag{2-3}$$

From Eq. (2-2) we know that $E_o = AE$ for any amplifier. Therefore, in place of *E* in this equation, let's substitute Eq. (2-3) to get:

$$E_o = A(E_{in} + BE_o) \tag{2-4}$$

$$= AE_{in} + ABE_o \tag{2-5}$$

Now let's arrange Eq. (2-5) so that it reads $E_o/E_{in} = f(A, E, B)$. (That is, E_o/E_{in} is a function of *A*, *E*, and *B*):

$$AE_{in} = E_o - ABE_o \tag{2-6}$$

$$E_{in} = \frac{E_o(1 - AB)}{A} \tag{2-7}$$

or
$$\frac{E_o}{E_{in}} = \frac{A}{1 - AB} \tag{2-8}$$

It's important to remember that this equation is for a system that uses *positive* feedback for its operation.

Let's now take a look at some what-if situations using Eq. (2-8):

1.   *What if the amplifier's gain, A, times the feedback control (FC) gain, B, is less than one?*

Answer: The gain of the system is increased over and above what it would have been with the open-loop system depending on the magnitude of $A \times B$.

2.   *What if $A \times B$ is exactly equal to one?*

Answer: The system's gain would be infinitely high causing the amplifier to just go into smooth oscillation provided that the AB product is maintained just at the value of one. Theoretically, one could then remove the signal generator at this point and have a perpetual motion device. However, like all good ideas, there are catches! In this case, because of system losses due to less than 100% efficiencies of component performances, it is necessary periodically to add additional make-up energy to the system to keep it oscillating smoothly. Or think of it this way: In order to keep a child's swing swinging at the same amplitude or height of swing, you must occasionally add an additional push, otherwise the swinging would eventually run down and cease.

3.   *What if $A \times B$ is greater than one?*

Answer: The system's gain calculation would result in a negative number, and as a result, can't be defined. The system's behavior that would result from this condition would be described as uncontrollable, wild, or unmanageable. Oscillations at this point could result ultimately in damage or self-destruction to the system.

Systems that employ positive feedback for their operation are extensively used today, but not for automatic control applications, as we will find out later on. Positive feedback has widespread use in electronic circuitry in the design of oscillators. Years ago, a circuit that was very popular in radio manufacturing was the regenerative radio receiver circuit that used a tickler coil circuit shown in Figure 1-38 in Section 1-15. This system was a positive feedback system (its operation was described in general in Chapter 1 but is worth repeating here). The circuit worked like this: The very weak radio signal to be amplified entered the grid of the vacuum tube by way of the antenna and step-up transformer coils *a* and *b*. The so-called grid-leak detector circuit aided the vacuum tube in detecting the radio signal while the vacuum tube also amplified the signal. The signal coming from the plate of the tube was then fed back to the antenna coil by means of the tickler coil to allow the amplified signal to be reamplified all over again. The amount of feedback was controlled by the radio's listener by adjusting the variable resistor, *R*. *R* was adjusted for maximum volume or just to the point before allowing the circuit to break into oscillation (i.e., just before $A \times B = 1$). If too much signal was fed back to the antenna, the radio's output at the headphones or speaker would emit loud and uncontrollable howls and screeching noises, indicating oscillations and too much feedback. The radio had amazing sensitivity to weak signals because of this circuit and was used for many years before ultimately being replaced by the modern-day superheterodyne receiver circuit. Even today, some people still argue that the regenerative receiver, because of the positive feedback circuit, still has very definite and beneficial applications in weak signal communications.

Now that we have explored the behavior patterns of a system with positive feedback, let's take a look at our same system but with a *negative feedback circuit* installed. What is interesting here is the modification necessary to convert the positive feedback system to a negative feedback system. All that is needed is to reverse the wire connections out of *FC* going to the input of the amplifier (Figure 2-10). With this wiring revision, the signal leaving *FC* will now be introduced to E_{in} 180° out of phase with E_{in}. In other words, when E_{in} is increasing in voltage, the output of *FC* is decreasing, and visa versa. Let's now see how the E_o/E_{in} gain equation is calculated for our system with this reversed behavior pattern.

Because of the fact that BE_o is now out of phase with E_{in} by 180°,

$$E = E_{in} - BE_o \quad \text{(note the sign change)} \tag{2-9}$$

Again, from Eq. (2-2), we know that $E_o = AE$ for any amplifier. Therefore, in place of *E* in this equation, let's substitute Eq. (2-9) to get:

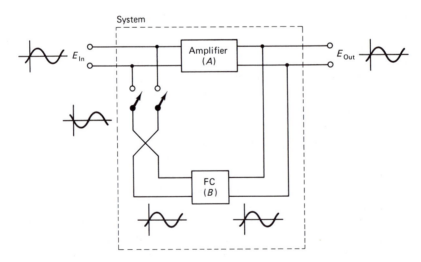

Figure 2-10   A negative feedback system setup.

$$E_o = A(E_{in} - BE_o) \qquad (2\text{-}10)$$

$$= AE_{in} - ABE_o \qquad (2\text{-}11)$$

Now let's arrange Eq. (2-11) so that it reads $E_o/E_{in} = f(A, E, B)$ as we did in Eq. (2-5):

$$AE_{in} = E_o + ABE_o \qquad (2\text{-}12)$$

$$E_{in} = \frac{E_o(1 + AB)}{A} \qquad (2\text{-}13)$$

or

$$\frac{E_o}{E_{in}} = \frac{A}{1 + AB} \qquad (2\text{-}14)$$

Equation (2-14) is used for calculating the system's gain *employing negative feedback*. Notice that the only difference between Eq. (2-14) and Eq. (2-9) is the sign in the denominator. Be sure not to confuse them.

Now that we have defined positive and negative feedback both verbally and mathematically, let's go back to our discussion of the audio amplifier in Section 2-5 concerning the use of feedback in controlling the phase angle shift between the amplifier's output and input signals. By using negative feedback, we can nullify or cancel out any phase shift tendencies in our amplifier by introducing this feedback at the proper instances. This is precisely what is done in hi-fi amplifiers to give them the broad frequency responses necessary to faithfully reproduce the sounds of a record, tape, or compact disk system. Figure 2-11 shows the results of our redesigned amplifier now using negative feedback. Compare these curves with those shown in Figure 2-7. A notable feature of Figure 2-11 is the two humps

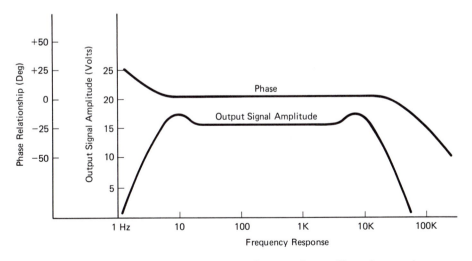

Figure 2-11 Frequency response curves for an audio amplifier using negative feedback.

occurring in the gain curve. Because of the negative feedback circuitry added to the amplifier, it operates as negative feedback circuitry only when the out-of-phase relationship is not too severe between the amplifier's input and output. This is certainly the case for the majority of the flat response region of the gain curve. However, near its ends, as we already discussed, the phase shifting becomes much greater. It becomes so much greater in fact that this circuitry behaves momentarily like a positive feedback system resulting in an actual gain increase for these frequencies. This characteristic may be more fully understood and better appreciated after studying Chapter 7.

2-7 THE BLOCK DIAGRAM

Because electronic and mechanical circuit systems can become rather complicated and difficult to draw, it is often necessary to use a shorthand method of depicting these systems in illustrations. Blocks are used to represent the various parts of the system and to simplify the system's description. Figure 2-12 shows the block diagram of a closed-loop system with all the necessary labeling needed to identify the system's operation. Notice that this type of drawing is easier to draw and to understand than if a mechanical drawing of a water valve with its flow valve being controlled by a float sensor were presented instead. And notice the similarity between the block diagram shown in this figure and those shown in Section 2-4. In general, block diagraming can be used to break down any control system into the more basic components of a system similar to the one depicted in Figure 2-12. Let's now take a look at these basic elements. These are shown in Figure 2-13.

   We can define the functions of the individual blocks of Figure 2-13 in the following statements:

1. *Input components:* These are the components of the overall system that have to deal with the system's input signal. More importantly, these are the components that produce some sort of *reference* signal against which the

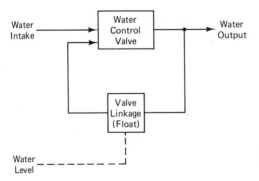

Figure 2-12   Block diagram of a simple water level controller.

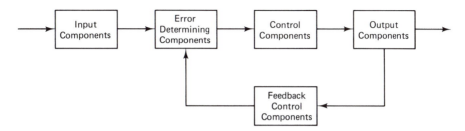

Figure 2-13 A generic block diagram for an automatic control system.

output may be compared in order to determine which way a correction must be made to correct the system's behavior.

2. *Error determining components:* These are the components that produce the *error* signal as the result of comparing the input signal to that of the system's output signal thereby resulting in either a difference signal or an additive signal that is then sent on to the control portion of the system. Don't assume that the error signal is an undesirable signal. On the contrary, the error signal is an absolute necessary constituent of an automatic control system.

3. *Control components:* These are the components that modify the error signal to put it into the desirable form for outputting. Typical control components would be voltage, current, or power amplifiers, mechanical linkages that control position, and control windings of servo devices.

4. *Output components:* These are the components that actually produce the output of the system.

5. *Feedback control components:* These are the components that produce a feedback signal that is of the same form as the input signal, but is 180° out of phase, and is a scaled version of the output signal such that the feedback signal may be combined with the input signal to produce the error signal.

2-8 THE SUMMING POINT

Refer back to Figure 2-13 for a moment. The block representing the error determining components is often referred to as *the summing point*. This is where the input signal and feedback output signals are added algebraically to produce the error signal. Figure 2-13 may be redrawn as shown in Figure 2-14 where the error determining components block has been replaced now with a summing point symbol. This is simply a circle with an *X* drawn through it with the input and output signal polarities all properly labeled. It can be generally assumed at this point that all signals marked negative are 180° out of phase with the positive-marked signals. Notice that the feedback signal entering the summing junction is negative. This is to indicate that the feedback being used in this system is nega-

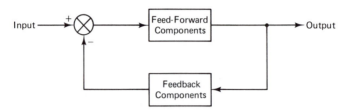

Figure 2-14   The summing point.

tive and not positive. Also, the input and output component blocks have all been consolidated into one block called the forward components block. This name is derived from the fact that the consolidated boxes all had signals flowing through them in a forward input-to-output fashion. Quite often, you will see this form of block diagramming used rather than the slightly more detailed form shown in Figure 2-13. Instead of noting in each block the type of components involved, you will often find the block's *system transfer function* written.

2-9 THE TRANSFER FUNCTION

The *transfer function* of a system is nothing more than the gain of that system. In other words, the transfer function is defined by the following equation:

$$\text{Transfer function} = \frac{\text{the system's output}}{\text{the system's input}} \tag{2-15}$$

Let's take a look at some examples. Let's determine the transfer function for an AC motor. To do this we must first identify the motor's input and output. The input would, of course, be an AC voltage. The output would most likely be measured in rpm. Therefore, its transfer function units would be rpm/AC volts. For another example, let's write the transfer function units for a water valve. The input would be an angular displacement, degrees. This would represent the angle through which the valve stem must be turned to control the desired water flow. The output would be the water flow itself, measured in gallons per minute, or gpm. Therefore, its transfer function units would be, gpm/deg.

In both cases, there most likely would be a numeric value associated with the units of the transfer function. For instance, in the case of the water valve, a 45° turn of the valve stem may produce a flowrate of 9 gpm. Consequently, its complete transfer function would be stated as 0.2 gpm/deg. We'll come back to this subject of transfer functions in a later section.

2-10 AN AUTOMATIC CONTROL SYSTEM

Up to this point, we have covered many of the basic concepts used in understanding automatic control systems. Let's now try our hand at analyzing a system just to make sure we have a good understanding of all the concepts covered so far. This is probably the best way to uncover questions and to present additional ideas that are covered later.

Assume that you and I are hot-air balloonists wanting to design a system that maintains our balloon at a constant altitude regardless of air temperature change or air draft conditions. We realize that our balloon's lift, and to a great extent its flight altitude, is determined by the difference in the balloon's air bag temperature and the temperature of the outside air. We need a system that will automatically modulate or turn the propane burner off and on when needed to maintain the proper lift or altitude. Obviously, there are other factors too that determine the balloon's altitude, such as wind direction and solar heating of the air bag, so we need a system that will compensate for all of these conditions. Figure 2-15 shows just such a system. The principle of operation is based on the aneroid barometer. This is a device that is sensitive to changes in air pressure and therefore responds to slight changes in altitude. If the aneroid mechanism senses a decrease in air pressure, it interprets this change as an increase in altitude. If it senses an increase in pressure, the altitude must, therefore, be decreasing. To get our system functioning properly, we must hook up the mechanism normally used to move the needle that produces the barometric pressure reading on its scale to the gas regulator on the propane burner. Assuming that we have done this, the system will work like this: We must first adjust the altitude input control for the desired altitude that we wish to maintain before taking off. Also, the aneroid barometer must be adjusted for the existing barometric pressure reading so that at ground level, the propane burner's valve is opened initially to create the needed take-off lift. As the balloon begins lifting off, the aneroid barometer responds to the increase in altitude by gradually shutting off the gas supply to the propane burner. The balloon's ascent is therefore slowed until it reaches the desired height. At that time, the gas supply has been completely shut off.

Let's analyze what would happen if a sudden updraft of air came along to disturb our nice, level flight path. The aneroid would respond to the sudden increase of the balloon's altitude by first noting that the burner is already shut off.

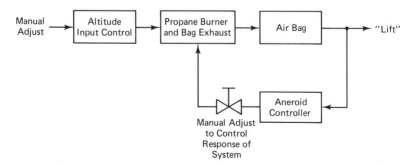

Figure 2-15 The automatic altitude controller for a hot air balloon.

Therefore, the aneroid would send a command signal to the gas bag to open its venting valve to allow hot air to escape. The balloon would then respond by losing some of its altitude. The aneroid, sensing this altitude loss and detecting that this altitude is less than the desired altitudes as set by the input control, would turn the gas burner on again to compensate for the balloon's decrease in altitude. Again, the aneroid, sensing an increase in altitude and sensing that the balloon's height is again greater than the desired preset altitude (but not as great as previously, since a partial correction was already made), opens the vent valve to let more hot air escape, and so on. Each correction would create a diminishing amount of the automatic correction needed until finally the balloon would again be at precisely the desired altitude.

Figure 2-16 is a diagram of the flight path of our balloon. After assuming level flight, notice the two possible flight response paths in the diagram as a result of an updraft. One is for our balloon if we had no automatic controls whereas the other path is for our balloon with automatic controls. Can you imagine what our flight path would look like if our automatic system had a much quicker response time to the updraft of air current? Or, what would our flight path look like if our automatic system responded very, very slowly? Figure 2-17 shows these two possibilities. As you study this figure, ask yourself these questions: What would happen to the response characteristics of our control system if we increased the

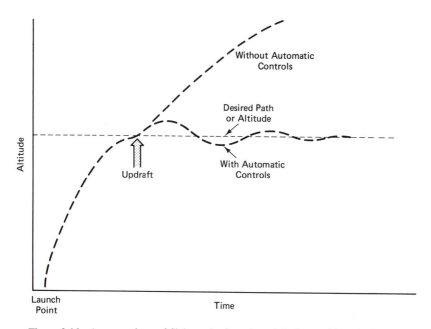

Figure 2-16 A comparison of flight paths for a hot air balloon with and without automatic altitude control.

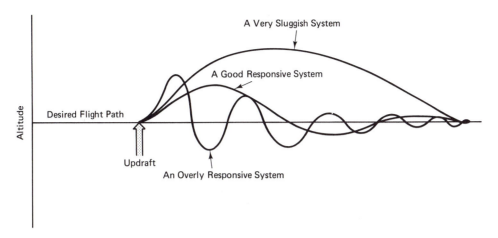

Figure 2-17   Various kinds of response characteristics for the altitude controller.

weight of our balloon or its payload? Is there a possibility that our control system would become so sluggish or so poorly coordinated due to slow response speeds and general massiveness of the balloon system itself that the gas burner and exhaust systems could become *out of phase* with their responses to updrafts or downdrafts? Obviously, it would do us no good to have a system that, because of cheap components or poor judgment in hardware design, responded so slowly to command signals that when the balloon calls for an increase in altitude, the venting valve on the air bag is still open.

As we soon find out in our study of automatic control systems, the behavior or response of the automatic altitude system of our hot air balloon is very typical of other automatic control systems in general. And to comment on the question raised in the previous paragraph, out-of-phase response is a serious problem in automatic control design.

SUMMARY

In studying the audio amplifier, such as the ones typically used in hi-fi systems, we developed an understanding of *frequency response, gain, bandwidth, and feedback.* Even though all of these terms were applied to the audio amplifier, they are common terms applied to automatic control systems in general.

Another important concept in automatic control design is *phase.* In the case of the amplifier, we found that by reducing or eliminating an out-of-phase signal response problem, we could increase the bandwidth of our amplifier. We found that this phasing problem could be affected through the introduction of negative feedback.

Positive feedback has little if any use in the design of an automatic control system; however, it is vital in the design of electronic oscillators. On the other hand, negative feedback is vital to the design of automatic control systems.

The following important equations were discussed in Chapter 2:

$$\text{Gain} = \frac{E_o}{E_{in}} \qquad \text{(Eq. 2-2)}$$

For positive feedback:
$$\frac{E_o}{E_{in}} = \frac{A}{1 - AB} \qquad \text{(Eq. 2-8)}$$

For negative feedback:
$$\frac{E_o}{E_{in}} = \frac{A}{1 + AB} \qquad \text{(Eq. 2-14)}$$

$$\text{Transfer Function} = \frac{\text{the system's output}}{\text{the system's input}} \qquad \text{(Eq. 2-15)}$$

EXERCISES

2-1. If an automatic control system has an input voltage of 15 volts, an amplifier gain of 3, and a *B* value of 0.25, find its output voltage.

2-2. Determine the amount of feedback needed to create a smoothly running oscillator having an amplifier whose gain is 25.

2-3. Design an automatic control system using a block diagram style (similar to Figure 2-15) showing how a temperature controller would be used to regulate the temperature inside a refrigerated chamber used for research applications. There are several possible methods that can be used for this design.

2-4. Write the transfer function for a gear box transmission whose output shaft rotates 100 revolutions for every 726 revolutions of its input shaft.

2-5. What is the transfer function for a hydraulic cylinder that produces a piston rod displacement of 49 inches with a supply of hydraulic fluid of 2.37 gallons?

2-6. Determine the gain of an amplifier needed in an automatic control system to produce an overall system gain of 0.9, assuming a feedback value of 0.37.

2-7. Calculate the decibel output of an amplifier having a gain of (a) 1,000 (b) −21.5 (c) 14.5 (d) 0.65.

2-8. Determine the output voltage resulting from a system having a gain of 135 dB and an input voltage of 7 VDC.

2-9. Show mathematically that a 3-dB down point is the same as a voltage gain of 0.707.

2-10. For a negative feedback system, how much feedback is needed in order to create an overall gain of 0.25 for a system assuming an amplifier gain of 20?

REFERENCES

DeRoy, Benjamin E., *Automatic Control Theory,* New York, N.Y.: John Wiley & Sons, Inc., 1966.

Miller, Richard W., *Servomechanism Devices and Fundamentals,* Reston, Va.: Reston Publishing Co., 1977.

Sante, Daniel P., *Automatic Control System Technology,* Englewood Cliffs, N.J.: Prentice-Hall, 1980.

3

Laplace Transforms

3-1 THE MATHEMATICS OF AN AUTOMATIC CONTROL SYSTEM

Perhaps a better title for this chapter would have been, "Laplace Transforms—The Painless Way of Avoiding Calculus Problems." In working with automatic control systems, it is often necessary to describe their operations using rather complicated differential equations and calculus expressions. Needless to say, for many engineers and technicians, this can be a rather somewhat laborious ordeal. Unfortunately, though, some of these math approaches defy any other methods of solving, and so in some cases no choice is left but to use calculus. For someone just starting to understand a new engineering concept, the beauty and enjoyment of understanding and working with that concept or system can be easily destroyed by becoming bogged down in the mathematical justifications. At one time or another we have all become victims of this problem.

3-2 A LITTLE BACKGROUND CONCERNING LAPLACE TRANSFORMS

About 200 years ago, Pierre Simon, Marquis de Laplace (pronounced Mar-kee Day Laplahz), developed a method of greatly simplifying the solutions of differential equations. He must have had the modern-day controls engineer in mind. (Actually, it was Oliver Heaviside who, some years later, really perfected the modern-day principles of Laplace transforms. But that's another story.)

74

3-3 HOW TRANSFORMS ARE USED

Let's see how Laplace transforms can make life so much easier for us. Think of performing the division problem, 477,123 ÷ 147, in your head. A difficult task, right? But what if, instead, we were to present the same problem in this form: $1 \times 10^{5.67863} \div 1 \times 10^{2.16732}$. Almost immediately, we see the answer as being $1 \times 10^{3.51131}$. Simply by using the laws of scientific notation, we arrive at our answer. However, to get our answer into a more familiar form, we have to determine the antilog of 3.51131, which turns out to be 3245.7. Admittedly, this job requires using log tables or a calculator.

In using Laplace transforms, the general idea is somewhat the same. You must first convert a given differential equation expression to its Laplace equivalent. Then, you would do whatever algebraic manipulations would be necessary to arrive at your solution, then reconvert the expression by converting the anti-Laplace of that expression back to its original form. Basically, you will have reduced what could have been a rather complicated calculus solution down to some very simple algebraic steps.

Laplace transforms are used in time-dependent types of expressions, that is, expressions that say $y = f(t)$. More specifically, they can be used only for solving linear differential equations whose variables are functions of time. In automatic control expressions these will, fortunately for us, occur more frequently than not.

We won't need differential equations for the work we do here. Instead, we look for certain time-related math expressions that crop up periodically in the course of performing math operations. When we spot these expressions, we go to a lookup table of Laplace transform equivalents and make the appropriate substitution. After doing that, and after performing what algebra is necessary to arrive at the form that we want, we then look up the inverse transform to get the expression back into the time domain.

Let's discuss an example here to see how this works. Look back at Figure 1-32 in Chapter 1. In this figure we see a capacitor, C, (assumed initially to be uncharged) in series with a resistor, R. Between terminals $A-B$, we apply a step voltage $e(t)$ (i.e., a voltage whose magnitude rises to a particular value *immediately* at $t = 0$). By using differential and integral calculus, we can arrive at the fact that the current, i, $= (E/R)(e^{-t/RC})$. Now note the expression $e^{-t/RC}$ in the expression just mentioned. This occurs time and again throughout discussions of systems that have transient conditions. Remember, transient conditions last only for a small period of time and then disappear. Many math operations in automatic controls involve this particular expression. As a result, things can get fairly messy during the course of performing these operations, especially when using calculus. Fortunately, this expression has been assigned a Laplace transform because of its frequent use.

The best way to appreciate all of this and to get a better foothold in understanding all that has been said so far is to work an example problem. Again, look at Figure 1-32. Let's say that we want to develop a mathematical expression that relates the capacitor's (C) charging current, $i(t)$, to the capacitor's time constant (RC). In other words, we want an expression that says: $i(t) = f(RC)$. To do this, we have to find what $f(RC)$ is.

First, we look at the involved solution. If you have no interest whatever in this solution, which will require integral and differential calculus, you can skip this part and go on to the simpler solution. Otherwise, here now is our involved solution:

EXAMPLE

Involved Solution:

The voltage,
$$e(t) = RC \frac{dv_c(t)}{dt} + v_c(t) \tag{3-1}$$

Let
$$E = e(t)$$

then
$$E = RC \frac{dv_c(t)}{dt} + v_c(t) \tag{3-2}$$

We want to find $v_c(t)$ because $v_c = E - iR$ from Kirchhoff's voltage law. And from this expression we can solve for i, which is:

$$i = \frac{E - v_c}{R}$$

which can be restated as

$$i(t) = \frac{E - v_c(t)}{R} \tag{3-3}$$

Now, to find $v_c(t)$ so that we can place it into Eq. (3-3) for a solution of $i(t)$, we want to set up Eq. (3-2) with $v_c(t)$ on one side.

$$E - v_c(t) = RC \frac{dv_c(t)}{dt}$$

And,
$$\frac{dv_c(t)}{E - v_c(t)} = \frac{1}{RC} dt$$

Integrating both sides:

$$\int \frac{dv_c(t)}{E - v_c(t)} = \int \frac{1}{RC} dt$$

or
$$Ln(E - v_c(t)) = \frac{-t}{RC} + K$$

where K is a constant of integration.

When $t = 0$, then $LnE = K$. Substituting $K = LnE$ into Eq. (3-3) will give:

$$Ln(E - v_c(t)) = \frac{-t}{RC} + LnE$$

or
$$Ln(E - v_c(t)) - LnE = \frac{-t}{RC}$$

or
$$Ln\left(\frac{E - v_c(t)}{E}\right) = \frac{-t}{RC}$$

Taking the natural antilogs of both sides results in:

$$\frac{E - v_c(t)}{E} = e^{-t/RC}$$

Solving for $v_c(t)$,

$$E - v_c(t) = Ee^{-t/RC}$$

Solving for $v_c(t)$,

$$v_c(t) = E - Ee^{-t/RC}$$
$$= E(1 - e^{-t/RC})$$

Since
$$i(t) = \frac{E - v_c(t)}{R} \quad \text{(Eq. 3-3),}$$

then
$$i(t) = \frac{E - E(1 - e^{-t/RC})}{R},$$

and finally,
$$i(t) = \frac{E(e^{-t/RC})}{R} \qquad (3\text{-}4)$$

Now, the simpler solution for the same problem:

EXAMPLE

Simpler Solution:

Remember, we are trying to find the expression for $i(t)$, the charging current for capacitor, C. First, we note that the reactance for C is:

$$X_c = \frac{1}{2\pi fC} \quad \text{or} \quad \frac{1}{j\omega C}$$

We also know that Z_T (the total circuit impedance as measured across terminals A–B in Figure 1-32) is

$$Z_T = R + \frac{1}{j\omega C}$$

Going to a table of Laplace transforms (Table 3-1), we see that $j\omega$ has a transform, s. Therefore,

$$Z_T = R + \frac{1}{sC} \qquad (3\text{-}5)$$

According to this same table, E (our voltage step) = E/s. And since $E = IZ$, we can therefore state that $E(s)$, which is E/s in our case, = $I(s) \times Z(s)$. This means that E/s = a Laplace transform expressing impedance as a function of s. In other words:

$$I(s) = \frac{E(s)}{Z(s)}$$

and since $E(s) = E/s$ and, referring to Eq. (3-5), $Z(s) = R + 1/sC$, then

$$I(s) = \cfrac{\dfrac{E}{s}}{R + \dfrac{1}{sC}} = \cfrac{\dfrac{E}{s}}{\dfrac{sRc + 1}{sC}} = \frac{E\,C}{(sRC + 1)}$$

Now, multiply the far right equation by R/R, which of course is 1, to get:

$$\frac{R}{R} \cdot \frac{E\,C}{sRC + 1} = \frac{E\,R\,C}{R(sRC + 1)}$$

Since $RC = \tau$, then

$$I(s) = \frac{E\,\tau}{R(s\tau + 1)} \tag{3-6}$$

Now, to find the time equivalent of Eq. (3-6), we have to determine its inverse transform. Referring again to Table 3-1, we find from pair 6 that the inverse of $1/(s\tau + 1)$ is $(1/\tau)e^{-t/\tau}$. Notice that $E\tau/R$ is a constant. Constants multiplied by time functions are not changed by transformations, and vice versa. And since $I(s)$ transforms back to $i(t)$, we finally have:

$$i(t) = \frac{E\,\tau}{R}\frac{1}{\tau}\,e^{-t/\tau}$$

or

$$i(t) = \frac{E}{R}\,(e^{-t/\tau}) \tag{3-7}$$

Equation (3-7) checks with Eq. (3-4), which was obtained using the involved method.

Admittedly, at this point it's probably difficult to see how using Laplace transforms is any less complicated or involved than using the calculus approach. But, once you become familiar with the Laplace method, you'll begin to see its benefits.

3-4 TRANSFORM TABLES

Now that we have demonstrated the using of transforms showing the possibility of making our math somewhat simpler, let's take a closer look at the other transforms listed in Table 3-1. You will probably find a number of familiar looking expressions that will have a not-so-familiar Laplace equivalent transform. After a

TABLE 3-1   LAPLACE TRANSFORM PAIRS

Pair no.	$f(t)$	$F(s)$	Comments
1.	I or E	$\dfrac{I}{s}, \dfrac{E}{s}$	A step input of magnitude I or E.
2.	1	$\dfrac{1}{s}$	A unit step input.
3.	At	$\dfrac{A}{s^2}$	A ramp function.
4.	At^2	$\dfrac{2A}{s^3}$	A parabola.
5.	Ae^{-at}	$\dfrac{A}{s+a}$	
6.	$\dfrac{Ae^{-t/\tau}}{\tau}$	$\dfrac{A}{\tau s + 1}$	General response of first-order system.
7.	$A(1 - e^{-t/\tau})$	$\dfrac{A}{s(\tau s + 1)}$	First-order response to a step input.
8.	$\dfrac{A(1 - e^{-at})}{a}$	$\dfrac{A}{s(s+a)}$	
9.	$A\sin\omega t$	$\dfrac{A\omega}{s^2 + \omega^2}$	A sine function.
10.	$A\cos\omega t$	$\dfrac{As}{s^2 + \omega^2}$	A cosine function.
11.	ω	$s\theta$	
12.	$j\omega$	s	
13.	α	$s^2\theta$	
14.	$\dfrac{A\omega_n e^{-z\omega_n t}}{\sqrt{1 - z^2}} \sin(\omega_n \sqrt{1 - z^2})$ $\dfrac{A\omega_n^2}{s^2 + 2z\omega_n s + \omega_n^2}$		Second-order system, general response where $z < 1$.
15.	$\dfrac{Ate^{-t/\tau}}{\tau^2}$	$\dfrac{A}{(\tau s + 1)^2}$	General response of second-order system ($z = 1$).
16.	Ate^{-at}	$\dfrac{A}{(s+a)^2}$	
17.	$1 + \dfrac{e^{-z\omega_n t}}{\sqrt{1 - z^2}} \sin\left(\omega_n \sqrt{1 - z^2}\, t - Arc\, tan\, \dfrac{\sqrt{1 - z^2}}{-z}\right)$ $\dfrac{\omega^2}{s(s^2 + 2z\omega_n s + \omega_n^2)}$		Second-order system response to a unit step input ($z < 1$).

$$\text{where} \quad 0° < Arc\, tan\, \frac{\sqrt{1 - z^2}}{-z} < 180°$$

while, however, you will feel just as at home with the Laplace equivalent as you will with the now more familiar time domain expressions.

Because automatic controls deal with system characteristics that can be short-lived or transient in nature, the mathematical expressions for these transient conditions are seen as decaying functions. The charging capacitor is a good example. The equation for this charging condition is seen in Figure 1-32. As you look through the transforms in Table 3-1 and their time domain counterparts, notice how often the expression, e^{-Kt}, occurs. This should tell you how important that particular exponential function is to automatic control systems and how often it occurs.

The derivations of all the Laplace transforms follow three basic steps:

1. Choose the time domain expression that you want to convert to a Laplace transform. In other words, this expression must be in the form of $y = f(t)$.
2. Multiply this expression by a factor, e^{-st}. This is a factor that decreases or decays to zero as time becomes infinitely large.
3. Integrate this new expression with respect to time between the limits of zero and infinity.

Mathematically, these three steps can be expressed in the following way:

$$F(s) \equiv \pounds[f(t)] \equiv \int_0^\infty f(t)e^{-st} \cdot dt$$

where $F(s)$ = the Laplace transform expression
$\pounds[f(t)]$ = to be read as "take the Laplace transform of the time domain expression. . . ."
$\equiv$ = to be read as "defined as. . . ." or "the same thing as. . . ."

But again, we really won't have to concern ourselves with the preceding three rules, or with the preceding mathematical expression for deriving our transforms. This has already been done for us. The information has been given here only to complete our discussion of transforms. Hopefully, you may become interested in the mathematical proofs at a later time, since there is no better way to understand this material. If so, the seed has been planted here for that study.

While we are defining the various Laplace math symbols that we use soon, another symbol is $\pounds^{-1}[F(s)]$. This means, "take the inverse transform of the Laplace expression. . . ." Therefore,

$$\pounds^{-1}[F(s)] \equiv f(t)$$

Table 3-2 shows the various Laplace mathematical operations and their time domain equivalents. Looking at this table, the only differences in the math operations in the two systems are differentiation and integration. Since we plan not to do either, we won't worry about them. These two operations are presented here only to offer a complete table of math operation equivalents.

TABLE 3-2   EQUIVALENT MATH OPERATIONS USING LAPLACE TRANSFORMS

Math operation	$f(t)$	$F(s)$
Addition of two functions	$f_1(t) + f_2(t)$	$F_1(s) + F_2(s)$
Subtraction of two functions	$f_1(t) - f_2(t)$	$F_1(s) - F_2(s)$
Multiplying a function by a constant	$f_1(t) \cdot K$	$F(s) \cdot K$
Multiplying two functions together	$f_1(t) \cdot f_2(t)$	$F_1(s) \cdot F_2(s)$
Dividing a function by a constant	$\dfrac{f_1(t)}{K}$	$\dfrac{f_1(s)}{K}$
Dividing two functions	$\dfrac{f_1(t)}{f_2(t)}$	$\dfrac{f_1(s)}{f_2(s)}$
Differentiation	$\dfrac{df(t)}{dt}$	$s \cdot F(s)$
Integration	$\int f(t)\, dt$	$\dfrac{1}{s} \cdot F(s)$

3-5 EXAMPLES USING TRANSFORMS

The only good way to fully understand and appreciate the power of the Laplace transform is to become involved in some controls-type problems. At this point, however, we still need to digest some additional theory. So, instead, let's look at some basic problems involving transforming an expression from one domain to the other.

EXAMPLE 3-1

Transform $f(t) = 25$ to a Laplace transform. Note that 25 could be a step input to a circuit having a magnitude of 25 volts or 25 amps, etc.

Solution:

Referring to Table 3-1, we see that pair 1 appears to fit the description of the given problem. Therefore,

$$F(s) = \frac{25}{s}$$

EXAMPLE 3-2

Transform $1/(s + 5)$ into an $f(t)$ expression.

Solution:

$$F(s) = \frac{1}{(s + 5)}$$

Referring to Table 3-1, we see that pair 5 appears to fit the wanted description, where $A = 1$ and $a = 5$. Therefore,

$$f(t) = (1)e^{-5t}$$

EXAMPLE 3-3

Find $\pounds[f(t)] = 10 + e^{-t}$

Solution:

Noting that this is an addition problem of two $F(s)$ expressions, we have to refer to Table 3-2 to see that $f_1(t) + f_2(t) = F_1(s) + F_2(s)$. Therefore, referring to pairs 1 and 5, we have:

$$\pounds[f(t)] = \frac{10}{s} + \frac{1}{s+1}$$

EXAMPLE 3-4

Find $\pounds^{-1}[4/(s^2 + 16)]$.

Solution:

Look at pair 9 in Table 3-1. (We discarded pair 10 as a possibility since there was no s-term in the given function's numerator.) Looking at the given function, we note that the ω-term in the denominator is a perfect square of the ω in the numerator. This is what pair 9 requires. Also, we can say that $A = 1$. Therefore,

$$\pounds^{-1}[F(s)] = sin4t$$

EXAMPLE 3-5

Find $F(s)$ for the function $f(t) = 12t$.

Solution:

Pair 3 shows that for $f(t) = A \cdot t$, $F(s) = A/s^2$. Therefore, with $A = 12$, $F(s) = 12/s^2$.

EXAMPLE 3-6

Find $F(s)$ for the function $f(t) = 50(1 - e^{-100t})$.

Solution:

Look at pair 8. Notice that A *appears* to be 50 in the given function, but we must have an a under the 50 (which is 100 in our function) in order to use this pair. So, our only recourse is to force a 100 under the 50. But to do this and not change the value of the function itself, we must multiply the function by 100. Therefore, the result is:

$$f(t) = 50(1 - e^{-100t}) = \frac{100}{100} \times 50(1 - e^{-100t})$$

$$= \frac{5{,}000}{100}(1 - e^{-100t})$$

We now see that *A* actually = 5,000 and *a* = 100, and we can now proceed with using the *F*(*s*) conversion given in pair 8, which is:

$$\frac{5{,}000}{s(s + 100)}$$

EXAMPLE 3-7

Given that $f(t) = (5t/36)e^{-t/6}$, find $F(s)$.

Solution:

Look at pair 15. Notice that the 6 in the given exponent is a square root of the figure in the denominator of *At*/36. Therefore, let 6 = τ. Also, let *A* = 5. Then, *F*(*s*) = $5/(6s + 1)^2$.

SUMMARY

Laplace transforms represent a straightforward alternative mathematical approach to automatic control system analysis. (We investigate this further in the coming chapters.) The Laplace method allows us to convert a time domain $f(t)$ expression into the so-called *s*-plane domain or $f(s)$ expression, and then back again. Table 3-1 lists the most commonly used conversions that are used in this book.

EXERCISES

Given the function $f(t)$, find $F(s)$ using Tables 3-1 and 3-2.

3-1. $f(t) = 107$

3-2. $f(t) = 20t^2$

3-3. $f(t) = 3t^2 + 9t - 2$

3-4. $f(t) = \frac{25}{10}(1 - e^{-10t})$

3-5. $f(t) = 2.5(1 - e^{-10t})$

3-6. $f(t) = 2t^2 + 6(1 - e^{-34t})$

3-7. $f(t) = 5e^{-t/4}$

3-8. $f(t) = 1.4e^{-14t}$

3-9. $f(t) = (1 - e^{-t/3})$

3-10. $f(t) = (2.9e^{-5t})(2e^{-t/4})$

3-11. $f(t) = 3.2\sin\omega t$

3-12. $f(t) = 5\cos 144t$ Hint: What does $\omega = $?

3-13. $f(t) = 2te^{-0.5t}$

3-14. $t(t) = 8(e^{-t/6} - e^{-t/12})$

In the following expressions, find $f(t)$ for the given $F(s)$ functions:

3-15. $F(s) = \dfrac{22}{(s + 1)^2}$

3-16. $F(s) = \dfrac{5}{s(0.02s + 1)}$ Hint: What is the value of τ?

3-17. $F(s) = \dfrac{6.2}{s + 3}$

3-18. $F(s) = \dfrac{23.5\omega}{s^2 + \omega^3}$

3-19. $F(s) = \dfrac{100}{s^2 + 1,000}$

3-20. Using pair 14, determine the value of ω_n and z in the following expression:

$$F(s) = \frac{3.96}{0.2s^2 + s + 3.96}$$

REFERENCES

KUHFITTIG, PETER K. F., *Basic Technical Mathematics with Calculus,* pp. 921–34. Monterey, Calif.: Brooks/Cole Publishing Co., 1984.

RICE, BERNARD J. and JERRY D. STRANGE, *Technical Mathematics and Calculus,* pp. 827–38. Boston, Mass. Prindle, Weber & Schmidt Pub., 1983.

WASHINGTON, ALLYN J., *Basic Technical Mathematics with Calculus,* (4th. ed), pp. 933–45. Menlo Park, Calif.: The Benjamin/Cummings Publishing Co., 1985.

4

Transducers and Control System Components

4-1 *DEFINING TRANSDUCERS*

In general, a transducer is a device that transforms one form of energy into another. A transducer is generally thought of as a device that is relatively small; that is, it can usually be held in the hand, and the amount of energy being transformed is usually quite small. In addition, the conversion process between the input and output is done quantitatively. That is, by knowing the input quantity, one can predict the amount of output based on a predetermined calibration factor, or set of factors. (A furnace is not usually considered a transducer even though it is an energy converter. This is because of its sheer size and because its output has not been precisely correlated with its input as in the case of a transducer.) The kinds of transducer devices that we discuss here can be hand-held for the most part. They can change quantities such as heat, light, motion, etc. into a form of electrical energy. More specifically, each transducer is specifically designed to change a particular quantity into a small current or voltage signal that will be proportional in magnitude in some way to the quantity being sensed.

In automatic control systems, transducers are used extensively to sense position, motion, or some other quantity. This information is then transmitted in the form of an electrically equivalent signal to some sort of control circuitry for the purpose of producing the desired response (Figure 4-1).

All transducers contain some sort of sensing element or device. This is the portion of a transducer that actually does the transducing. Figure 4-2 shows, schematically, the difference between the sensing element of a transducer and the transducer itself. It is important to understand the difference between these two

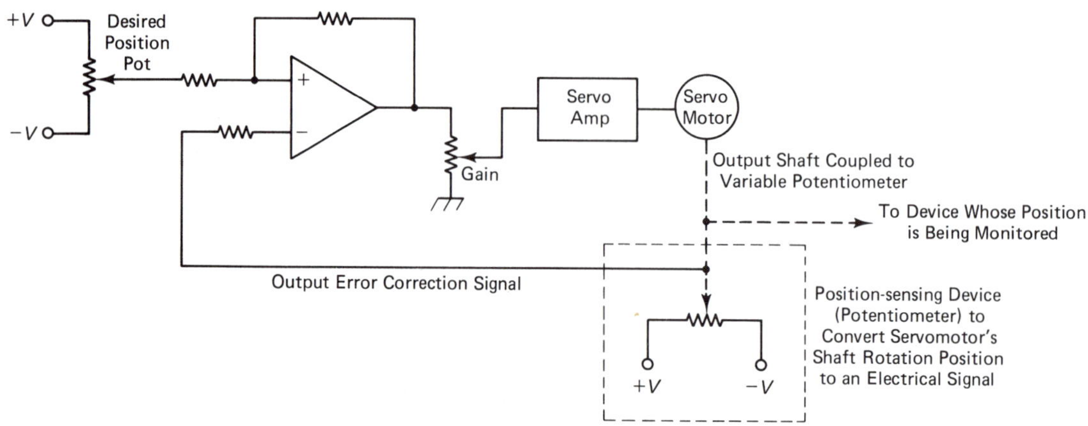

Figure 4-1 A position controller.

terms. Each has its own particular transfer function. The units of measurement may be the same for both, but the magnitude of the transfer function may be radically different. And it is the transducer's transfer function that we are primarily interested in. Admittedly, up to this point we have developed little understanding of the importance of transfer functions in automatic controls, much less developed an understanding of what an automatic control circuit is. This will come later. For the present, however, we must become familiar with the transfer functions of the various devices discussed in this chapter, and then, as we get into the heart of automatic control theory, the purpose of this type of function will become clear.

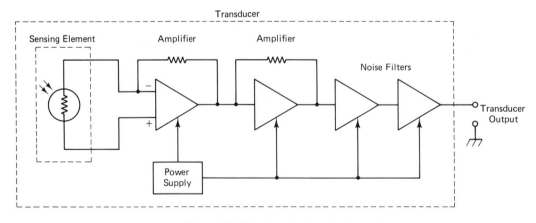

Figure 4-2 The circuit of a typical transducer.

4-2 POSITION SENSING

An example of position sensing in an automatic control system is found in the application of positioning a radar antenna. The operator, sitting inside a control bunker, has the ability to remotely control the positioning of his or her radar dish. If a gust of wind comes along to disturb that position, the dish's rotation control circuitry (containing positional transducers in the dish's mount) will sense the disturbance and counteract the movement by causing the dish to move back to its original position. This is done automatically without the operator's intervention (Figure 4-3).

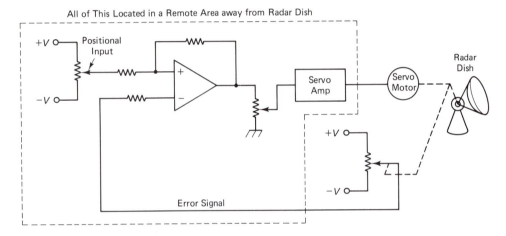

Figure 4-3   A radar's antenna positional control system.

4-2.1 The Potentiometer

One of the most frequently used devices used for sensing position is the potentiometer. Potentiometers, or "pots," are also called variable resistors. When used for automatic control system applications, these pots are generally of the precision type. That is, the resistive element inside the pot is constructed of a durable temperature-stable material that has been formulated to create the same resistive values over a long period of time. The resistances measured at the pot's output terminals are usually very repeatable to within a very small percentage.

There are two major types of pot configurations used in automatic control circuits. These are shown in schematic form in Figure 4-4, (A) and (B). Schematics (C) and (D) in that same figure are the actual schematic representations used for either configuration shown in (A) and (B). However, (D) is preferred. Notice that (C) doesn't show any output connection for the wiper. This information is

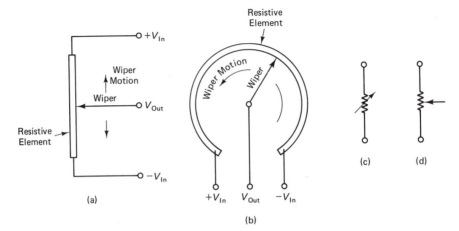

Figure 4-4 Potentiometer configurations.
A. Linear pot
B. Rotary pot
C. Electrical symbol
D. Alternate electrical symbol

vital in some automatic control system circuitry. Figures 4-5 and 4-6 illustrate typical linear-constructed and circular-constructed pots.

Referring again to Figure 4-4, let's now analyze the pot in order to determine its transfer function. Looking at (A), we can see that the input quantity to the wiper is a displacement. This is depicted in the wiper's drawing by the label *wiper*

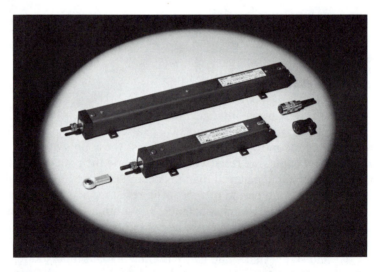

Figure 4-5 A linear precision pot. (*Courtesy of Waters Manufacturing, Inc., Wayland, Mass.*)

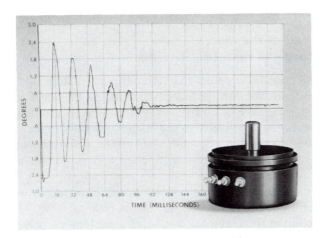

Figure 4-6 A circular precision pot and transducer. (*Courtesy of Waters Manufacturing, Inc., Wayland, Mass.*)

motion. The output from the pot is a variable voltage, labeled E_{out}, and is measured between the wiper and one of the two ends of the resistive element. Consequently, this pot's transfer function would be:

$$\frac{E_{out}}{l} = K_p \frac{volts}{inch} \tag{4-1}$$

where K_p = gain or transfer function value of pot (volts/in.)
$\quad E_{out}$ = output voltage (AC or DC volts)
$\quad\quad l$ = displacement of wiper (inches, or any other convenient unit of displacement)

EXAMPLE 4-1

Find the value of K_p for a linear pot whose output shaft must be moved a distance of 5.87 inches in order to create an output voltage of 12 volts DC for a positioning circuit.

Solution:

According to Eq. (4-1), the pot's transfer function value is calculated by the following method:

$$K_p = \frac{12 \text{ VDC}}{5.87 \text{ in.}}$$
$$= 2.044 \text{ volts/in.}$$

The transfer function for the circular pot (Figure 4-4, Schematic B) is calculated similarly. In this case, however, since the input displacement to the wiper is circular, the K_p value would be:

$$\frac{E_{out}}{\theta} = K_p \frac{volts}{deg} \tag{4-2}$$

where K_p = gain or transfer function value of pot (volts/θ)

$\quad$ E_{out} = output voltage (AC or DC volts)

$\quad\quad$ θ = angular displacement of wiper (deg. or rad.)

There are other types of transducers other than the pot that can be used for sensing displacement or position. The pot, however, is probably the most frequently used and is the least expensive of all devices. As you become familiar with other types of transducers, you will be able to see other methods of detecting position other than just the one mentioned here. Refer also to the discussions on electromagnetic sensing (Section 4-7), light sensing (Section 4-9), and resolvers (Section 4-13).

One of the greatest disadvantages in using pots for position sensing is that their mechanical construction frequently prohibits them from rotating a full 360°. This can be a serious restriction in some applications using an automatic control design. In addition, pots tend to wear out more quickly than do some other types of position transducers.

4-2.2 The LVDT

LVDT stands for *linear variable differential transformer.* This type of transducer is used for sensing position and is noted for its precision and accuracy for performing this function.

The LVDT works on the principle of transformer action. The location of a moveable plunger, acting as a transformer core, determines the amount or magnitude of signal coupled to the two secondary windings of the LVDT. Also, the plunger's position determines the phase of the secondary signal. This phase is determined relative to the primary signal. The secondary signal may either be in phase for one direction of travel, or it may be shifted 180° out of phase for an opposite travel direction, depending on the core's relative position with its center of travel. The transfer function for an LVDT is:

$$\frac{E_{\text{out}}}{l_{\text{in}}} = K_{LVDT} \frac{\text{volts}}{\text{inch}} \tag{4-3}$$

where K_{LVDT} = the sensor's gain or transfer function value for LVDT (volts/in.)

$\quad$ E_{out} = output voltage (either AC or DC volts)

$\quad$ l_{in} = input displacement of core (inches)

4-2.3 The Linear and Rotary Photocell Encoder

The linear photocell encoder consists of a photocell and light source that has a transparent strip or disk with opaque closely spaced bars which are passed between the photocell and light source. Displacements as small as 0.0001 inch can be detected. The transfer functions for these devices are similar to those of the linear and rotary pots described in Section 4-2. (See equations 4-1 and 4-2.)

4-3 VELOCITY SENSING

Many control systems require having to know the linear or rotational velocity of certain components for the system's proper operation. An example might be the cruise control system on an automobile as it travels down a highway. Once the desired speed of the car is set by the driver, the system must be able to measure the car's forward velocity in order to make corrections due to wind velocity changes and changes in the slope of the highway.

There are numerous methods available for measuring velocity. We discuss a few methods here and, after having done this, you will undoubtedly see other ways of making this kind of determination.

Quite often, it is possible to use a positional type of transducer, such as the precision pot discussed earlier, and by making a positional measurement during a measured length of time, the velocity becomes known. Let's investigate an example here. Refer to Figure 4-7. We see a linear measuring pot being used, similar to the ones shown in Figure 4-5, to measure the velocity of a rod and piston on an hydraulic cylinder. The change of DC voltage per unit of time can be directly correlated to the input velocity of the cylinder's follower attached to the pot's wiper arm. This is the purpose of the time base in Figure 4-7. This is an electronic circuit that produces a signal that is proportional to the distance traveled per unit time. A clocking circuit periodically samples the distance voltage coming from the linear pot, each sample being measured over a precisely known length of time. The output of the time base circuitry may be in the form of a variable amplitude DC voltage, or it could be a variable frequency AC waveform. The transfer function for this transducer device would then be:

$$\frac{E_{\text{out}}}{\text{vel}} = K_p \frac{\text{volts}}{\text{ft/sec}} \tag{4-4}$$

where K_p = sensor constant or transfer function value for resistive-type velocity sensor. Must be obtained from manufacturer or experimentally derived.

E_{out} = output voltage of transducer (AC or DC volts)

vel = input velocity of wiper arm (ft/sec)

Figure 4-8 shows a typical velocity measuring transducer used for determining linear velocity. It can also be used for rotational velocity when used with the proper mechanical rotary-to-linear motion conversion techniques, as is often done. The sensor that it uses is a photocell device that detects the passage of light- and dark-spaced marks on a rotating disk. The photocell in turn produces either a voltage or no voltage signal, depending on the light level that it senses, and sends this pulsating voltage level to a signal conditioning circuit. This circuit then converts the frequency of the pulses into a proportionally varying DC voltage level whose magnitude represents the velocity amount. The transfer function for this transducer would be determined by:

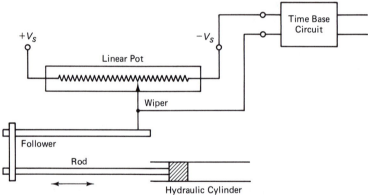

Figure 4-7 Measuring velocity of a hydraulic cylinder using a linear pot. (*Photo Courtesy of Waters Manufacturing, Inc., Wayland, Mass.*)

$$\frac{E_{\text{out}}}{\text{vel}} = \frac{K_{\text{photo}}}{\tau s + 1} \frac{\text{volts}}{\text{ft/sec}} \text{ (for a linear transducer)}$$

$$\frac{\text{volts}}{\text{rad/sec}} \text{ or } \frac{\text{volts}}{\text{deg/sec}} \text{ (for a rotary encoder)} \qquad (4\text{-}5)$$

where K_{photo} = sensor gain or transfer function value for velocity photocell trans-
        ducer. (volts/ft/sec, volts/rad/sec, or volts/deg/sec)
E_{out} = voltage output (DCV)
        vel = input velocity (ft/sec, rad/sec, or volts/deg/sec)

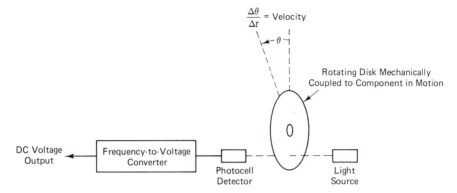

Figure 4-8   A velocity sensing transducer.

s = Laplace transform for $j\omega$

τ = time constant of photocell (sec)

Note that in the preceding transfer function you see the Laplace transformation expression, $(\tau s + 1)$. This is a transform associated with many sensing elements or transducers whose outputs are not instantaneous, but rather gradual in that they require time to build up their outputs to their final values. The characteristic of this response is similar to the R–C time constant discussed in Chapter 1. The mathematical expression of this characteristic is the familiar $e^{-t/\tau}$ function seen repeatedly in the Laplace transform Table 3-1 in Chapter 3. The theory of operation of the photocell is explained later in Section 4-9.

4-3.1 The Rotary Magnetic Encoder

The circuit in Figure 4-8 can be modified to use a magnetic sensing device to detect rotation of the disk rather than using a photocell. The disk itself would of course have to be modified so that the magnetic sensor could sense its degree of rotation. Often, what is done is to use a many-toothed gear made of ferromagnetic material such as iron. The magnetic sensor, sensing a change in the magnetic field that it creates, as the result of a tooth or any other protrusion sweeping past it, generates a pulse of DC voltage. The frequency of this pulse would of course depend on the rotational velocity of the disk or gear. As in the case of the photodetector circuit in Figure 4-8, the frequency of the pulses could then be converted to a variable DC voltage output. The transfer function for this system would be identical to that of the photo-detector device, except perhaps for the magnitude of the constant, K.

Another common method used for measuring velocity is the rate generator. This is basically nothing more than a DC voltage generator whose output voltage is directly proportional to the input velocity of its rotating shaft. The principle of

operation and the method of calculating its transfer function is discussed in detail
in Section 4-11.

4-4 PRESSURE SENSING

The sensing of fluid pressure, whether that fluid is a liquid or a gas, can be done
using similar transducer methods. Perhaps the most commonly used method for
sensing pressure is through the usage of strain gages. The principle of operation
of a strain gage rests in the fact that if one were to stretch an electrical conductor
such as a wire, that wire's DC resistance would increase proportionally to the
applied stress. Very small changes in the wire's resistance can be detected by
arranging the wire in a Wheatstone Bridge configuration, seen in Figure 4-9. The
resistor, R_3, represents the stressed wire, whereas the other three resistors, R_1,
R_2, and R_4 are precisely known and temperature-stable resistors. Often, R_4 is
another strain gage identical to the strain gage at R_1. However, unlike R_1, R_4 does
not receive any stress. Its sole purpose is to compensate, that is, nullify, any
changes in temperature occurring to the stressed strain gage. It can do this because of the current-balancing characteristics of the Wheatstone Bridge and the
locations within the bridge of R_1 and R_4. Being sensitive to temperature is one of
the unfortunate characteristics of a strain gage. The unstressed strain gage at R_4 is
called a dummy gage.

A voltage source, V, supplies a voltage to the bridge. By design, when R_3 is
unstressed by no pressure being present on the transducer's pressure diaphragm,

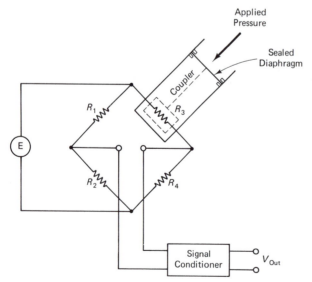

Figure 4-9   A pressure sensing transducer.

the opposing currents flowing through the bridge's central leg are equal and are therefore nullified. Consequently, no current at all flows under this condition. However, as soon as a pressure occurs at the transducer's diaphragm, R_3 becomes stressed, thereby upsetting the balanced condition of the bridge and causing current to flow through the indicator in the bridge's central leg. The amount of current will be proportional to the amount of pressure being applied. This resulting imbalance of current is then transmitted to a signal conditioning circuit that converts the current into a designed signal form such as a variable DC voltage. The transducer's transfer function becomes:

$$\frac{E_{\text{out}}}{P_{\text{in}}} = \frac{K_{pr}}{1 + s\tau} \frac{\text{volts}}{\text{lbs/in}^2} \qquad (4\text{-}6)$$

where K_{pr} = sensor constant or transfer function value for pressure transducer. Must be obtained from the manufacturer or experimentally obtained.

$\quad\quad E_{\text{out}}$ = output voltage of transducer (DCV)
$\quad\quad P_{\text{in}}$ = input pressure to transducer (lbs/in²)
$\quad\quad\quad s$ = Laplace transform for $j\omega$
$\quad\quad\quad \tau$ = system time constant (sec)

4-5 SOUND SENSING

The sensing of sound involves using one of the more frequently used transducers, as in the use of the microphone. Admittedly, this form of transducing is not all that often used in automatic control applications, but nevertheless, it is discussed here to round out the discussion on transducers.

Before we begin this discussion, however, we first must understand what sound waves are and how their intensities are measured. Sound waves are nothing more than pressure fronts emitted from a sound source. The pitch of the sound is determined by the frequency of the sound waves, that is, the number of pressure fronts passing a particular point each second of time. The sound's intensity is measured in units of the *decibel*.

There are many ways to convert sounds into electrical signals, so we start with one of the simplest and earliest methods used in sound sensing. Figure 4-10 shows the construction of the carbon-type microphone. This was the earliest type of device used to convert sound into electrical signals. The principle of operation is quite simple and ingenious. A diaphragm was fastened rigidly to a flexible container holding dried carbon granules. The diaphragm was free to vibrate whenever struck with sound waves. This vibration was passed on to the carbon granule container where the granules would be alternately compressed and

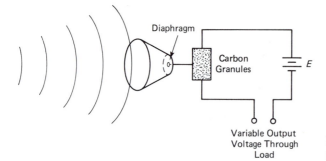

Figure 4-10   The carbon button micro-phone.

allowed to decompress with each impinging wave front on the diaphragm. A small electrical signal was passed through the granules so that the electrical current would experience a proportional increase and decrease in electrical resistance depending on whether the granules were being compressed or decompressed at that instant. This would result in a varying voltage or current that would result in a fair representation of the frequency and intensity of the sound picked up by the microphone.

Figure 4-11 shows another type of microphone. This one uses a *piezoelectric* substance (see Table 4-1) to convert sound waves into a varying output volt-

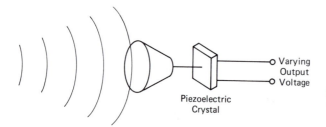

Figure 4-11   The piezoelectric or "crystal" microphone.

age. When a piezoelectric substance experiences an applied force, an output voltage is generated. The sensitivity (i.e., the ability of the substance to generate a voltage for a given input of force) is often stated in coulombs/newton, the coulomb being a unit of electrical charge, whereas the newton is a unit of force. In the case of a piezoelectric microphone, sound waves are allowed to strike a diaphragm that alternately compresses and decompresses an attached piezoelectric crystal. Table 4-1 shows the varying output sensitivities of various and often-used materials in microphone construction.

Figures 4-12 and 4-13 show two additional microphone types that both per-form very much the same way as the one just described, but utilizing different

TABLE 4-1   COMMONLY USED PIEZOELECTRIC MATERIALS[1]

Material type	Sensitivity (coulomb/newton)
Quartz	2.3×10^{-12}
Tourmaline	1.9×10^{-12}
Rochelle salts	550.0×10^{-12}
Ammonium dihydrogen phosphate	48.0×10^{-12}
Lithium sulfate	16.0×10^{-12}

[1] Richard L. Allen and Bob R. Hunter, *Transducers,* (Albany, N.Y.: Delmar Publishers, 1972), p. 35.

processes. The microphone in Figure 4-12 contains a capacitor whose one plate is attached to the vibrating diaphragm. This creates a varying capacitance between the two plates. A voltage is supplied to the capacitor to allow the charge to vary.

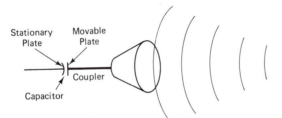

Figure 4-12   The capacitive microphone.

On the other hand, Figure 4-13 shows a microphone that generates a proportional voltage because of a magnet and coil combination. A magnet is attached to the diaphragm and is allowed to vibrate back and forth inside a coil. The lines of magnetic flux sweeping past the coiled turns then generate a voltage that is proportional to the amount of vibration. This voltage, in turn, will be proportional to the arriving sound waves at the microphone.

One of the more recent innovations utilizing sound waves in an automatic control circuit is the automatic focusing unit used in certain instant cameras.

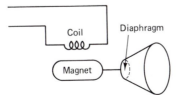

Figure 4-13   The "dynamic" or electromagnetic microphone.

Figure 4-14 shows this device along with its circuitry. An ultra-high-frequency sound pulse is radiated from the camera's sound transmitting circuitry toward the subject. An echo is detected by a sensitive microphone located back on the camera and, after timing the interval between pulses, the camera's circuit computes the subject's distance and then automatically adjusts the camera's optics for that distance.

The one thing that all these microphones have in common with each other, aside from the fact that they all have the capability to sense sound, is that their transfer functions are all alike. Their transfer function is:

$$\frac{E_{\text{out}}}{P_{dB}} = \frac{K_{\text{sonic}}}{1 + s\tau} \tag{4-7}$$

where K_{sonic} = sound detector constant or transfer function value. Must be obtained from manufacturer or experimentally obtained.

Figure 4-14 Camera with automatic sonic focusing. (*Courtesy of Polaroid Corporation, Cambridge, Mass.*)

E_{out} = DC output voltage (volts)
P_{dB} = input sound pressure level (dB)
s = Laplace transform for $j\omega$
τ = Sound detector's time constant (sec)

4-6 FLOWRATE SENSING

There are many schemes that are used for sensing the flow-rates of gases and liquids. A common method is to immerse a paddle-wheel-like mechanism into the fluid and to allow the flowing liquid to turn the paddle wheel much like a water wheel on a mill house or a windmill. These devices are surprisingly accurate. Figure 4-15 shows one such device along with its schematic. Notice that it uses a

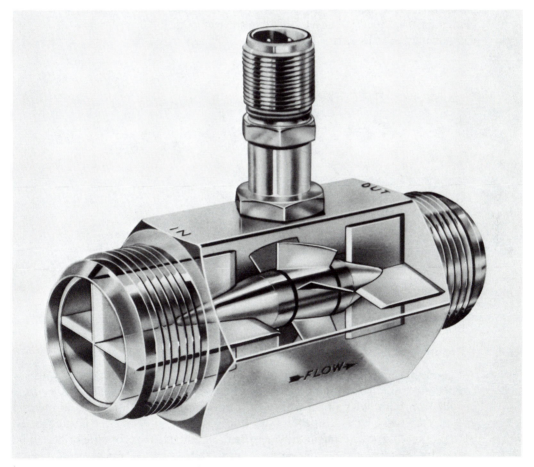

Figure 4-15 A flowrate transducer for sensing liquid flows in a pipe. (*Courtesy of Flow Technology, Inc., Phoenix, Ariz.*)

magnetic pick-up device that senses the passage of an individual paddle as it sweeps past the pick-up. The pick-up device is comprised of two components. One is a magnet whose magnetic field becomes distorted by any nearby metallic movement (a paddle). In the process of becoming distorted, this field sweeps past a second component, a pick-up coil, causing a voltage to be induced inside the coil. Therefore, for every passage of a paddle wheel past the pick-up, there is a voltage pulse that is generated by the pick-up itself. Then it becomes a simple matter of correlating the number of pulses per second to the flowrate of the passing fluid in the pipe or tube in which the transducer is installed.

Another common method for the sensing of gases uses what is referred to as the *hot-wire anemometer principle*. In this process (Figure 4-16) a heated wire is allowed to be cooled by the flowing gas. The amount of cooling is determined by the amount of decrease in the wire's electrical resistance. (Remember that most pure substances decrease their electrical resistance when cooled.) Once this change in resistance is known, this can then be correlated to the flowrate of the gas flowing past the sensor itself.

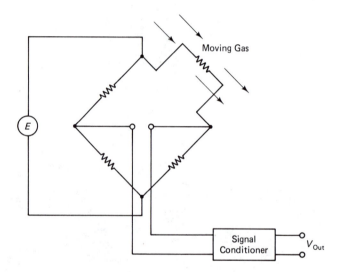

Figure 4-16   Circuit for a hot-wire anemometer.

Another more indirect method for measuring the flowrates of fluids uses a created pressure drop for its flowrate detection. As fluids flow past *pressure taps* installed on both sides of an orifice in a pipe, a pressure difference is created between the two taps as a result of this orifice. A *manometer* device is attached to both pressure taps resulting in the pressure drop that exists across the orifice to be sensed by the manometer. The manometer usually contains some sort of electrical sensing device, such as an LVDT, to detect the height or pressure head of the manometer's reading fluid. Figure 4-17 shows such a system.

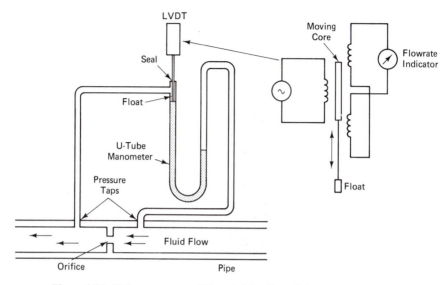

Figure 4-17 Using a pressure differential for determining liquid flowrates.

The transfer functions for all of the aforementioned flow-rate-sensing units are the same. Namely,

$$\frac{E_{out}}{Q} = \frac{K_{flowrate}}{1 + s\tau}$$

(4-8)

where $K_{flowrate}$ = sensor constant or transfer function value for flowrate sensor. Must be obtained from manufacturer or obtained experimentally.

Q = fluid flowrate expressed either as weight flow rate (lbs/sec) or volumetric flow rate (ft³/sec)

s = Laplace transform for $j\omega$

τ = time constant of sensor (sec)

4-7 ELECTROMAGNETIC SENSING

Electromagnetic sensing covers a very large range of sensor types. This is due to the very broad nature of the electromagnetic spectrum. This is the broad illustrated band of radio and light frequency designations, called the *electromagnetic spectrum,* often seen illustrated in physics textbooks and illustrated in Figure 4-18. But we limit our electromagnetic sensing to just the radio communications frequencies, since the other portions of this spectrum already have their own unique sensor types. For example, look at the visible light portion of the spectrum. There are special light-sensitive sensors that have the task of responding

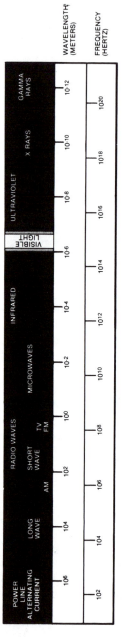

Figure 4-18 The Electromagnetic Spectrum. (*Courtesy of American Radio Relay League, Newington, Conn.*)

only to visible, or nearly visible, light radiation. And in the X-ray portion of the spectrum, there are sensors that are devoted solely to the detection of X-rays. We concentrate only on the radio frequency sensing devices for this particular discussion.

Basically, a radio frequency sensing device (or rf sensing, as it is usually called) is nothing more than a radio receiver that has been designed to tune a desired range of radio frequencies. There are a wide variety of radio receiver circuits available, and many have been constructed out of a single integrated circuit similar to the one depicted in Figure 4-19. In general, an rf receiver's circuitry can be subdivided into the following individual circuits: a radio frequency amplifier section, which is usually tunable for a range of frequencies; a detector section that "strips" the audio off the carrier; and an audio amplification section. The only input adjustments needed to the IC are for tuning and for adjusting the receiver's sensitivity. In many cases, the receiver is purposely designed to be broadbanded or tuned so that no tuning is necessary at all. Wide ranges of frequencies can then be received simultaneously without having to tune, as long as it isn't necessary to receive information from more than one frequency carrier at a time that might be transmitted simultaneously on any of the many received frequencies.

Typical applications requiring the use of rf sensing in an automatic control system would be the automatic tracking of a moving radio-wave-emitting object such as an airplane, missile, or satellite. These are fairly sophisticated systems and are not discussed in this book.

In order to develop the transfer function for an rf receiver, we must first look at the description for the incoming radio signal to the receiver. The radio signal's strength or quantity is usually given in units of the *microvolt*. The output of the receiver is usually given in volts. Consequently, the receiver's transfer function would then be:

$$\frac{\text{output volts}}{\text{input volts}} = K_{rf} \tag{4-9}$$

where K_{rf} = gain or transfer function value for an rf receiver (volts/μvolt)
output volts = output voltage of receiver (DCV)
  input volts = input signal strength (ACμV)

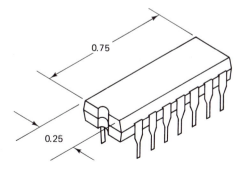

Figure 4-19 A typical IC package (a DIP or dual in-line pin configuration) for housing an entire FM radio receiver.

4-8 TEMPERATURE SENSING

There are a number of methods available for converting temperature into an electrical signal. One method is through the application of a solid state device called a *thermistor*. Most pure elements in nature increase their electrical resistance when exposed to an increase in temperature, with carbon being the most notable exception. Many compounds, however, react oppositely. And in particular, the compounds referred to as semiconductors, which are used in the production of semiconductor devices, react with this opposite behavior. That is, their electrical resistances decrease with an increase in temperature.

The thermistor is one such device (Figure 4-20). Thermistors are relatively inexpensive and can be placed into very small recesses for temperature measuring. A typical calibration curve correlating a thermistor's DC resistance to tem-

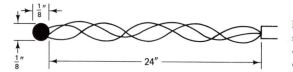

Figure 4-20 Thermistors used for sensing temperature. (*Courtesy of Omega Engineering, Inc., Stamford, Conn.*)

perature would look like the ones in Figure 4-21. It would be rare, however, to use a thermistor's resistance measurement for a direct temperature reading. Instead, a current flow or a voltage drop (most likely a voltage drop across a series resistance) would be used for a temperature indication. A thermistor's transfer function would then be determined in the following manner:

$$\frac{E_{out}}{°F} = \frac{K_{thrm}}{1 + s\tau} \tag{4-10}$$

where K_{thrm} = sensor constant or transfer function of thermistor. Must be obtained from manufacturer or obtained experimentally.

  E_{out} = AC or DC voltage output
  °F = the temperature amount being sensed (°F)
  s = Laplace transform for $j\omega$
  τ = time constant (sec)

Another commonly used temperature sensing device is the *thermocouple*. This device depends on a phenomenon called the *Seebeck Effect*. This is where one can take two dissimilar metal wires and, by heating the ends of these wires where they have been physically joined either through twisting or welding, produce an electromotive force (emf) at the wire's opposite ends. The emf, or voltage, being generated is quite proportional to the amount of heat being applied to the junction. Since this voltage is quite small, in the millivolt range typically, this voltage is usually amplified to a larger value and represented either by a larger DC

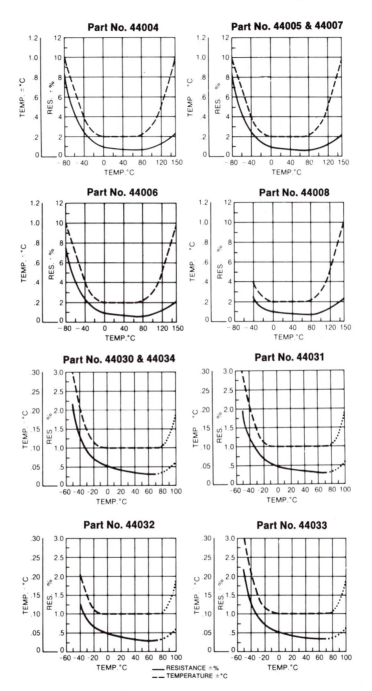

Figure 4-21   Response curves for thermistors. (*Courtesy of Omega Engineering, Inc., Stamford, Conn.*)

or AC voltage. Consequently, the transfer function units of the thermocouple would be identical to the thermistor's transfer function. Namely,

$$\frac{E_{out}}{°F} = \frac{K_{tcpl}}{1 + s\tau} \tag{4-11}$$

where K_{tcpl} = sensor constant or transfer function value. Must be obtained from manufacturer or obtained experimentally.

E_{out} = DC voltage output (mV)

$°F$ = the temperature amount being sensed (°F)

s = Laplace transform for $j\omega$

τ = time constant (sec)

There are other ways of detecting temperature, some of which are rather sophisticated, but we consider only the two aforementioned methods. These cover a fair percentage of all temperature detection systems.

4-9 LIGHT SENSING

In the case of light sensing, there are several means available for the detection of its intensity. As a result, light sensing may be broken down into roughly three different categories:

1. the photoresistive sensor,
2. the photoemissive sensor, and
3. the photovoltaic sensor.

We discuss each without getting too deeply involved with the physics or chemistry of how each works. However, before we do this, we first define the unit of light intensity, since this is going to be the input quantity to all of our transducers that we discuss in this section.

The unit of illumination can be quite confusing and is not too well understood by some people. For the purpose of our discussion, we define the unit of illumination as *the flowrate of light.* In this context, think of light as being comprised of lines of flux, just as you thought of magnetic lines of force as flux when dealing with the discussion of magnetic fields in electricity. The input of this light flowrate is the *lumen,* a unit of measurement in the Meter-Kilogram-Second (MKS) system of units. Having now defined the unit of illumination, let's continue now with our discussion of light sensing transducers.

4-9.1 The Photoresistive Sensor

This photoresistive device (Figure 4-22) varies its internal DC resistance according to the light intensity falling on the sensor. Its chemistry is somewhat similar to

Figure 4-22   The photoresistive sensor. (*Courtesy of Motorola Semiconductor Products, Inc., Phoenix, Ariz.*)

that of the thermistor except that the wavelengths of electromagnetic energy response are shorter. (Refer back to the electromagnetic spectrum chart in Figure 4-18.) Like the thermistor, however, the change in resistance response is usually transformed to a varying voltage for the correlation to light intensity.

As a result, the sensing transducer containing the photoresistor has the following transfer function:

$$\frac{E_{out}}{Q} = \frac{K_{photo}}{1 + s\tau} \tag{4-12}$$

where K_{photo} = constant or transfer function value for photoresistor transducer. Must be obtained from manufacturer or obtained experimentally.

$\quad Q$ = the input flowrate of light (lumens)

$\quad E_{out}$ = the output voltage of the transducer in either AC or DC volts

$\quad s$ = Laplace transform for $j\omega$

$\quad \tau$ = time constant (sec)

The substance the light sensitive material is made from will determine that sensor's time constant, τ. Photodiodes and phototransistors have time constants less than 1 μs. Lead selenide sensors have constants in the 10 μs range, whereas lead sulfide cells are in the 100 to 1,000 μs range. Cadmium selenide cells are typically as high as 10 μs, whereas cadmium sulfide cells typically run as high as 100 μs.

4-9.2 The Photoemissive Sensor

The photoemissive transducer uses a light sensing device that emits electrons in some proportion to the incoming flowrate of light. These transducers have the unique characteristic of being extremely sensitive to weak levels of light. Other than that however, the transfer function is identical to that of the photoresistive sensor.

4-9.3 The Photovoltaic Sensor

The photovoltaic transducer uses a light sensing material that has the capability of producing an emf that is proportional to the incoming flowrate of light. A unique characteristic of this material is that the generated voltages are extremely linear to the lightflow amounts. However, as in the case of the photoemissive sensor, the transfer function is identical to that of the photoresistive sensor. Typical time constants are 20 μs for this type of cell.

4-10 OTHER CONTROL SYSTEM COMPONENTS: SERVOMECHANISM DEVICES

Up to this point, we have been concentrating our discussions on sensing devices and their transfer functions. Now, we consider a group of devices that are used extensively in automatic control systems, namely, the *servomechanism*. A servomechanism is an electromechanical system comprised of devices designed specifically for accurate positioning or the controlling of motion. These devices are manufactured to close mechanical and electrical tolerances and may be considered to be the very heart of the automatic control system.

4-11 THE SERVOMOTOR

The first servomechanism device that we study here is the servomotor. The servomotor may be thought of as a precision electric motor whose function is to cause motion in the form of rotation or linear motion in proportion to a supplied electrical command signal. There are two general types, the AC servo and the DC servo. DC servos are used generally for high power applications requiring large amounts of torque capability. However, in addition, DC servos are noted for generating radio frequency (rf) interference and requiring a certain amount of maintenance, such as the periodic changing of their brushes. And since these motors require DC for their operation, the DC amplifiers needed to run these motors have had a tendency in the past to *drift,* thereby suffering from a reliability problem. Recent developments in DC amplifier design, however, have produced highly stable amplifiers. AC servos, on the other hand, tend to be more stable in their operation. They also tend to be lighter in weight, somewhat more rugged, and require less maintenance. However, they do lack the large torque capabilities of the DC servo.

Figure 4-23 shows a view of a typical servomotor used in small automatic control systems. We now derive the transfer function for the AC servomotor and try to gain some insight into how this motor actually works.

Figure 4-23   Servomotor. (*Courtesy of The Singer Company, Kearfott Guidance and Navigation Division, Little Falls, N.J.*)

To begin with, the AC servomotor is considered to be a two-phase induction-type motor. It's an induction motor because the rotor is inductively coupled (similar to the coupling that exists between the windings of a transformer) to the stator rather than through brushes (as in the case of a DC motor), and it's a two-phase motor because of the two separate voltage source windings that it possesses (Figure 4-24). In reality, the servo has more than just the two windings. Typically, four windings are used. Figure 4-25 shows how they are all wired together.

The servo's speed of rotation is determined by the amplitude of the incoming control signal being supplied from the servo's amplifier. Its speed can also be dependent on the control signal's frequency. An amplifier is usually needed, preceding the motor, to supply the control signal since a sizeable voltage *and* current signal is needed to produce the power to drive the servo. It is rare to find a control signal being supplied directly from a transducer or other nonamplifying device having this necessary power capability.

The servo's direction of rotation is determined by the phase relationship existing between the control signal current and the power line current signal, in this case, called the fixed phase current. Figure 4-26 shows the waveforms of the

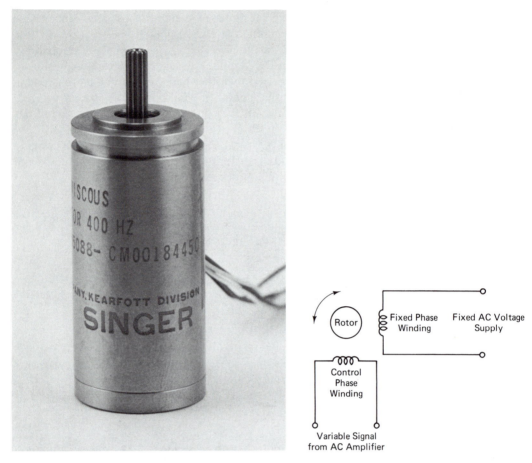

Figure 4-24 Schematic of an AC servomotor. (*Photo courtesy of The Singer Company, Kearfott Guidance and Navigation Division, Little Falls, N.J.*)

two currents being supplied to the servo's windings. Notice that the control phase current *leads* the fixed phase current by 90°. This particular relationship is normal and is created by the servo amplifier that is feeding the control phase signals to the servomotor. This also causes the servomotor's output shaft to rotate in a particular direction. Now, if the control phase current is shifted 180°, the resultant waveforms would look like Figure 4-27. The control phase current signal will now *lag* the fixed phase current by 90°. This new relationship will cause the servomotor's output shaft to rotate in the opposite direction from the previous direction. The reason for this reversal in rotation has to do with the revolving fields created by the changing stator currents and the 90° control phase differences within the servomotor. A discussion of this is beyond the purpose of this text but may be found in any good small induction motor theory text.

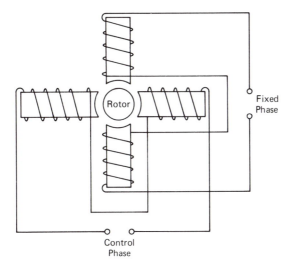

Figure 4-25 How the four windings of an AC servomotor are connected.

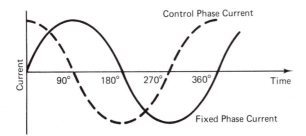

Figure 4-26 The control phase current leads the fixed phase current by 90°.

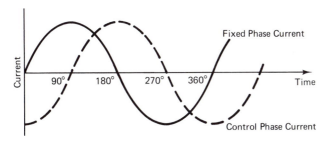

Figure 4-27 The control phase current lags the fixed phase current by 90°.

4-11.1 Calculating the Transfer Function for a Servomotor

In order to develop the transfer function for a servomotor, we must first under-stand the concept of torque. *Torque* is defined as the product resulting from multiplying a force being applied at a right angle to a *moment arm* by the length of that moment arm. For a motor, its output torque for any given speed of shaft rotation would be equal to the force that would be produced at the end of an arm attached to the motor's shaft, multiplied by the length of that arm. This is the length, *l*, indicated in Figure 4-28.

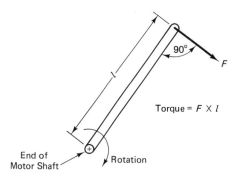

Figure 4-28   Calculating torque.

The torque curve for a motor is the curve relating that motor's torque output with the speeds at which the torque measurements were made. A typical curve for an ordinary induction motor is seen in Figure 4-29. Also shown is the torque curve for an induction servomotor. The stator winding's DC resistance is typically several times more than the stator of an ordinary induction motor. The effect of this is to cause the shape of the servomotor's torque curve to straighten out considerably, as can be seen in Figure 4-29. This will make mathematical approximations somewhat easier for us when we have to use this information later on, because we can approximate these curves with straight lines. In fact, many of

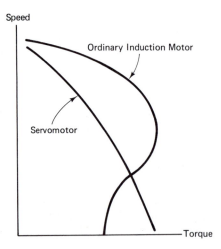

Figure 4-29   Torque curves for an ordinary induction motor and an AC servomotor.

the torque curves seen in catalogs published by servomotor manufacturers are already straight lines. Figure 4-30 is a typical example.

Let's take a further look at Figure 4-30. Notice that there are several speed-torque curves, each curve representing the speed-torque for a particular control voltage value. In reality, servomotors operate at fairly low voltage values. In other words, the signal voltages supplied to the servo are usually quite small, certainly less than 25% of the servo's rated control voltage. The reason for this is

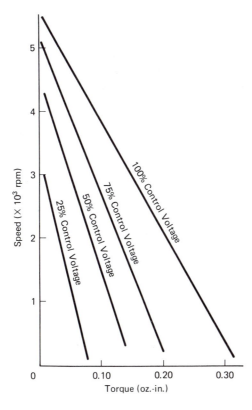

Figure 4-30   Typical servomotor straight-line speed-torque curves found in many manufacturers' catalogs.

because most of any automatic control system's time is spent making minor touch-up adjustments to maintain a particular position, pressure, temperature, etc. In the case of the systems we study here, these small adjustment voltages are continuously and automatically being fed to our servomotor. These voltages will be found to be switching back and forth in polarity. So what the servomotor experiences are small nudging voltages causing the servo to jockey back and forth, performing what closely resembles a balancing act, in order to maintain a certain desired shaft output location or position.

Looking at the 25% curve in Figure 4-30, note where this straight line meets the 0 rpm level at approximately the 0.8 torque value. This torque value is called the motor's *stall torque*. If you were to supply this motor with 25% of 26 VAC for its control signal voltage, and hold down its shaft rigidly, you would experience its stall torque for that particular control voltage.

Looking at Figure 4-30 once again, notice the intersection of the upper portion of each curve with the left-hand vertical axis. This intersection point, as read on the speed axis, is the *free-wheeling speed* (i.e., the torque is zero) for that particular control voltage.

Now that we have analyzed a servomotor's speed-torque curve, its free-wheeling speed, and its stall torque value, we now must define yet another important characteristic before deriving our transfer function. This characteristic is called the motor's *viscous damping*. Viscous damping is a resistance that the servomotor's shaft experiences when it rotates. This resistance is created by an interaction that takes place between the magnetic fields of the rotor and stator. (Viscous damping can be observed by applying full fixed phase voltage to the proper set of windings, and shorting out the control phase windings. Next, try rotating the motor's shaft. The resistance felt is the servomotor's viscous damping. The faster you try rotating the shaft, the greater the resistance that you feel becomes.)

Viscous damping could be considered good or bad depending on your viewpoint and application. It may be considered bad from the standpoint that it robs the motor's output of power; it may be considered good because it reduces *overshoot,* a condition that we discuss in a later chapter. For now, though, we want to be able to calculate the magnitude of this damping effect, because it has a profound effect on the motor's transfer function.

The amount of viscous damping, D_v, is calculated by the following equation:

$$D_v = \frac{L_s}{\omega_f} \tag{4-13}$$

where D_v = viscous damping (oz-in/rev/min or dyne-cm/rad/sec)
$\quad\ L_s$ = stall torque (torque at $\omega = 0$) at 25% rated control voltage (oz-in or dyne-cm)
$\quad\ \omega_f$ = free-wheeling speed (torque = 0) at 25% rated control voltage (rad/sec)

Note: oz-in/rev/min = ounce-inch per revolution per minute.
$\quad$ dyne-cm/rad/sec = dyne-centimeter per radian per second.

Notice that both English and Centimeter-Gram-Second (CGS) units have been listed for the terms in Eq. (4-13). To convert oz-in/rev/min to dyne-cm/rad/sec, multiply by the conversion factor, 6.742×10^5.

We now must calculate the amount of torque *lost* due to the viscous damping. This is done by letting:

$$L_v = \omega D_v \tag{4-14}$$

where L_v = torque lost due to viscous damping (oz-in or dyne-cm)

ω = rotational velocity of shaft (rev/min or rad/sec)

D_v = viscous damping from Eq. (4-12) (oz-in/rev/min or dyne-cm/rad/sec)

Now, we must define the servomotor's *torque constant*, K_{LM}. This is done with the following equation:

$$K_{LM} = \frac{L_s}{e_{cv}} \tag{4-15}$$

where K_{LM} = the motor's torque constant (oz-in/volt or dyne-cm/volt)

L_s = the motor's stall torque at 100% control voltage (oz-in or dyne-cm)

e_{cv} = 100% control voltage (volts)

In general, therefore:

$$L_{\text{produced}} = L_s - L_v$$

or
$$L_{\text{produced}} = (K_{LM}) \cdot (e_{cv}) - \omega V \tag{4-16}$$

Now the produced torque, L_{produced}, must equal the absorbed torque, L_{absorbed}. The absorbed torque is comprised of the inertial torque of the rotor, J_α, and the bearing friction torque (which is negligible, unless the bearings are gone). We are also assuming that the load connected to the motor's shaft is zero. This is an assumption that is not too far from the truth. Most servomotors work through a gear transmission, and because of this the motor "sees" very little torque load. We can now write:

$$(K_{LM})(e_{cv}) - \omega D_v = J\alpha$$

or
$$(K_{LM})(e_{cv}) = J\alpha + \omega D_v \tag{4-17}$$

where J = moment of inertia of motor rotor (slugs-ft^2 or gm-cm^2)

α = angular acceleration (rad/sec^2)

Before going on, let's identify what our goal is so that we know what it is that we want. To begin with, we want a transfer function that says:

$$\frac{\text{output}}{\text{input}} = \frac{\theta}{e_{cv}} = K_{SM} \tag{4-18}$$

where θ = angular displacement of shaft (deg)

e_{cv} = control phase voltage (volts)

K_{SM} = transfer function value (deg/volt or rad/volt)

We now resort to Laplace transforms to write the final equation for our transfer function. Referring to Table 3.1, we find the equivalent Laplace transform expressions for the following terms:

$$\alpha = s^2\theta$$

$$\omega = s\theta$$

$$e_{cv} = E_{cv}$$

Therefore, $\qquad K_{LM}E_{cv} = Js^2\theta + D_v\theta$

or $\qquad\qquad\qquad K_{LM}E_{cv} = \theta(Js^2 + D_v)$

Solving for θ/E_{cv}:

$$\frac{\theta}{E_{cv}} = \frac{K_{LM}}{D_v s + Js^2}$$

There is some additional cleaning up that we can do to make the final transfer function of a servomotor appear in a more standard form. Let τ be the motor's time constant. This is the time it takes for the motor to reach 63.2% of its intended speed after receiving 100% of rated control voltage signal. Then,

$$\tau = \frac{J}{D_v} \qquad (4\text{-}19)$$

where τ = servomotor's time constant based on 100% rated control voltage signal (sec)

We also define now a new term, *velocity constant*, K_V:

$$K_V = \frac{K_{LM}}{D_v} \qquad (4\text{-}20)$$

or

$$K_{LM} = K_V \cdot D_v$$

therefore

$$\frac{\theta}{E_{cv}} = \frac{K_V \cdot D_v}{D_v s + Js^2}$$

Note: The velocity constant may also be calculated using the following equation:

$$K_V = \frac{\omega_f}{E_{cv}} \qquad (4\text{-}21)$$

This equation is not used in this particular transfer function development, but is used later.

Now, multiplying the θ/E_{cv} right-hand expression by $1/D_v$, (both the numerator and the denominator), we get:

$$\frac{\theta}{E_{cv}} = \frac{K_V}{s + \dfrac{Js^2}{D_v}}$$

Because $J/D_v = \tau$ (Eq. 4-19),

$$\frac{\theta}{E_{cv}} = \frac{K_V}{s + \tau s^2}$$

Finally,
$$K_{VDSM} = \frac{\theta}{E_{cv}} = \frac{K_V}{s(1 + \tau s)} \frac{\text{rad}}{\text{volt}} \qquad (4\text{-}22)$$

where K_{VDSM} = transfer function value for viscous damped servomotor (rad/volt)

Equation 4-22 represents the transfer function for a servomotor assuming that you know its time constant, τ, and its velocity constant, K_V. These values can be obtained from the manufacturer's catalog data, except for one problem. Remember when we based all of our operating conditions for a servomotor on its 25% of rated control voltage line rather than using the 100% line as the manufacturer does? Unfortunately, much of the published data in a catalog is based on this 100% line. We have to make adjustments to the data to place it into the more realistic 25% of rated control voltage operating area. To do this, we make the following adjustments to any catalog data that we see:

$$\text{Actual } D_v = \frac{1}{2} D_v ! \qquad (4\text{-}23)$$

$$\text{Actual } \tau = 2\tau ! \qquad (4\text{-}24)$$

$$\text{Actual } K_{VM} = 2K_{VM} ! \qquad (4\text{-}25)$$

$$\text{Actual } J = J ! \qquad (4\text{-}26)$$

The exclamation mark (!) indicates that the information is unconverted catalog data. We use this convention from now on.

Bear in mind that the time constant that we are talking about here is a *mechanical* time constant, not an electrical time constant. Both, however, have similar characteristics. That is, one time constant equals 63.2% of the full reaction time. It takes approximately 5τ to equal the full reaction time. For a complete transfer function that takes into account both the mechanical and electrical time constants, the function would look like this:

$$\frac{\theta}{E_{cv}} = \frac{K_V}{s(1 + \tau s)(1 + \tau_e s)} \frac{\text{rad}}{\text{volt}} \qquad (4\text{-}27)$$

where τ_e = servomotor's *electrical* time constant (sec)

EXAMPLE 4-2

Derive the transfer function for a servomotor given the following catalog information: $\tau! = 0.0147$ sec., the no-load speed ($\omega_f!$) is 4,900 rpm, and the rated control voltage ($E_{cv}!$) is 36 VDC.

Solution:

First, calculate $K_V!$. Do this by first converting the 4,900 rpm figure to rad/sec so that we can use Eq. (4-21):

$$\omega_f = \frac{(4,900)(2\pi)}{60} \text{ rad/sec}$$

$$= 513 \text{ rad/sec}$$

Then, using Eq. (4-21): $K_V! = \dfrac{513}{36} \dfrac{\text{rad/sec}}{\text{volt}}$

$$= 14.25 \text{ rad/sec/volt}$$

And since

$$K_V = 2K_V! \text{ (Eq. 4-25)}$$

then

$$K_V = 2(14.25) \text{ rad/sec/volt}$$
$$= 28.5 \text{ rad/sec/volt}$$

Next, we have to convert the time constant for the servomotor into actual data. We do this by using

$$\tau = 2\tau! \text{ (Eq. 4-21)}$$

Therefore,

$$\tau = 2(0.0147) \text{ sec}$$

$$= 0.0294 \text{ sec}$$

Finally, using Eq. (4-22):

$$\frac{\theta}{E_{cv}} = \frac{28.5}{s(1 + 0.029s)}$$

The type of servomotor that we have been describing here uses viscous damping that is applied continuously throughout its operation. This means that the motor has to work continuously against this magnetic damping whether it wants to experience damping or not. There is another form of this motor which has viscous damping applied only when there is a change in speed. From the standpoint of efficiency, this is certainly more desirable. This type of servomotor is called an *inertially damped* servomotor. Figure 4-31 shows a diagram of this device. It works like this: Surrounding the motor's output shaft is a series of magnets that is free to rotate on a circular raceway. The magnetic field from this arrangement interacts with a similar set of magnets rigidly mounted on the shaft itself. As long as the shaft rotates at a constant speed, the outer raceway magnets will rotate at that same speed. Under this condition, few, if any, magnetic lines of force are being cut. However, should the shaft's speed suddenly change, the magnetic fields become cut, causing an opposing viscous damping torque to be generated. This in turn causes the shaft's rotation to slow down.

The transfer function for this type of servomotor becomes somewhat more complicated. There is more than just one time constant that becomes envolved here. The transfer function now becomes:

$$K_{IDSM} = \frac{\theta_{out}}{E_{cv}} = \frac{K_V(1 + s\tau_2)}{s(1 + s\tau_1)(1 + s\tau_3)} \frac{\text{rad}}{\text{volt}} \qquad (4\text{-}28)$$

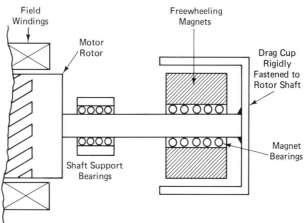

Figure 4-31 An inertially damped servomotor (IDSM). (*Photo courtesy of The Singer Company, Kearfott Guidance and Navigation Division, Little Falls, N.J.*)

where τ_1, τ_2, and τ_3 = mechanical time constants for the servomotor (secs.)

K_{IDSM} = transfer function value for inertially damped servomotor (rad/volt)

4-12 THE GEAR TRAIN

It perhaps may seem a little strange to introduce the subject of gear trains at this point, since gear trains appear to have nothing to do with a discussion on precision electromechanical devices. However, precision gear box transmissions play an extremely important role in the design of automatic control systems. Rarely do you hook up or couple a servomotor directly to some mechanical device to be driven by the servo directly. The servo's shaft speed would be entirely too high. Typically, these shaft speeds can exceed 10,000 rpm. For a motor to operate at that speed and perform precision responses to command signals would be virtually impossible. Consequently, servomotors generally operate through geared-down transmissions, or gear trains, whose purposes are to furnish the more desirable slower shaft speeds and to create an increase in needed output torque. Whenever you purchase a servomotor, you will most likely be asked to specify your preference in gear ratios to be supplied with the servo. The geared transmissions can be easily attached or exchanged on these servomotors to satisfy the controls hardware design when needed.

Let's look more closely at a gear train. In Figure 4-32 we see several gears that are in mesh forming a simple gear train mechanism. Gear 1 is the input gear and gear 4 is the output gear. Gears 2 and 3 couple the rotation of gear 1 to gear 4 but really don't enter into the transfer function that we develop here. Again, using the relationship that a mechanism's transfer function equals input/output, and because we are interested in the displacement, θ, that takes place between the gear train's input and output, the transfer function is going to depend on the gear

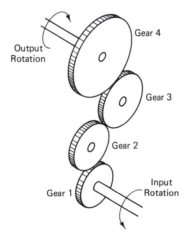

Figure 4-32   A gear train.

train's gearing ratio. The gear ratio, in turn, is going to be dependent on the gear tooth numbers existing between the driver gear and driven gear in our gear train. In other words:

$$\frac{\text{output}_2}{\text{input}_1} = \frac{\theta_2}{\theta_1} = \frac{N_1}{N_2}$$

Also,

$$\frac{\text{output}_3}{\text{input}_2} = \frac{\theta_3}{\theta_2} = \frac{N_2}{N_3}$$

Also,

$$\frac{\text{output}_4}{\text{input}_3} = \frac{\theta_4}{\theta_3} = \frac{N_3}{N_4}$$

Then, the overall output/input transfer function will be equal to the products of the individual output/input functions. Or,

$$\frac{\text{output}}{\text{input}} = \frac{\text{output}_2}{\text{input}_1} = \frac{\text{output}_3}{\text{input}_2} = \frac{\text{output}_4}{\text{input}_3}$$

Or,

$$\frac{\text{output}}{\text{input}} = \frac{\text{output}_4}{\text{input}_1} = \frac{\theta_4}{\theta_1}$$

Then,

$$K_{\text{gear}} = \frac{\theta_4}{\theta_1}$$

In general, for any gear train: $K_{\text{gear}} = \dfrac{\theta_{\text{output}}}{\theta_{\text{input}}}$ (4-29)

Equation (4-29) tells us that regardless of the number of gears existing inside a gear box or transmission, its transfer function is dependent only on the ratio of its output gear displacement versus its input gear displacement. We have to be careful here, though. We don't want to confuse a gear train's transfer function with its gear ratio. The *gear ratio,* sometimes called the *speed ratio* of a gear train, is equal to its input gear teeth number divided by its output gear teeth number. In many instances, this ratio is expressed as the letter N. Also, N represents the ratio of the output speed divided by the input speed; this ratio, as it turns out, is the *reciprocal* of the gear train's transfer function. Summarizing,

$$K_{\text{gear}} = \frac{\theta_{\text{out}}}{\theta_{\text{in}}} = \frac{T_{\text{in}}}{T_{\text{out}}} = \frac{\omega_{\text{out}}}{\omega_{\text{in}}} = \frac{1}{N}$$ (4-30)

where K_{gear} = the transfer function value for a gear train (no units)
  θ_{out} = output gear's displacement (rad)
  θ_{in} = input gear's displacement (rad)
  ω_{out} = output gear's angular velocity (rad/sec)
  ω_{in} = input gear's angular velocity (rad/sec)
  T_{out} = output gear's tooth number
  T_{in} = input gear's tooth number
  N = gear train's speed ratio or tooth ratio

Another often used gearing system in automatic controls applications is the rack and pinion. Its construction is shown in Figure 4-33. The rack and pinion is used to change the direction of motion and to change rotary motion or displacement to linear motion or displacement. Its transfer function is:

$$K_{rp} = \frac{\text{output}}{\text{input}} = \frac{\text{linear displacement}}{\text{rotary displacement}} \frac{\text{inches}}{\text{rad}} \qquad (4\text{-}31)$$

where K_{rp} = transfer function value of rack and pinion combination (in/rad)

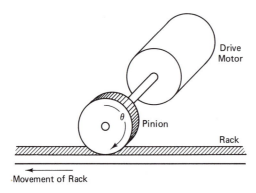

Movement of Rack

Figure 4-33   The rack-and-pinion gear drive system.

4-13 RATE GENERATORS

A rate generator is nothing more than a small precision-made electrical generator whose function is to produce an electrical voltage in proportion to its rpm. Other common names for this device are the *tachometer, servo-generator,* and *rate generator.* There are three basic applications for this device:

1. as a speed controller device,
2. as a stabilizing device for position control, and
3. as a mathematical integrator.

The most common application for this device is probably as a speed controller. This application requires the generator to control and maintain a particular rotational velocity of a mechanism. The output voltage of the rate generator is fed back to the mechanism in such a way to nullify or cancel a portion of that mechanism's supply voltage. Any increase in supply voltage would then be cancelled, thereby serving to hold the same velocity. This, if you will recall, is an application of negative feedback, which we discuss in detail in a later chapter.

Rate generators are manufactured for the purpose of generating both DC and AC voltages. The AC generator is the more commonly used of the two types. Figure 4-34 shows a simple wiring layout for an AC generator. Because of its

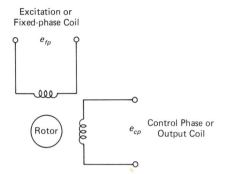

Figure 4-34   An AC two-phase rate generator.

frequent use as a speed controller and as a positional controller, the rate generator is often built as an integral part of a servomotor.

4-13.1 The Rate Generator's Transfer Function

The transfer function value for the rate generator, K_g, is determined in the following way:

$$K_g = \frac{E_{\text{out}}}{\omega_{\text{in}}} \tag{4-32}$$

where K_g = transfer function value for rate generator (volts/rad/sec or volts-sec/rad)

E_{out} = output voltage (volts)

ω_{in} = input rotational velocity (rad/sec)

4-14 SYNCHROMECHANISMS

This group of devices is somewhat off into its own area of special automatic control devices. This is a hardware group comprised of those electromechanical devices that have the capability to transduce position into an electrical signal. In these devices are found a set of phasing coils for a stator, a supply voltage coil for a rotor, and an output shaft. Synchromechanisms usually operate in pairs, one unit referred to as the transmitter while the other is called the receiver. Their theory of operation can be somewhat complicated, so we discuss the theory only superficially here.

To begin with, synchros can be divided into two general groups depending on their application. These groups are:

1. *control synchros,* for indicating readings of position or location from a remote location; and

2. *torque synchros,* for performing work based on remotely transmitted signals. Torque synchros usually use servomotors for their operation.

The only difference between these two groups is in their internal construction. The torque synchro uses heavier wire in its windings and is generally more ruggedly built. It is used for the transmission of power. The control synchro is usually more lightly constructed and, consequently, smaller in physical size because it doesn't require the carrying of much current in its windings.

Figure 4-35 shows the schematic of a synchro. Note that there is a control transmitter (CX) and a control receiver (CR). (The discussion here would be the same for a torque transmitter, TX, and a torque receiver, TR). Basically, the theory of operation for this system is based on the action of a transformer. It is important to realize when considering Figure 4-35 that the stator voltage's magnitude and phase are determined by the angle of rotation of the rotor relative to the stator. It is this relationship that permits the CX (or TX) to accurately transmit its rotor position to the matching CR (or TR). The receiver's rotor will respond to the voltages transmitted to its three windings and will automatically rotate, due to motor action. It will continue to do this until it finds a spot in its rotation where the three transmitted voltages and the supply voltage produce a torque that is essentially zero. This will cause the CR synchro's shaft to come to rest at precisely the same location as the CX's shaft.

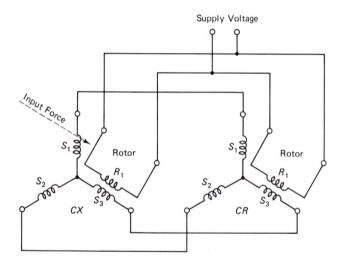

Figure 4-35 How the synchromechanism operates.

Figure 4-36 shows a typical application of a remote reading synchromechanism. The CX is attached to the dial mechanism of a Bourdon-tube pressure gage. The CR is located at a remote data site and is attached to a dial that would have normally been installed in the pressure gage. As the pressure gage responds to some pressure, the CX transmits the amount of rotation of the Bourdon mechanism to the CR where the actual pressure reading can then be read.

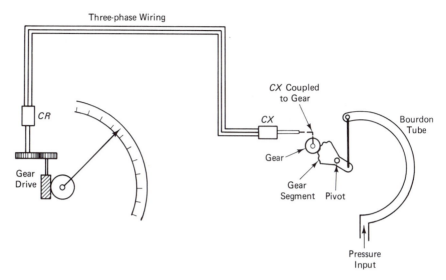

Figure 4-36 A remote reading pressure gage using a CX–CR synchromechanism pair.

4-14.1 The Transfer Function for Synchros

Because the input to a CX or TX is rotational motion and the output of these two devices is voltage, the transfer function for an individual CX or TX synchro becomes:

$$\frac{E_{\text{out}}}{\theta_{\text{in}}} = K_{CX/TX} \frac{\text{volts}}{\text{rad}} \tag{4-33}$$

where $K_{CX/TX}$ = transfer function value for either a CX or a TX synchro (volts/rad)

θ_{in} = input displacement of rotor shaft (rad)

E_{out} = output voltage of synchro (volts)

The transfer function for a CR or TR synchro becomes the inverted equivalent of the units' transfer function described in (Eq. 4-33), since the input now becomes a voltage and the output becomes a rotational motion. Namely:

$$\frac{\theta_{\text{out}}}{E_{\text{in}}} = K_{CR/TR} \frac{\text{rad}}{\text{volts}} \tag{4-34}$$

It's a little unusual to see a CX or CR synchro unit being used by itself. And it's just as unusual to see a CR or TR synchro being used by itself. Normally, these units are used in pairs, since one unit depends on the other units to complete a remote reading system. However, their individual transfer functions are given here as more of an exercise than anything else. Of course, what has been said of the CX–CR systems is also true for the TX–TR systems.

4-14.2 *The Synchro Control Transformer*

A *synchro control transformer,* or CT, is electrically quite similar to the synchro control transmitter (CX) and receiver (CR). However, mechanically it operates quite differently. It has an input shaft and winding construction much like the CX, to indicate position. However, on its other shaft is a winding whose output is an AC voltage that is proportional to the difference between the rotational displacements of the input and output shafts (see Figure 4-37). It may help to think of the CT synchro as being similar to a CX–CR synchro all in one unit with the common rotor connections removed. The CX rotor winding, instead, becomes the input

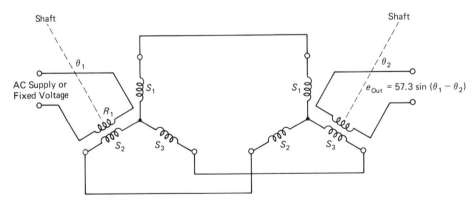

Figure 4-37 Using a CT to generate an error voltage. (*Photo courtesy of The Singer Company, Kearfott Guidance and Navigation Division, Little Falls, N.J.*)

connections for the CT where an input AC voltage is applied, while the CR windings are the output connections. The output AC voltage is then proportional to the relative displacements of the two input shafts, the CX rotor shaft and the CR rotor shaft. This output voltage is called an *error voltage* and is frequently used in automatic control systems for controlling position.

The transfer function for a CT system, K_{CT}, is somewhat similar to the CX or TX system. The only difference is that the rotational input for the synchro is an expression of a rotational difference existing between its two input shafts.

4-15 RESOLVERS

Resolvers are another group of synchros that can be classified under the heading of synchromechanisms. They are used for the purpose of performing trigonometric calculations and may be thought of as analog electromechanical computers. They can transform coordinates and perform conversions between polar and rectangular coordinates.

A resolver is like a variable transformer, similar in construction to that of the synchros just discussed. However, unlike the CX–CR systems requiring a three-phase coil system for a stator, the resolver can have as few as two coils for its stator winding and the same for its rotor winding. The coil pairs are wound at right angles to each other, as seen in Figure 4-38.

Resolvers, like the CX–CR or TX–TR systems, operate in pairs. However, generally no distinction is made between which unit is a transmitter and which is the receiver. Looking at Figure 4-38, we see how a typical resolver system might be wired. In this case, two resolvers are being used to determine the sine and cosine of two angles, α and β. The system would operate like this:

1. A fixed supply voltage would be applied at $E_{S\ 1\text{-}2}$.
2. Rotor 1 would be rotated through an angle, α.
3. Rotor 2 would be rotated through an angle, β.
4. The voltage resulting at $E_{R\ 1\text{-}2}$ would then be $= E_{S\ 1\text{-}2}\cos(\alpha + \beta)$.
5. The voltage resulting at $E_{R\ 3\text{-}4}$ would then be $= E_{S\ 1\text{-}2}\sin(\alpha + \beta)$.

If the wires were crossed over at the two spots marked with Xs in Figure 4-38, the two resolvers would then produce entirely different results. The resultant voltage at $E_{R\ 1\text{-}2}$ would then become $E_{S\ 1\text{-}2}\cos(\alpha - \beta)$; the resultant voltage at $E_{R\ 3\text{-}4}$ would become $E_{S\ 1\text{-}2}\sin(\alpha - \beta)$.

4-16 AUXILIARY SERVO CIRCUITS

Up to this point, we have discussed the major servomechanical components used in automatic control systems. However, there are several electronic circuits that are frequently used to modify or to create certain characteristics within portions

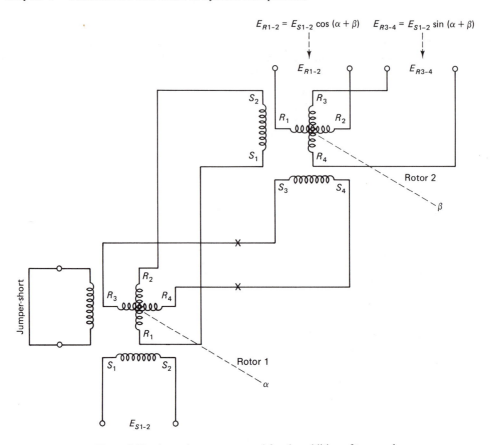

Figure 4-38 A resolver system used for the addition of two angles.

of the control system, just as there are electromechanical devices used for the same purpose. In the next few sections we discuss these circuits in some detail to see what their characteristics are and how their transfer functions are derived.

4-16.1 The Modulator

There are control systems designed to operate on either DC or AC signals. There are also control systems that operate on AC signals in certain portions of their systems and DC signals in other portions. Consequently, these systems must have the capability of converting one form of signal to the other. This is where the modulator comes in. *A modulator circuit is used for converting DC type signals to AC signals.* Circuits that do this conversion are sometimes called *inverters.*

Unfortunately, the name *modulator* is somewhat misleading if you happen to have some radio theory background. You know then that a modulator circuit is used for combining a lower frequency signal, usually an audio-rate signal, with a much higher frequency carrier signal to prepare the lower frequency signal for radio transmission. This is somewhat similar to what the modulator does in a control circuit. In this case, the modulator converts a DC signal whose amplitude varies quite slowly (due to varying command levels and responses) as compared to a true formal AC signal, to an AC signal whose amplitude or frequency varies in accordance to the amplitude of the DC signal.

There are a number of circuit designs that can be used for the DC-to-AC conversion process, all of which work very well. Years ago, these circuits were electromechanical in their operation and used what were called *vibrator reed circuits* (see Figure 4-39). They were popular in the automotive industry for converting the then 6 VDC car battery voltage to a much higher AC voltage signal. This higher voltage was then rectified, that is, converted to DC, to supply the high DC voltage needs of the car radio's vacuum tubes.

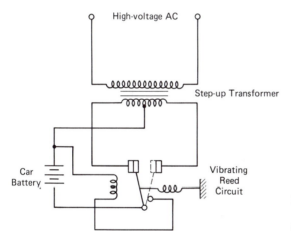

Figure 4-39   Vibrating reed circuit for "chopping" a DC voltage for conversion to an AC voltage.

The modern-day control circuit modulator operates on roughly the same principle as the old-fashioned vibrator circuit. Figure 4-40 shows a solid-state chopper that does the same thing as the mechanical vibrator. The DC voltage signal is alternately chopped by the two transistors, Q_1 and Q_2, called *switching transistors*. They replace the mechanical switching that was performed by the vibrating reed in our old vibrator circuit. The transistors are far more efficient, and there are no moving parts to wear out as was the case in the older electromechanical circuits. In the solid-state chopper, or inverter as it is more frequently called, the two transistors are alternately turned off and on allowing the input DC voltage to flow first in one direction through the center-tapped winding, N_1, and then the other. This produces a square-wave signal (voltage V_T in Figure 4-40) which, in turn, produces an AC voltage across the output winding, N_2. The amplitude of this transformed signal will be directly proportional to the amplitude

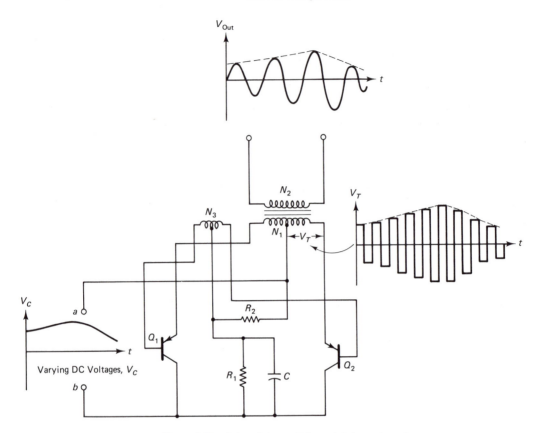

Figure 4-40   A transistor switch modulator or inverter.

of the input DC voltage supplied to the inverter. The alternate switching action of the two transistors is produced and controlled by the alternate saturation of each half of winding N_1. This, in turn, causes the center-tapped feedback transformer winding, N_3, to alternately bias the bases of Q_1 and Q_2 allowing one or the other transistor to conduct. R_2 and C_1 encourage the circuit to begin oscillation at startup of the system.

The transfer function for the modulator is the following:

$$K_{\text{mod}} = \frac{E_{\text{out}}}{E_{\text{in}}} \frac{\text{ACV}}{\text{DCV}} \tag{4-35}$$

4-16.2 The Demodulator

The *demodulator* performs the opposite function of a modulator. Again, the radio enthusiast, when he or she hears the term *demodulator,* probably thinks of a receiver circuit that removes the low frequency audio signals from the incoming high frequency carrier radio signal. In the control circuit demodulator, somewhat of the same process takes place. However, in this case, the amplitude or frequency of the AC signal containing the command level information is transformed into a DC voltage whose amplitude is a representation of this command information.

If the demodulator converts the *amplitude* of the AC signal to a DC signal, that circuit is called an *amplitude demodulator* or simply *AM demodulator*. If the demodulator converts the *frequency* of the AC signal to DC signal, that circuit is called a *discriminator*. The AM demodulator can be a simple diode that rectifies the AC signal into a DC level signal. The filtering out of any remaining AC components is done by R–L or R–L–C networks, as shown in Figure 4-41(a). Figure 4-41(b) shows a full-wave demodulator that is more efficient than the

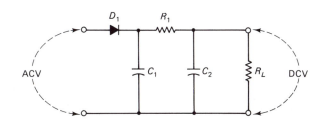

D_1 = Diode
$R_1 C_1 C_2$ = PI Filtering Network for Reducing AC Components Occurring at Output
R_L = Output Load

(a) Half-wave Demodulator

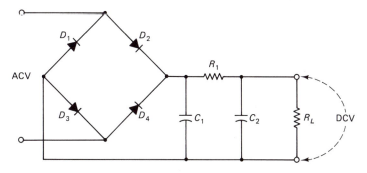

$D_1 D_2 D_3 D_4$ = Diodes Arranged in Full-wave Bridge Configuration
$R_1 C_1 C_2$ = PI Filtering Network
R_L = Output Load

(b) Full-wave Bridge-type Demodulator

Figure 4-41   Two popular demodulator circuits.
   A. Half-wave demodulator
   B. Full-wave bridge-type demodulator

circuit in Figure 4-41(a), producing a larger output signal as a result. FM discriminator circuits dealing with servo applications are rather complicated in their circuit design and are not discussed here.

The two circuits in Figures 4-41(a) and 4-41(b) represent the very simplest of circuits that can be used. There are many other more sophisticated circuits that won't be covered here and would be superior in performance to the ones shown. Nevertheless, the circuits presented here represent the very basic circuit principles used in these applications.

The transfer function for a demodulator (regardless of whether it is designed for converting AM or FM signals to DC) may be stated as:

$$K_{\text{demod}} = \frac{E_{\text{out}}}{E_{\text{in}}} \frac{\text{DCV}}{\text{ACV}} \tag{4-36}$$

4-16.3 The Phase-Lead Network

The phase-lead network, as the name implies, is used to cause an existing phase angle to become leading. This type of circuit is often referred to as a *compensating circuit*. In other words, the circuit compensates for a lagging phase angle by reducing the phase-lag which may be present between the output signal and the input signal of a control system by shifting the output signal into a more leading phase angle. As we find out in a later chapter, system stability is determined by the phase relationship existing between the output and input signals. By shifting this phase, we can stabilize or destabilize our control circuits at will. Figure 4-42 shows a typical phase-lead network. As you can see, the circuit is comprised solely of passive components. No external power is required to operate the circuit. The transfer function is derived by the following process:

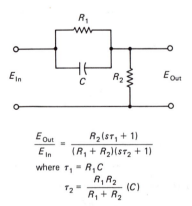

$$\frac{E_{\text{Out}}}{E_{\text{In}}} = \frac{R_2(s\tau_1 + 1)}{(R_1 + R_2)(s\tau_2 + 1)}$$

where $\tau_1 = R_1 C$

$$\tau_2 = \frac{R_1 R_2}{R_1 + R_2} (C)$$

Figure 4-42 Phase-lead network.

Looking at Figure 4-42 and using the voltage divider method for impedances:

$$\frac{Z_{out}}{Z_{in}} = \frac{R_2}{\dfrac{R_1/sC}{R_1 + \dfrac{1}{sC}} + R_2} = \frac{R_2}{\dfrac{\dfrac{R_1}{sC} + \dfrac{R_2(sCR_1 + 1)}{sC}}{\dfrac{sCR_1 + 1}{sC}}}$$

$$= \frac{R_2}{\dfrac{R_1 + sR_2CR_1 + R_2}{sCR_1 + 1}} = \frac{R_2(sCR_1 + 1)}{R_1 + R_2 + R_1R_2sC}$$

$$= \frac{R_2(sCR_1 + 1)}{(R_1 + R_2)\left(\dfrac{R_1R_2}{R_1 + R_2}sC + 1\right)} = \frac{R_2}{R_1 + R_2}\frac{s\tau_1 + 1}{s\tau_2 + 1} \qquad (4\text{-}37)$$

where $\tau_1 = R_1C$

$$\tau_2 = \frac{R_1R_2}{(R_1 + R_2)C}$$

Because of the resistive component, $R_2/(R_1 + R_2)$ in Eq. (4-37), there is an insertion loss as a result of using this circuit. Usually, this can be nullified by increasing the gain of an existing amplifier in the control circuit.

4-16.4 The Phase-Lag Network

The phase-lag network is used to do just the opposite from the phase-lead network. The purpose of the phase-lag network is to cause the existing phase angle of a system to increasingly lag, thus compensating for a too leading phase angle. Again, this network is comprised of passive components and therefore requires no external power supply. Its circuit is shown in Figure 4-43. The derivation of its transfer function is the following:

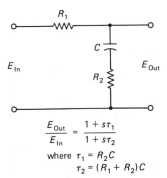

$$\frac{E_{Out}}{E_{In}} = \frac{1 + s\tau_1}{1 + s\tau_2}$$

where $\tau_1 = R_2C$

$\tau_2 = (R_1 + R_2)C$

Figure 4-43   Phase-lag network.

Again, using the voltage divider rule for impedances:

$$\frac{Z_{\text{out}}}{Z_{\text{in}}} = \frac{\dfrac{1}{sC} + R_2}{R_1 + \dfrac{1}{sC} + R_2} = \frac{\dfrac{1 + sCR_2}{sC}}{\dfrac{sCR_1 + 1 + sCR_2}{sC}}$$

$$= \frac{1 + sCR_2}{1 + sC(R_1 + R_2)} = \frac{1 + s\tau_1}{1 + s\tau_2} \qquad (4\text{-}38)$$

where $\tau_1 = R_2 C$
   $\tau_2 = (R_1 + R_2)C$

Servomotors and phase-lag networks have certain similar characteristics. When the frequency response curves are plotted for both devices, there are close similarities in the appearances of their response curves. As a result, servomotors are sometimes referred to as phase-delay devices.

4-17 THE STEPPER MOTOR

The stepper motor is a type of motor that lends itself very nicely to the digital controlling of automatic control systems. The stepper motor has found use in many low torque applications where computer control is used. Good examples are found in the design of printers and plotters.

Figure 4-44 represents a very basic depiction of what goes on inside a stepper motor. Shown is a four-pole motor having alternate pole polarities. The polarity of each pole is determined by a set of digital bits, each bit determining the polarity of its particular pole. In the example shown in Figure 4-44, a 1 bit causes a north pole to appear at the pole's end adjacent to the armature, while a 0 bit causes a south pole to appear. By systematically switching the bit patterns in the sequence shown in Table 4-2, it's possible to create rotation in the stepper motor's armature, as seen in Figure 4-45.

Having established the method of rotation, it then becomes a matter of developing the proper sequence of bits to cause the stepper to step in the desired directions. This would become the function of the software program.

Stepper motors are manufactured to index in various amounts measured in degrees. A stepper having a stepping angle of 45° would obviously have only 8 discrete positions in a full 360° angle of rotation. On the other hand, a stepper that can step as little as 5° per index has 72 discrete positions, which means finer resolution for positioning purposes.

The transfer function for a stepper motor, for the purposes of this text at least, is similar to those for the servomotor. The same dynamic and transient

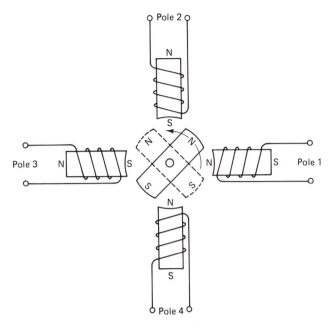

Figure 4-44 The "stepping" of a stepper motor.

TABLE 4-2   A STEPPER MOTOR'S CONTROL BIT PATTERN FOR ROTATION

Counterclockwise rotation			
Pole 1	Pole 2	Pole 3	Pole 4
1	0	0	1
1	1	0	0
0	1	1	0
0	0	1	1
Clockwise rotation			
0	1	1	0
1	1	0	0
1	0	0	1
0	0	1	1

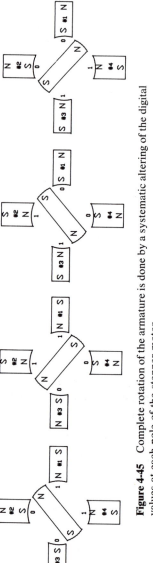

Figure 4-45   Complete rotation of the armature is done by a systematic altering of the digital values at each pole of the stepper motor.

problems of motion are involved with the stepper as there are with the servomotor. The transfer function for a stepper is:

$$K_{StM} = \frac{K_V}{s(1 + \tau_s)} \text{ rad/pulse} \tag{4-39}$$

For additional information on the stepper motor and how it operates, refer to Section 12-4 in the last chapter.

SUMMARY

Transducers are essential components that make up a control system. Because of their importance, we investigated their characteristics and determined the transfer functions for many of the more frequently used sensors. We also discussed the characteristics of the many servomechanisms commonly used in control systems. We found that the servomechanism is a precision electromechanical device used in feedback systems for creating the output commands of position and velocity. We also investigated the synchromechanism and discussed its transfer functions. The synchromechanism is also a precision electromechanical device but is used primarily in positional and positional measurement applications. It can sometimes be used without amplifiers and is designed for analog applications.

Table 4-3 summarizes the transfer functions that were discussed in Chapter 4.

TABLE 4-3 SUMMARY OF TRANSFER FUNCTIONS

Device	Transfer function	Typical units
1. Potentiometer displacement transducer	$\dfrac{E_{out}}{l}$ or $\dfrac{E_{out}}{\theta} = K_p$	volts/inch, volts/deg, or volts/rad
2. Velocity xducr	$\dfrac{E_{out}}{\text{vel}} = K_p$	volts/in/sec
3. Photocell vel. transducer	$\dfrac{E_{out}}{\text{vel}} = \dfrac{K_{photo}}{1 + s\tau}$	volts/ft/sec
4. Pressure xducr	$\dfrac{E_{out}}{P_{in}} = \dfrac{K_{or}}{1 + s\tau}$	volts/lbs/in²
5. Microphone or sonic xducr	$\dfrac{E_{out}}{P_{dB}} = \dfrac{K_{sonic}}{1 + s\tau}$	volts/lb/in²
6. Flowrate xducr	$\dfrac{E_{out}}{Q} = \dfrac{K_{flowrate}}{1 + s\tau}$	volts/lb/sec or volts/ft³/sec
7. RF receiver	$K_{rf} = \dfrac{\text{output volts}}{\text{input volts}}$	volts/μvolt
8. Thermo xducr	$\dfrac{E_{out}}{°\text{F}} = \dfrac{K_{thrm}}{1 + s\tau}$	volts/°F

TABLE 4-3   *Cont.*

Device	Transfer function	Typical units
9. Thermocouple	$\dfrac{E_{out}}{°F} = \dfrac{K_{tcpl}}{1 + s\tau}$	mV/°F
10. Photocell transducer	$\dfrac{E_{out}}{Q} = \dfrac{K_{photo}}{1 + s\tau}$	volts/lumen
11. Servomotor (VDSM)	$K_{VDSM} = \dfrac{K_V}{s(1 + \tau s)}$	rad/volt
12. Servomotor (IDSM)	$K_{IDSM} = \dfrac{K_V(1 + \tau_2)}{s(1 + s\tau_1)(1 + s\tau_3)}$	rad/volt
13. General TF for stepper motor	$K_{StM} = \dfrac{K_V}{s(1 + \tau s)}$	rad/pulse
14. Gear train	$\dfrac{\theta_{output}}{\theta_{input}} = K_{gear}$	deg/deg
15. Rate generator	$\dfrac{E_{out}}{\omega_{in}} = K_g$	volt/rad/sec
16. CX or TX synchro	$\dfrac{E_{out}}{\theta_{in}} = K_{CX/TX}$	volt/rad
17. CR or TR synchro	$\dfrac{\theta_{out}}{E_{in}} = K_{CR/TR}$	rad/volt
18. Phase-lead network	$\dfrac{R_2}{R_1 + R_2} \cdot \dfrac{s\tau_1 + 1}{s\tau_2 + 1}$	
19. Phase-lag network	$\dfrac{1 + s\tau_1}{1 + s\tau_2}$	

EXERCISES

4-1. Determine the transfer function for a potentiometer transducer system that produces a total of 12.8 volts DC output for a total displacement of 42 inches of its sensing arm.

4-2. Find the transfer function of a gear transmission system for an antenna tracking system having the following specifications for the individual gears (refer to Figure 4-32):

$$\text{input gear } 1 = 57 \text{ teeth}$$

$$\text{gear } 2 = 108 \text{ teeth}$$

$$\text{gear } 3 = 63 \text{ teeth}$$

$$\text{output gear } 4 = 211 \text{ teeth}$$

4-3. During the testing of a servomotor to determine its time constant, it was found that it required 0.090 seconds for the motor to respond fully to a command signal given it. Measurements were made with a recording oscilloscope. What would be its actual time constant?

4-4. Referring to Figure 4-42, develop the transfer function for a phase-lead network having the following component values:

$$R_1 = 22 \text{ ohms}$$

$$R_2 = 1,000 \text{ ohms}$$

$$C = 1,000 \text{ pF}$$

4-5. In order to create a particular frequency response curve for a servo system, a phase-lag network (see Figure 4-43) must be installed in which τ_2 must be four times the amount of τ_1. If C is 1.3 μF and R_1 is 100 ohms, what is the phase-lag network's complete transfer function?

4-6. Find the velocity constant for a servomotor whose no-load speed is 6,500 rpm and whose rated control voltage is listed as 32 VAC.

4-7. A certain manufacturer of VDSMs lists the following information for one of its servomotor models:

$$\text{time constant} = 0.016 \text{ sec}$$

$$\text{no-load speed} = 5500 \text{ rpm}$$

$$\text{rated control voltage} = 24 \text{ VDC}$$

Derive the servomotor's transfer function based on the foregoing information.

4-8. A certain VDSM has a no-load speed of 377 rad/sec. Using Figure 4-30 as its speed-torque curve, find its actual viscous damping amount in dyne-cm/rad/sec.

4-9. A servomotor manufacturer lists the following transfer function for one of its models:

$$K_{IDSM} = \frac{25(1 + 0.03j\omega)}{j\omega(1 + 0.05j\omega)(1 + 0.004j\omega)}$$

Correct this transfer function to show *actual* values.

4-10. In the following transfer function, calculate the actual phase angle and magnitude of the output in degrees and rad/volt at a frequency of 12 rad/sec.

$$K_{IDSM} = \frac{67(1 + 0.009s)}{s(1 + 0.007s)(1 + 0.067s)}$$

REFERENCES

ALLOCCA, JOHN A. and ALLEN STUART, *Transducers, Theory & Applications*, Reston, Va.: Reston Publishing Co., 1984.

HONEYCUTT, RICHARD A., *Electromechanical Devices,* Englewood Cliffs, N.J.: Prentice-Hall, Inc., 1986.

MILLER, RICHARD W., *Servomechanisms, Devices & Fundamentals,* Reston, Va.: Reston Publishing Co., 1977.

O'HIGGINS, PATRICK, *Basic Instrumentation,* New York, N.Y.: McGraw–Hill Book Co., 1966.

PATRICK, DALE R. AND STEPHAN W. FARDO, *Industrial Process Control Systems,* Englewood Cliffs, N.J.: Prentice-Hall, Inc., 1979.

5

Block Diagram Mathematics

5-1 *BREAKING UP SYSTEMS INTO BLOCKS*

Admittedly, discussions of block diagrams or flow charts are not the most dynamic portions of any textbook. These figures are typically used for explaining the operation of some kind of system. Or they may be used to show the flow of information within a system, showing how it gets from point *A* to point *B*. In automatic control systems, block diagrams are used for many of the same kinds of applications. However, there is an added dimension to their usage. There exists a rigorous, math-disciplined method for changing and moving the locations of these blocks within a system; this is usually done for the purpose of making system simplifications.

In an automatic control system, each block represents a system component having a particular transfer function. And, it's these transfer functions that must work with the other transfer functions in an orderly systematic fashion so as to create the desired overall system output. Look at Figure 5-1. Here we see a block diagram of a somewhat simplified version of an automobile's power plant system. Study it for a moment. Not the inputs. These are the driver's inputs needed to properly control the car's various systems. Remember too, that like the simple block diagram systems that we studied back in Chapter 2, each of the system components seen in Figure 5-1 has its own unique transfer function.

For example, look at the block labeled Transmission in Figure 5-1. The transfer function for this system component would be output/input = θ_{out}/θ_{in}, where θ_{out} represents the angular displacement of the output shaft and θ_{in} is the angular displacement of the input drive coming from the engine. If we were to

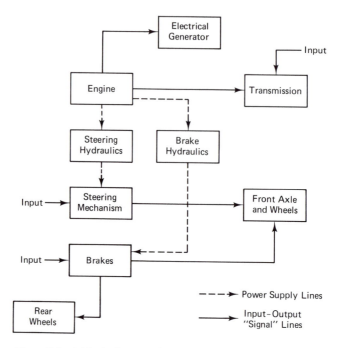

Figure 5-1   A block diagram of the power plant in an automobile.

swap the transmission for, say, the brakes system, our intuition would tell us that we would have a disaster on our hands if we attempted to drive that car. But let's look at this problem mathematically. The transfer function for the braking system would look something like this: pressure/displacement. Pressure is the output hydraulic pressure needed to operate the wheels' brakes; displacement is the brake pedal movement needed to create the pressure. The transfer function is obviously radically different from that of the transmission's. If we investigated the other system components in our car, we would find a similar lack of input–output compatibility. In other words, if we were to swap components around in our car, we would find that each component has been specifically designed to operate in that, and only that, location or sequence of systems. And if we were to substitute or combine components, we would want to make certain that the inputs and outputs of the changed components match exactly with the systems they must work with. This is because we want our car to have the same output performance that we've been accustomed to experiencing.

An automotive design engineer, when designing a new car system, has a very good idea what transfer functions are required for each component, in theory, to make the car perform properly. If he or she decides to combine or swap these components for the sake of simplifying the system, the output versus input results must remain the same. These must not change.

Let's look at another system now. Figure 5-2 shows a drawing of a system that can be used for automatically tracking the trajectory of a satellite-launching rocket, for photographic purposes. A series of infrared photocell sensors are used for following the heat exhaust of the rocket. The sensors cause servomotors to continuously adjust the telescope and camera assembly for proper attitude. Figure 5-3 is the block diagram of this system. Each system component block has its transfer function written inside.

As in the case of the automotive engineer, the design engineer on this particular project would want the freedom to replace blocks, combine blocks, add and remove blocks, while still maintaining a desired overall transfer function for that portion of the system being worked on. Bear in mind too, that merely changing a single wiring or piping connection to one of these blocks can greatly affect the system's overall transfer function. The point is this: In order to have this freedom of design change, a thorough knowledge of block diagram mathematics is necessary.

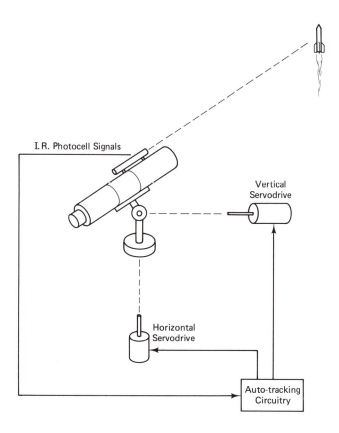

Figure 5-2 An automatic telescopic camera for tracking rocket trajectories.

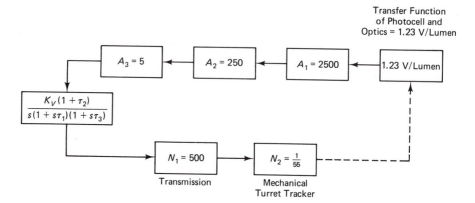

Figure 5-3   Block diagram for the automatic telescopic camera described in Figure 5-2.

5-2 *REVIEWING OLD TERMS AND LEARNING NEW ONES*

In order to understand how block diagram mathematics work, we must first understand the symbols used in block diagrams. Refer to Figure 5-4. The four symbols shown are defined as follows:

1. *The block:* The block represents a component, or a combination of several components, of a system. The block has a signal input and an output. Associated with this block is its transfer function. Usually, all power connections to this block are omitted simply for clarity reasons.

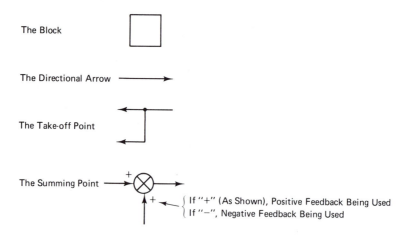

Figure 5-4.   The four major symbols used in automatic controls block diagraming.

2. *The directional arrow:* The directional arrow indicates the direction of flow of a signal or data into and out of each block or along a circuit line.

3. *The take-off point:* This is a point where two or more circuit lines are physically joined together allowing their separate signals to combine into one signal, or to allow a single signal to split into the separate lines. In the second case, it is assumed that the divided signal is of the same amplitude and phase as the original signal.

4. *The summing point:* This is a point where two signals are either added together or subtracted from each other. Note that despite the fact that two signals may be subtracted from each other, it is still called a summing point. Note that if bottom sign is $(+)$, positive feedback is being used in the system; if bottom sign is $(-)$, negative feedback is being used.

Let's review a few of the basic points mentioned in Chapter 2 concerning system gain, positive feedback, and negative feedback:

1. *System gain:* May be considered another way of saying transfer function; used primarily in electronics to describe the amount of amplification of an amplifier or amplifier system. It is the ratio of the voltage or current output divided by the system's voltage or current input.

2. *Positive feedback:* A condition found in certain control systems where the output is coupled back to the input, in phase, so that there is continuous reamplification of the original input signal. Mathematically, this is described as:

$$\frac{E_{\text{out}}}{E_{\text{in}}} = \frac{A}{1 - AB} \quad \text{(see Eq. 2-8)}$$

3. *Negative feedback:* A condition used frequently in control systems where the output is coupled back, out of phase, to the input in order to cancel a portion of the input signal. Mathematically, this is described as:

$$\frac{E_{\text{out}}}{E_{\text{in}}} = \frac{A}{1 + AB} \quad \text{(see Eq. 2-14)}$$

The following is a list of eight basic rules or *identities* for changing block diagrams for the purpose of simplifying or modifying them. These rules are stated in the form of block diagrams and should be studied closely for their understanding.

1. Cascading blocks: Refer to Figure 5-5.

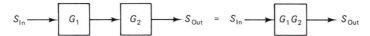

Figure 5-5   Cascading blocks.

2. Eliminating a forward loop: Refer to Figure 5-6.

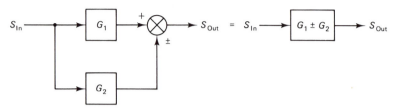

Figure 5-6 Eliminating a forward loop.

3. Moving a take-off point from an input to an output of a block: Refer to Figure 5-7.

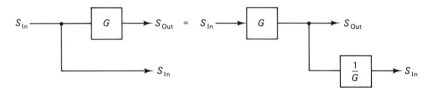

Figure 5-7 Moving a take-off point from an input to an output of a block.

4. Moving a take-off point to a block ahead: Refer to Figure 5-8.

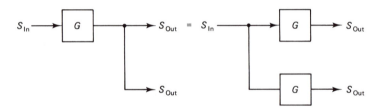

Figure 5-8 Moving a take-off point to a block ahead.

5. Moving a summing point beyond a block: Refer to Figure 5-9.

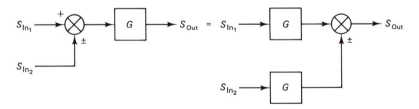

Figure 5-9 Moving a summing point beyond a block.

6. Moving a summing point from an output to an input of a block: Refer to Figure 5-10.

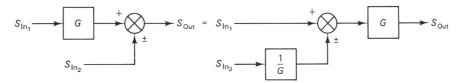

Figure 5-10   Moving a summing point from an output to an input of a block.

7. Eliminating a feedback loop: Refer to Figure 5-11.

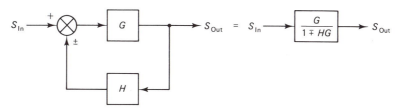

Figure 5-11   Eliminating a feedback loop.

8. Switching blocks: Refer to Figure 5-12.

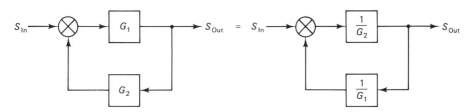

Figure 5-12   Switching blocks.

With these rules, you can combine blocks into single blocks, move pick-off points, or shift summing points ahead or behind blocks. Let's try our hand now at using these rules by going through some examples.

5-3 CASCADING BLOCKS

Probably the most often encountered system configuration in any automatic controls circuit is the one in which the output of one system is fed directly into the input of the next system, and so on. This is called a *cascading system*. Rule 1 shows how this type of configuration is simplified or reduced to a single block. As shown by this rule, the transfer function of the simplified version is nothing more than the product of all the individual blocks' transfer functions. The following example illustrates this point.

EXAMPLE 5-1

Simplify the system in Figure 5-13 by using rule 1.

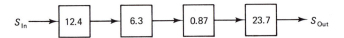

Figure 5-13   Example 5-1. The figures in each block are the individual transfer functions and S_{in} and S_{out} are the input and output signals for the total system.

Solution:

According to rule 1, the equivalent simplified system would have a transfer function equivalent to

$$12.4 \times 6.3 \times 0.87 \times 23.7 = 1610.76$$

5-4 ELIMINATING LOOPS

Often, it's desirable to get rid of feedback loops for the sake of simplification and cost reduction. This can be done as long as the feedback loop doesn't need any adjustments. If it can remain fixed in value, then most likely it should be simplified. The following example shows how this can be done.

EXAMPLE 5-2

Simplify the system in Figure 5-14 by finding the equivalent transfer function.

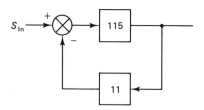

Figure 5-14   Example 5-2.

Solution:

The system shown in Figure 5-14 matches the configuration in rule 7. Note that Figure 5-14 is a negative feedback circuit. Therefore,

$$\frac{S_{\text{out}}}{S_{\text{in}}} = \frac{G}{1 + GH}$$

Since $G = 115$ and $H = 11$, then

$$\frac{S_{\text{out}}}{S_{\text{in}}} = \frac{115}{1 + (11)(115)}$$
$$= 0.091$$

Let's see what happens now if we change the circuit in Figure 5-14 to a *positive* feedback circuit as shown in Figure 5-15 and try to simplify it.

EXAMPLE 5-3

Simplify Figure 5-15.

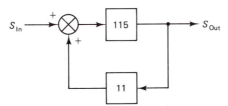

Figure 5-15   Example 5-3.

Solution:

Notice that the polarity sign at the bottom of the summing junction is now reversed to (+). Again, using rule 7, but being careful now to use the proper sign in the transfer equation:

$$\frac{S_{\text{out}}}{S_{\text{in}}} = \frac{G}{1 \quad GH}$$

Since $G = 115$ and $H = 11$, then

$$\frac{S_{\text{out}}}{S_{\text{in}}} = \frac{115}{1 - (11)(115)}$$
$$= -0.091$$

Since there is no such thing as a negative gain or transfer function, the preceding results must be discarded. The system, in other words, is unworkable and has to be redesigned.

5-5 *MOVING SUMMING POINTS*

At times it is advantageous to move summing points from one location to another in a circuit because of physical space requirements or because of other design limitations. The following two examples show how this can be done.

EXAMPLE 5-4

In Figure 5-16, the summing point is to be moved to the output side of the block. Find the new equivalent system.

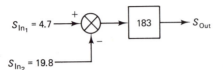

Figure 5-16 Example 5-4.

Solution:

This type of configuration is described by rule 5. Figure 5-17 shows the solution. Notice the addition of an identical block to the original circuit block in the S_{in2} line going to the newly moved summing point.

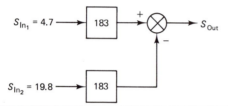

Figure 5-17 Solution to Example 5-4.

EXAMPLE 5-5

Refer to Figure 5-18. Assume that we want to move the given summing point to the input side of the given block. What is the resulting new equivalent circuit?

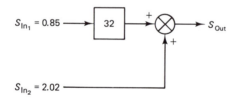

Figure 5-18 Example 5-5.

Solution:

The solution to this problem is seen in Figure 5-19. Notice the addition of a block in the newly relocated S_{in2} line whose transfer function is the reciprocal of the originally given block. The magnitude of $S_{out} = (0.85)(32) + 2.02 = 29.22$.

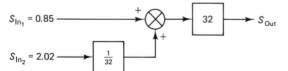

Figure 5-19 Solution to Example 5-5.

5-6 INTERCHANGING BLOCKS

The following example shows how you can swap the positions of two blocks with each other without changing the overall system transfer function.

EXAMPLE 5-6

Referring to Figure 5-20, we would like to interchange the two blocks without changing S_{out}/S_{in}.

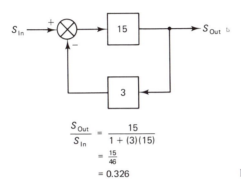

$$\frac{S_{Out}}{S_{In}} = \frac{15}{1 + (3)(15)}$$

$$= \frac{15}{46}$$

$$= 0.326$$

Figure 5-20 Example 5-6.

Solution:

The solution is shown in Figure 5-21. Notice that the system transfer function for the interchanged system did not change.

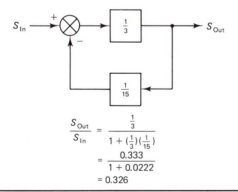

$$\frac{S_{Out}}{S_{In}} = \frac{\frac{1}{3}}{1 + (\frac{1}{3})(\frac{1}{15})}$$

$$= \frac{0.333}{1 + 0.0222}$$

$$= 0.326$$

Figure 5-21 Solution to Example 5-6.

5-7 COMBINED BLOCK REDUCTIONS

We now look at some complex block systems requiring several combined changes and reductions in order to eventually wind up with a much simplified system. There are many instances in which we won't want to reduce the number of blocks in a given system. There are times when it's more desirable to have individual blocks of systems rather than having them combined into one or just a few. From the standpoint of servicing a complex system, many times it's far cheaper to be able to remove an individual block for replacement rather than replacing a more complex circuit or block that contains many subcircuits or assemblies. The design decision must be made with some marketing knowledge for that product design.

Let's try our hand at reducing the system shown in Figure 5-22. The idea here is to reduce the system down to just one block and its transfer function. Again, bear in mind that from a practical standpoint, this may not be the most desirable thing to do. This is simply an exercise in block diagram mathematics.

EXAMPLE 5-7

Reduce the system in Figure 5-22(a) to a single block and its transfer function using the block reduction math method.

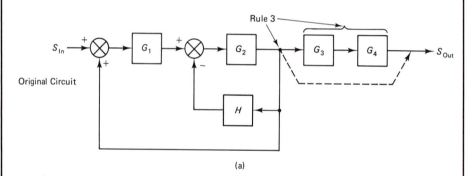

(a)

Figure 5-22   Example 5-7.

Solution:

The solution is shown in Figure 5-22(b) through (g) and summarized in the following step-by-step instructions. These steps can be applied to a majority of all block diagram math problems:

1. Move any blocks that are outside of any loops to the inside of these loops; Figure 5-20(b).
2. Combine any cascaded blocks; Figure 5-20(c).
3. Combine loops within loops where possible, when they share the same summing points and take-off points; Figure 5-20(d).
4. Combine any cascaded terms again where necessary; Figure 5-20(e).
5. Combine any remaining loops into one loop; Figure 5-20(f).
6. Finally, eliminate the remaining loop; Figure 5-20(g).

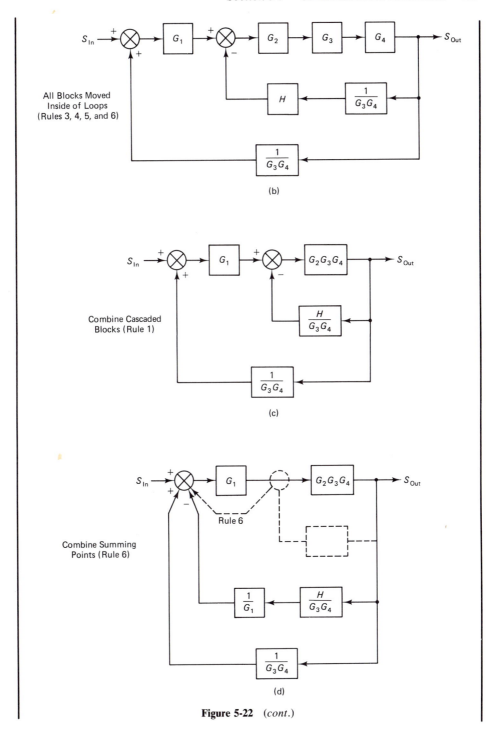

Figure 5-22   *(cont.)*

Combine Cascaded
Terms Once Again
Where Necessary
(Rule 1)

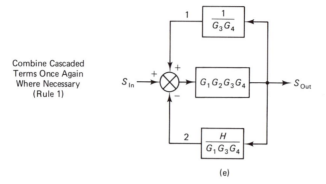

(e)

Combine Any Remaining Loops
into One Loop. (This Can Be
Done by Noting That Loops 1
and 2 in (e) Merely Subtract from
Each Other. Subtract Loop 2
from Loop 1 to Keep Expression
Positive Rather than Negative.)

$$\frac{1}{G_3 G_4} - \frac{H}{G_1 G_3 G_4} = \frac{G_1 - H}{G_1 G_3 G_4}$$

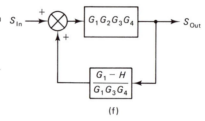

(f)

Eliminate Remaining Loop by
Using Rule 7. Since This is a
Positive Feedback Loop, Use

$$\frac{G}{1 - HG}$$

$$S_{In} \longrightarrow \boxed{\dfrac{G_1 G_2 G_3 G_4}{1 - (G_1 - H)}} \longrightarrow S_{Out}$$

Note: The G Term in the Positive Feedback Loop is $G_1 G_2 G_3 G_4$ and the H Term is $\dfrac{G_1 - H}{G_1 G_3 G_4}$

Therefore:

$$\frac{G_1 G_2 G_3 G_4}{1 - \left(\dfrac{G_1 - H}{G_1 G_3 G_4}\right)(G_1 G_2 G_3 G_4)} = \frac{G_1 G_2 G_3 G_4}{1 - (G_1 - H)G_2}$$

(g)

Figure 5-22   (*cont.*)

EXAMPLE 5-8

Determine the magnitude of the output signal, S_{out}, in Figure 5-23(a), using the block
reduction math method.

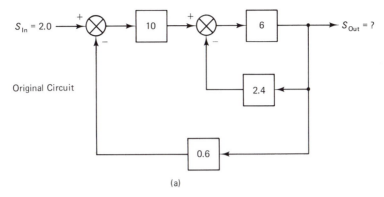

Original Circuit

(a)

Figure 5-23   Example 5-8.

Solution:

The solution is presented in Figure 5-23(b) through (e).

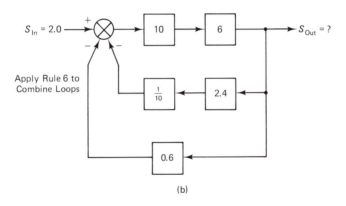

Apply Rule 6 to
Combine Loops

(b)

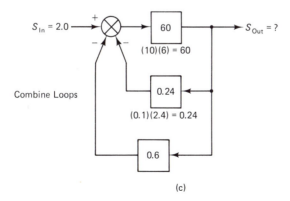

Combine Loops

(c)

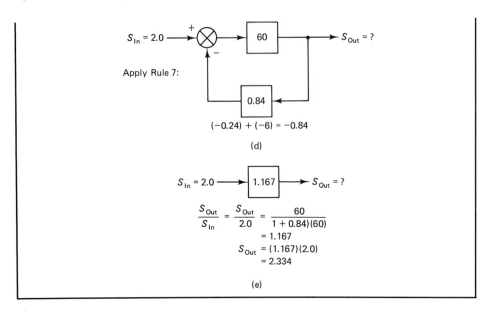

Apply Rule 7:

$(-0.24) + (-6) = -0.84$

(d)

$S_{In} = 2.0 \longrightarrow \boxed{1.167} \longrightarrow S_{Out} = ?$

$$\frac{S_{Out}}{S_{In}} = \frac{S_{Out}}{2.0} = \frac{60}{1 + 0.84)(60)}$$
$$= 1.167$$
$$S_{Out} = (1.167)(2.0)$$
$$= 2.334$$

(e)

After working a number of block reduction problems, you will undoubtedly find numerous shortcuts that can be taken in solving these kinds of problems. Rather than presenting these shortcuts here, it's better that you find them on your own. That way, you're less likely to forget them.

SUMMARY

Block diagram mathematics can be a very worthy time-saving exercise to go through for system simplification. The amount or degree of simplification depends on what system features are needed. The rules for block diagram math were summarized in Figures 5-5 through 5-12.

EXERCISES

The following exercises are strictly exercises in theory, applying the information discussed in this chapter. These are not problems that represent actual functioning systems.

5-1. Reduce the block diagram in Figure 5-24 to a single transfer function using the techniques outlined in this chapter.

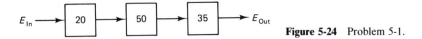

Figure 5-24 Problem 5-1.

5-2. Reduce the block diagram in Figure 5-25 to a single transfer function, again, using the techniques that were discussed.

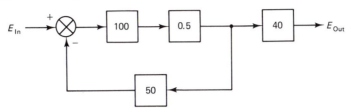

Figure 5-25 Problem 5-2.

5-3. Do the same for Figure 5-26.

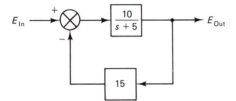

Figure 5-26 Problem 5-3.

5-4. Do the same for Figure 5-27.

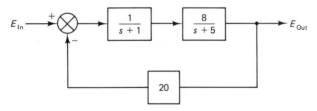

Figure 5-27 Problem 5-4.

5-5. Do the same for Figure 5-28.

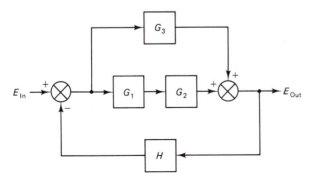

Figure 5-28 Problem 5-5.

5-6. Do the same for Figure 5-29.

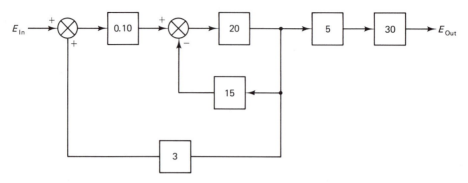

Figure 5-29 Problem 5-6.

5-7. Do the same for Figure 5-30.

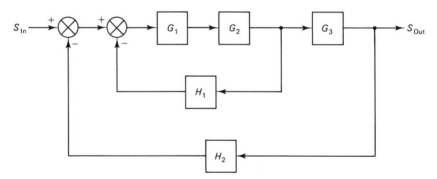

Figure 5-30 Problem 5-7.

5-8. Do the same for Figure 5-31.

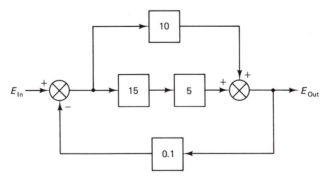

Figure 5-31 Problem 5-8.

5-9. Do the same for Figure 5-32.

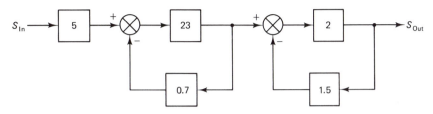

Figure 5-32   Problem 5-9.

5-10. Do the same for Figure 5-33.

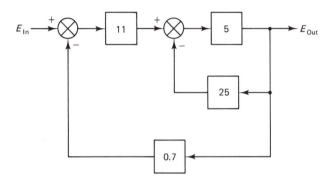

Figure 5-33   Problem 5-10.

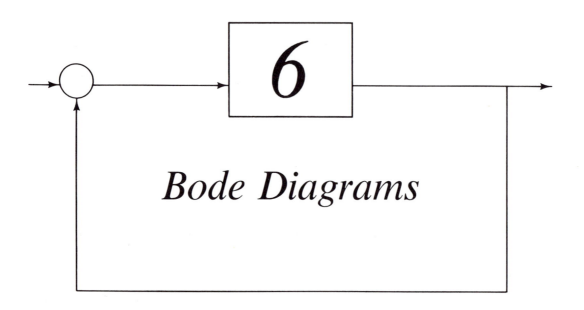

Bode Diagrams

6-1 INTRODUCTION

Hopefully, this chapter will open up a whole new perspective for you for understanding automatic control systems. Up to this point, most of what has been said on this subject has been ground work material. That is, with the possible exception of Chapter 2, virtually all that's been discussed up to now has had to do with preparing you for understanding automatic control concepts. To help us venture into this new territory, we first study the concept of the Bode plot. The *Bode* plot is to the automatic controls engineer what the frequency response curve is to the stereo enthusiast. Both plots tell the person what kind of gain can be expected out of his or her system for a particular given frequency input.

6-2 BODE DIAGRAM CONSTRUCTION

H. W. Bode (pronounced, Bo-dee) worked for the Bell Laboratories following World War II. Bode developed a systematic method of mathematically analyzing a controls system and plotting the results on graph paper. The method of presentation on the graph paper is what makes the Bode method so unique and easy to understand. In order to appreciate the Bode plot, let's first review the discussion in Chapter 1 concerning frequency response.

We described a method in Section 2-4 on how to observe the linearity of an audio amplifier. In place of the amplifier, however, we could have used virtually any kind of audio circuit to see how the output varied with its input as the input frequency is varied. As a matter of fact, if we knew the transfer function of the audio circuit being tested, we could have plotted the results ahead of time. We wouldn't have had to have gone through the trouble of setting up the hardware at all! And this is exactly what we want to be able to do with the circuits we study here. We want to be able to plot the transfer function characteristics based solely on the transfer function itself. We don't want to have to set up a circuit and record measurements. That's too time-consuming.

Let's take another look at our test circuit in Figure 2-3. In place of the audio amplifier, we substitute a resistor network, shown in Figure 6-1. Let's assume that this circuit has an input, *a–b,* and an output, *c–d.* Our dual-trace oscilloscope is also hooked up across these connections as shown. What we want to do now at this point is to calculate the transfer function of this circuit.

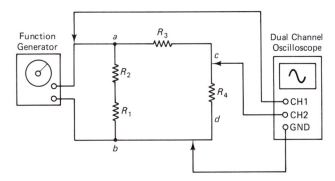

Figure 6-1 A test setup for determining the transfer function of a resistor network.

Look now at Figure 6-2. This is the very same circuit as seen in Figure 6-1, but the resistors in the network have been shifted around slightly so that the input and output voltage relationships can be seen a little more clearly. We can also see that resistors R_1 and R_2 will have no affect on the transfer function relationship since they don't determine the value of e_{in}. In other words, e_{in} has already been fixed in value, regardless of the input resistance. Of course, we are assuming that

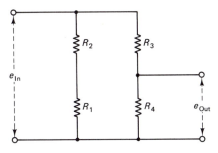

Figure 6-2 The circuit in Figure 6-1, redrawn in schematic form.

e_{in} is being supplied from a power supply capable of supplying any voltage at any current demand placed on it by the input resistance of the test circuit. Now, using the voltage divider rule (see Section 1-5), we see that

$$\frac{e_{out}}{e_{in}} = K_{network} = \frac{R_4}{R_3 + R_4}$$

which now becomes our transfer function for this circuit.

The next thing we want to do is to plot the frequency response curve for our circuit. This will help us to understand the mechanics of plotting when we attempt to plot the more complex circuits associated with a controls system. And this brings us to our first rule in plotting frequency responses in a Bode plot: All amplitudes are expressed in decibels rather than in v_{out}/v_{in} terms. In other words, rather than plotting frequency versus a voltage ratio, we will, instead, plot frequency versus decibels. The reason for this is: In many cases components are cascaded, such as in the case of two or more amplifiers connected in series with each other. The overall frequency response curve of this system will be equal to the *product* of the individual frequency response curves, assuming that we were plotting our results on a voltage ratio versus frequency graph. However, multiplying curves together is not the easiest thing to do graphically. But if we were to use the decibel system instead, all that is needed is to *add* the individual curves together. This will then give us the final overall curve for the system. Then, if we wanted, we could convert the final curve results back into the voltage ratio domain by taking the *antilog* of our results.

This brings us to our second rule in Bode plotting, which concerns frequency. Many times, but not always, the frequency is expressed in radians/sec rather than in hertz. The conversion from the one system to the other is simple: $\omega = 2\pi f$. Unfortunately, the ω is still referred to as frequency by many controls engineers, which may make a few non-controls engineers cringe. Obviously, the ω is an angular velocity term. However, custom takes precedence here. The reason for the frequency-to-rad/sec conversion is due to the frequent occurrence of the $j\omega$ term in controls mathematics. In this book we use ω for our frequencies.

Looking at the transfer function that we obtained for our foregoing resistive network, you have probably guessed by now that we are dealing with nothing more than a pure number, a constant. And when we take the log of this number we will still come up with a pure number. Therefore, we have a transfer function that is frequency independent. That is, regardless of frequency, we will always have the same value of amplitude. So we must keep this in mind when we eventually plot this information. And speaking of plotting, since we're dealing with logs here, all of our plotting will be done on semilog paper. That is, the x axis, or abscissa, will be comprised of logarithmic-spaced lines. On the other hand, the y axis, or ordinate, will be formed by evenly spaced or rectilinear lines. The units associated with this axis will be decibels. As a matter of fact, you will often see the designation dBA being used. This means, "Gain (A) converted to decibels." Recalling the purpose of log paper, any exponential curves existing in our transfer functions will generally stand a better chance plotting as a straight line, or nearly so, than if they were plotted on rectilinear paper. The one thing we must keep in mind though is the fact that log paper comes in various cycles to suit different x axis data ranges. To select the most suitable paper, determine the number of decades or powers of 10 that your x axis data covers. If, for instance, your data ranges from a value of 0.7 to, say, 3,400, then you must use five-cycle paper. This is because you will be plotting data from 0.1 to 1, 1 to 10, 10 to 100, 100 to 1,000, and 1,000 to 10,000.

The third rule in developing Bode plots is the often-done practice of plotting phase angle information on the same graph paper as the frequency versus amplitude information. (To refresh your memory on phase angles, refer back to Section 2-5.) Since our transfer function in Figure 6-2 is not frequency dependent in any way, the phase shift between the output and the input of our resistor circuit will be zero.

We are ready to make a Bode plot of our resistor network. But before we can do this, we must assign some resistor values to the resistors in our circuit in Figure 6-2. Let $R_3 = 400\Omega$ and $R_4 = 200\Omega$. Because K_{network} equals $e_{\text{out}}/e_{\text{in}}$ which equals $200/(200 + 400)$, or 0.33, and this value will be the same regardless of frequency (i.e., regardless of ω), and since $20 \log_{10}(K_{\text{network}})$ equals -0.48 dB, our frequency response, when plotted, will look like the curve in Figure 6-3. And because the phase angle is the same ($0°$) regardless of frequency, the phase angle curve of our Bode plot will also appear as shown in the upper half of our graph

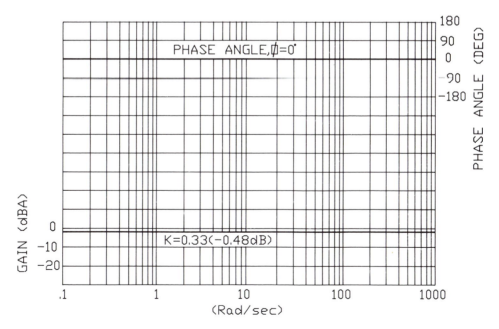

Figure 6-3   The Bode plot of Figure 6-2 where $R_3 = 400$ Ω and $R_4 = 200$ Ω.

paper. The results seen in Figure 6-3 are telling us that for *any* frequency independent component (i.e., there are no ω terms in the transfer function), the Bode plot of the gain has a slope of zero. Or, to state it in other terms, the slope on the Bode plot is 0 dB per each decade as seen and plotted on logarithmic graph paper. Also, the characteristic of the phase angle curve for this same kind of component is also a straight line drawn at 0°. Figure 6-4 shows the Bode plots for several other values of *K*.

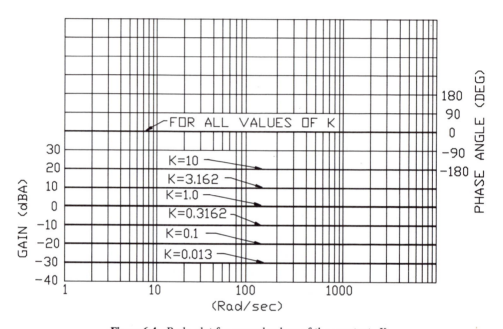

Figure 6-4   Bode plot for several values of the constant, *K*.

Now let's explore a different component configuration. Again, we use electrical components for our experiments, but keep in mind that we could just as well be using electromechanical components such as servomotors, generators, etc. It's just that electrical components such as resistors, capacitors, or inductors are easier to set up mathematically and to envision what is happening. Later, we investigate applications involving the electromechanical devices, but for now, let's take a look at Figure 6-5. Here we see another circuit but with a frequency dependent component added, a capacitor. The transfer function for this circuit is:

$$K_{rc} = \frac{e_{\text{out}}}{e_{\text{in}}} = \frac{Z_{\text{out}}}{Z_{\text{in}}}$$

Again, we have used the voltage divider rule to come up with our solution, but notice that in place of *R* we have now used impedance, *Z*. This, of course, is

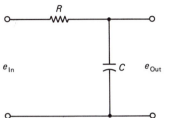

Figure 6-5

necessary since we are now dealing with AC voltages with frequency sensitive components (i.e., C). Using C to determine Z, our transfer function now becomes

$$K_{rc} = \frac{Z_{out}}{Z_{in}}$$

$$= \frac{\dfrac{1}{j\omega C}}{R + \dfrac{1}{j\omega C}}$$

When deriving transfer functions for Bode plotting, the numerator of that function is usually forcibly set to 1 to make the plotting easier. We consider the reason for this later. Therefore, we now multiply the numerator by $j\omega C$, and of course, in order not to change the value of the expression, we do the same thing to the denominator:

$$= \frac{\dfrac{1}{j\omega C} \cdot j\omega C}{R + \dfrac{1}{j\omega C} \cdot j\omega C}$$

$$= \frac{1}{1 + j\omega RC} \tag{6-1}$$

Or, more generally,

$$= \frac{1}{1 + j\omega \tau}, \text{ since } \tau = RC \tag{6-2}$$

We refer to the form of Eq. (6-2) from now on as *the standard form of a Bode transfer function.*

Now let's assign some values to the variables to Eq. (6-1). Let $R = 1$ Meg and $C = 0.05$ μF. This will make $RC = \tau = 0.05$ sec. Equation (6-1) now becomes:

$$\frac{1}{1 + j\omega 0.05} \tag{6-3}$$

Our next chore is to plot Eq. (6-3) on our semilog paper. But to do this we must first assign a wide range of values to ω, and for each assigned value calculate the gain, A, in dB. To do all of this and not lose track of what it is we are doing, we construct a table to record all our values. Our calculations will look something like this (refer to Appendix A to review converting rectangular complex number forms, i.e., j-operators, to polar forms):

$$\frac{1}{1 + \dfrac{j\omega}{20}} = \frac{1}{1 + j\omega\tau} = \frac{1}{\sqrt{1^2 + (\omega\tau)^2}} = \frac{1}{A \angle \arctan \dfrac{\omega\tau}{1}}$$

$$= \frac{1}{A \angle \theta} = \frac{1 \angle -\theta}{A}$$

Table 6-1 shows the results of our labor. We selected the values for ω that would give us a fair distribution of points across our graph paper.

Now, let's plot our figures, once again following the rules that were discussed earlier regarding making Bode diagrams. Figure 6-6 shows the results. Even though our table lists frequencies only in the range of 1 to 1,000 rad/sec for ω, we can assume that the curves resulting from our graphing extend to infinity in both directions. In other words, there are no additional significant features that occur beyond either curve end.

TABLE 6-1   RESULTS OF PLOTTING THE FUNCTION: $\dfrac{1}{1 + j\omega0.05}$

ω		$1 + j\omega0.05$		$1/(A \angle \theta)$		$\text{dBA} = 20\log(1/A)$
rad/sec	result. exp.	A	$\angle\theta$	$1/A$	$1/\angle\theta$	dBA
1	$1 + j0.05$	1.001	2.9°	.999	$\angle -2.9°$	−0.009
2	$1 + j0.1$	1.005	5.7°	.995	$\angle -5.7°$	−0.04
5	$1 + j0.25$	1.031	14.0°	.970	$\angle -14.0°$	−0.26
8	$1 + j0.4$	1.077	21.8°	.929	$\angle -21.8°$	−0.64
10	$1 + j0.5$	1.118	26.6°	.894	$\angle -26.6°$	−0.97
15	$1 + j0.75$	1.250	36.9°	.800	$\angle -36.9°$	−1.94
20	$1 + j1$	1.414	45.0°	.707	$\angle -45.0°$	−3.01
30	$1 + j1.5$	1.803	56.3°	.555	$\angle -56.3°$	−5.11
40	$1 + j2$	2.236	63.4°	.447	$\angle -63.4°$	−6.99
50	$1 + j2.5$	2.693	68.2°	.371	$\angle -68.2°$	−8.61
80	$1 + j4$	4.123	76.0°	.243	$\angle -76.0°$	−12.3
100	$1 + j5$	5.099	78.7°	.196	$\angle -78.7°$	−14.2
200	$1 + j10$	10.050	84.3°	.100	$\angle -84.3°$	−20.0
400	$1 + j20$	20.020	87.1°	.050	$\angle -87.1°$	−26.0
600	$1 + j30$	30.020	88.1°	.033	$\angle -88.1°$	−29.6
800	$1 + j40$	40.010	88.6°	.025	$\angle -88.6°$	−32.0
1,000	$1 + j50$	50.010	88.9°	.020	$\angle -88.9°$	−34.0

Let's take a closer look at Figure 6-6 now that we have constructed our first Bode plot containing a frequency dependent component. Figure 6-6 is the Bode plot for our *RC* network described in Figure 6-5. Looking at the phase angle curve in Figure 6-6, we are actually looking at the phase shift relationship existing between the output voltage, e_{out}, and the input voltage, e_{in}. It's important at this

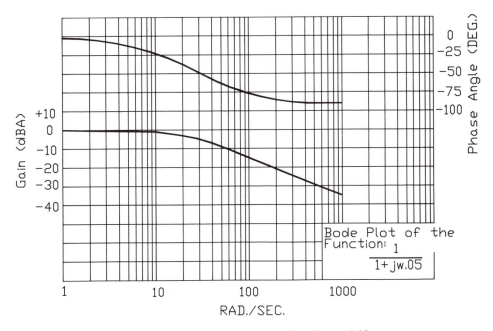

Figure 6-6   Bode plot for the function $1/1 + j\omega 0.05$.

point to remember not to confuse this phase shift relationship with the current and voltage phase shift relationship that exists in a purely capacitive reactance circuit. We already know that regardless of capacitor size and frequency being used, the current *always* leads the voltage by *exactly* 90° for a capacitor. However, in our situation described in Figure 6-5 and Figure 6-6, we are dealing with impedance, the combined effects of resistors and a capacitor in a circuit. And furthermore, we are comparing a circuit's output voltage to its input voltage, not merely a capacitor's output voltage to it's output current. It's very easy to confuse these phase relationships.

There are three significant observations that we make as we inspect the two curves in our Bode plot. First, notice the rather sudden bend in the gain curve that takes place around the 20 rad/sec segment of the curve (Figure 6-7). The bending is in the range of 10 to 30 rad/sec, but the center occurs at approximately 20 rad/sec. This sudden bending is an event that occurs in many Bode gain plots, and it can occur anywhere along the gain plot.

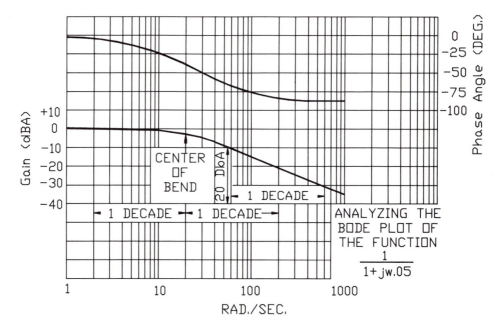

Figure 6-7 · Analyzing the Bode plot of the function $1/1 + j\omega 0.05$.

The second observation has to do with the phase angle curve. Looking again at the bending of the gain curve at the 20 rad/sec point, draw a straight line up to where it intersects the phase angle curve. Notice that it meets the phase angle curve at a value of $-45°$. From this point, look *one decade* to either side of this intersection point. Notice that we encompass most of the S-shaped portion of the phase angle curve. On either side of our one-decade zone we notice that the phase angle curve is relatively flat and is close to $0°$ on the left side of the zone, and is close to $-90°$ on the zone's right-hand side. This observation furnishes us with a valuable tool for helping us to construct our Bode plots later.

The third observation again concerns the gain curve. Preceding the bend in our curve, we notice the levelness of the curve. However, following the curve's sudden downward bend, we notice that the slope remains at a relatively constant value. If we measure this slope, we find it to be about -20 dBA/decade (again, refer to Figure 6-7).

Let's go back now to our first observation concerning the bending point in our gain curve. We had found that this occurred at 20 rad/sec. Look again at our equation (Eq. 6-3), and notice the value of our time constant, τ. By taking the reciprocal of τ, (1/0.05), we find that we come up with the value of 20. This is the value of our gain curve's bend center. When working with Bode plots, this bend, and its value, is referred to as the gain curve's *corner frequency*. This is the point beyond which the frequency response of the system ceases to be one value and gradually begins assuming another value.

Let's now see if we can simplify our plotting chores in Figure 6-6 so that we don't have to construct a table like Table 6-1 each time we want to construct a Bode plot. Instead of plotting individual points as we did in Figure 6-6, we can now make these observations:

1. The corner frequency of our gain plot will occur at 20 rad/sec (1/0.05 = 20).
2. Our gain curve will be flat at a value of 0 dBA from this corner frequency extending to its left.
3. At the corner frequency and to its right, the gain will slope downward at a constant rate of −20 dBA/decade.
4. Looking at our phase angle curve, we can approximate the left-hand portion of the curve with a straight flat line at 0° that extends from the far left end of the plot to within one decade of the corner frequency found on the gain curve.
5. Now, going to the extreme right-hand side of the phase angle plot, we can draw another straight flat line at 90° that extends toward the left to, again, within one decade of the gain curve's corner frequency.
6. The central portion of the phase angle curve can be filled in with a straight line that joins the two ends of the flat lines forming the two ends of the phase angle curves. The straight-line approximations of both the phase angle curve and the gain curve for our transfer function are shown in Figure 6-8.

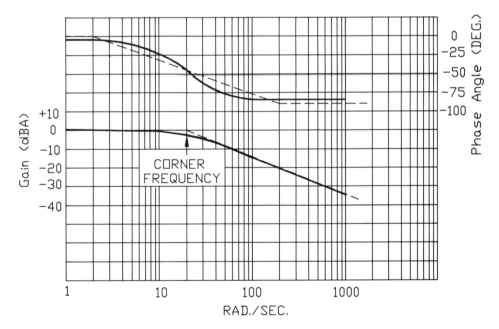

Figure 6-8   The corner frequency.

Let's now try another example to help us become familiar with this straight-line construction process:

EXAMPLE 6-1

Construct a Bode plot, using straight-line approximation, for the function,

$$\frac{1}{1 + j\omega 0.015}$$

Solution:

Step 1. We note that the transfer function is already in the form of Eq. (6-2).

Step 2. Determine the function's corner frequency. This is done by finding the reciprocal of the function's time constant, τ. Since $\tau = 0.015$ sec, the corner frequency is $= 1/0.015 = 66.7$.

Step 3. Lay out coordinates on three- or four-cycle semilog paper similar to the ones used in Figure 6-6.

Step 4. Locate the corner frequency on the graph and extend a straight line from this point to the extreme left of the graph. To the right of this point extend a straight line sloping downward having a slope of −20 dBA/decade and extend it to the far right of the graph. This completes the gain portion of the Bode plot.

Step 5. Draw a straight line at 0° from the extreme left side of the graph over toward the center of the graph. Draw this line lightly. Draw a straight line at −90° from the extreme right side of the graph over towards the center. Extend both lines so that each comes within one decade of the corner frequency. It may help to extend a lightly drawn vertical construction line upwards from the corner frequency on the gain plot to help in locating the decade widths on the phase angle curve construction.

Step 6. After determining the proper lengths of the ends of the phase angle lines, one being higher than the other, both horizontal, and having the center span missing, draw in the central missing portion. This is done by merely connecting the two inside line ends with a sloping straight line. The final results should look like Figure 6-9.

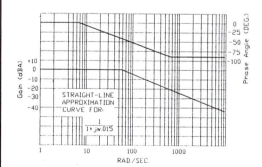

Figure 6-9 Straight-line approximation curve for $1/1 + j\omega 0.05$.

6-3 THE ELEMENTS OF BODE DIAGRAMS

So far, we have managed to plot the Bode diagram of an expression having the form, $1/(1 + j\omega\tau)$. As it turns out, there are other commonly encountered forms of transfer functions, and we want to know how to plot these also. We refer to these various forms as *elements*. Earlier, we discussed another element form, the constant. We found that when plotted, a constant makes a horizontal straight line for its gain plot, and its phase angle plot was another straight line that occurred at 0° regardless of the value of the constant (refer to Figure 6-4).

We now look at yet a third element type. This element is of the form:

$$K_{tf} = 1 + j\omega\tau \tag{6-5}$$

Let's look at an example:

EXAMPLE 6-3

Make a Bode plot of the function, $1 + j\omega 0.1$.

Solution:

Noting that we are no longer plotting a fraction, but instead a whole number expression, we again construct a table similar to Table 6-1, Table 6-2.

TABLE 6-2 RESULTS OF PLOTTING THE FUNCTION: $1 + j\omega 0.1$

ω		$1 + j\omega 0.1$		$dBA = 20\log(A)$
rad/sec	result. exp.	A	$\angle\theta$	dBA
0.1	$1 + j0.01$	1.000	0.6°	0.00
0.5	$1 + j0.05$	1.001	2.9°	0.01
1	$1 + j0.1$	1.005	5.7°	+.04
2	$1 + j0.2$	1.020	11.3°	+.17
5	$1 + j0.5$	1.118	26.6°	+.97
8	$1 + j0.8$	1.281	38.7°	+2.15
10	$1 + j1$	1.414	45.0°	+3.01
15	$1 + j1.5$	1.803	56.3°	+5.12
20	$1 + j2$	2.236	63.4°	+6.99
30	$1 + j3$	3.162	71.6°	+10.00
40	$1 + j4$	4.123	76.0°	+12.30
50	$1 + j5$	5.099	78.7°	+14.20
80	$1 + j8$	8.062	82.9°	+18.10
100	$1 + j10$	10.050	84.3°	+20.00
200	$1 + j20$	20.020	87.1°	+26.00
400	$1 + j40$	40.010	88.6°	+32.00
600	$1 + j60$	60.010	89.0°	+35.60
800	$1 + j80$	80.010	89.3°	+38.10
1,000	$1 + j100$	100.000	89.4°	+40.00

Plotting this type of expression follows the same analysis as was used for $1/(1 + j\omega\tau)$. The results of plotting individual points from a table and of using the straight-line approximation method are shown in Figure 6-10. Notice that our curves are simply upside-down and reversed from what you would have normally expected after having plotted a fractional-type transfer function. All the phase

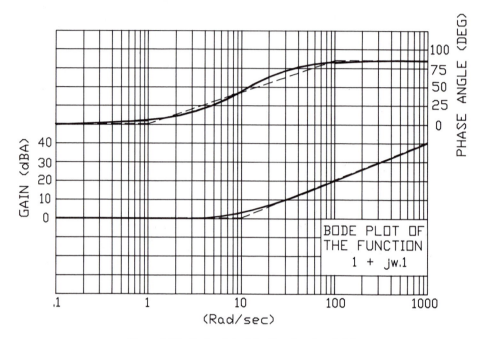

Figure 6-10   Bode plot of the function $1 + j\omega 0.1$.

angles are positive now, as are all the gain values. The slope of the gain curve is $+20$ dBA/decade to the right of the corner frequency point instead of -20 dBA/ decade. The straight-line curves are drawn from the corner frequency point just as they were done in the previous example, to form the gain curve. And the phase angle curve is constructed similarly. That is, the curve is flat $\pm$ one decade either side of the corner frequency.

A fourth type of transfer function element is of the form:

$$K_{tf} = (j\omega)^n \tag{6-6}$$

In this expression the value of n can be any positive or negative integer. The Bode plots for a few positive n values are seen in Figure 6-11. Figure 6-12 shows the Bode plots for some negative values of n.

After studying these curves for a few minutes, we can see an interesting correlation between all these curves. To begin with, let's look at the gain curves in Figure 6-11. We see that for every singular increase in the value of n, the slope

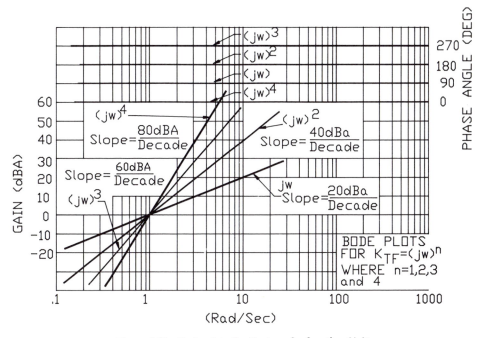

Figure 6-11 Bode plots for the transfer function $(j\omega)^n$.

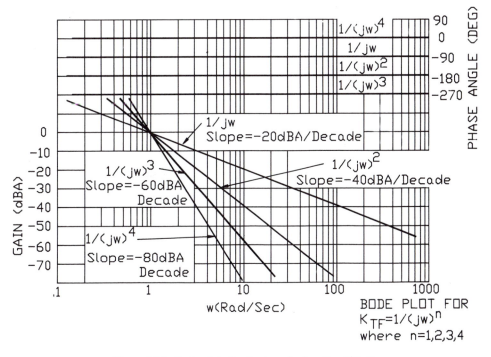

Figure 6-12 Bode plots for the transfer function $1/(j\omega)^n$.

of the gain curve increases in an increment of $+20$ dBA/decade. In addition, the phase angle curve increases by a $+90°$ increment for each of these changes in n. Looking at Figure 6-12, we see that as n is incremented, the negative slope of the gain curve is incremented similarly by an additional 20 dBA/decade. The phase angle curve is incremented by an additional $-90°$ for each change of n.

Let's now take a look at yet another element type found in transfer functions. This is of the form:

$$K_{tf} = \frac{1}{(1 + j\omega\tau)^n} \tag{6-7}$$

Figure 6-13 shows the results of plotting expressions of the form expressed in Eq. 6-7 for several values of n. Notice the similarities between plotting these expressions and plotting the expressions of $(j\omega)^n$.

We can conclude from Figure 6-13 that for every increase by 1 in the value of n, beginning with 1, the negative slope of the gain curve to the right of the corner frequency increases in increments of 20 dBA per decade. This is similar to the observations we made for the $(j\omega)^n$ function. We can also see that the phase angle curve increments by $-90°$ for each increase of n. Notice also the labeling of the x axis for this graph. Since the corner frequency, ω_c, always occurs at $1/\tau$, it stands to reason that for each decade increase or decrease to either side of $1/\tau$, you would obtain the values of ω shown.

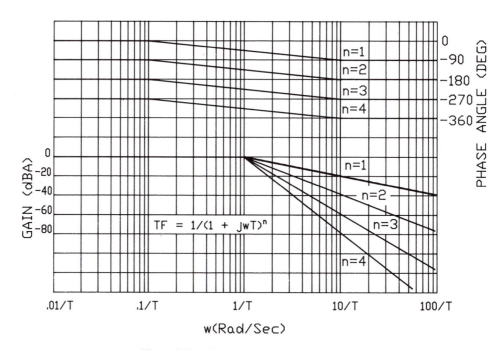

Figure 6-13   Plotting the function $1/(1 + j\omega\tau)^n$.

6-4 HOW ACCURATE ARE THE STRAIGHT-LINE APPROXIMATIONS?

Just how accurate are the straight-line approximation curves when compared to the true curves? Table 6-3 shows the relative comparison of gain plot values depending on how close or how far away you are from ω_c, the corner frequency. In other words, these are the corrections to apply to the function $1/(1 + j\omega\tau)$.

If corrections are to be applied to the straight-line approximation curves for $1 + j\omega\tau$, *add* instead of subtract the values listed in Table 6-3.

Table 6-4 shows the relative comparison of the phase angle curves for the straight-line approximation method versus the true curve plot, again depending on how far or how close you are from ω_c.

TABLE 6-3 GAIN CURVE CORRECTIONS FOR $1/(1 + j\omega\tau)$	
Corrections to subtract from the straight-line approximation curve to obtain the true dBA gain curve:	
ω	Correction
$0.2\omega_c$	0 dB
$0.5\omega_c$	1 dB
$0.7\omega_c$	1.5 dB
$1.0\omega_c$	3 dB
$1.3\omega_c$	2 dB
$1.5\omega_c$	1.5 dB
$2.0\omega_c$	1 dB
$3.0\omega_c$	0 dB

TABLE 6-4 PHASE ANGLE CURVE CORRECTIONS FOR $1/(1 + j\omega\tau)$	
Corrections to add to the straight-line approximation curve to obtain the true phase angle curve:	
ω	Correction
$0.05\omega_c$	$-3°$
$0.10\omega_c$	$-6°$
$0.15\omega_c$	$0°$
$0.20\omega_c$	$+2°$
$0.50\omega_c$	$+5°$
$1.00\omega_c$	$0°$
$3.00\omega_c$	$-5°$
$5.00\omega_c$	$-2°$
$6.00\omega_c$	$0°$
$10.00\omega_c$	$+6°$
$20.00\omega_c$	$+3°$

Again, if you are working with the function $1 + j\omega\tau$, merely *reverse* the signs associated with the correction values listed in Table 6-4.

As you go over Tables 6-3 and 6-4, refer back to Figures 6-8 and 6-10 to verify these corrections in your mind. It will help you to better understand where they came from.

6-5 GRAPHING COMPLEX FUNCTIONS

We have already looked at the characteristics of the four elements of transfer functions. Again, these elements are:

1. K_{tf} = a constant;
2. $K_{tf} = (j\omega)^n$, (where n is a positive or negative integer);
3. $K_{tf} = 1/(j\omega + \tau)^n$, (where n is a positive integer); and
4. $K_{tf} = 1 + j\omega\tau$.

Unfortunately, most transfer functions don't occur in just these simple elementary forms. Instead, they are found in more complex forms. That is, they occur in forms that are in combinations of these four elements. An example of a complex function would look something like:

$$K_{tf} = \frac{0.09}{(j\omega)^2(1 + j\omega0.2)}$$

We now have to discuss a method for making the Bode plot for this more complicated and more usual form of transfer function.

Let's use the preceding transfer function example to demonstrate a method for constructing its Bode diagram or plot. The first thing we should notice is that the preceding expression can be broken down into our smaller, less complex, elements. In other words, we could write the expression as:

$$K_{tf} = \frac{0.09}{1} \times \frac{1}{(j\omega)^2} \times \frac{1}{1 + j\omega0.2} \tag{6-8}$$

It now becomes obvious that we are dealing with three different elementary forms of the transfer function: the constant K, $(j\omega)^n$, and $1/(1 + j\omega\tau)$. It's a relatively simple matter to construct the Bode plot of each of these elements by themselves. Notice that the final transfer function is nothing more than its individual elements all multiplied together. Since we are plotting the logarithms of the expressions, we can merely add all the individual curves together to obtain the equivalent final curve since, in reality, this is the same thing as multiplying all of these expressions together.

In order to plot Eq. (6-8), you must first plot the individual elements that make up the expression. This has been done in Figure 6-14. The next step is to pick several values of ω and begin adding the values of the individual gain amounts for each curve. Also, add the individual phase angle amounts for each curve at each ω. Pick ω values that are prominently located, such as at the ends of the curves, at the corner frequencies, and at either side of the corner frequencies. Table 6-5 shows the results of doing just that for Eq. (6-7). The individual element plots have been numbered in Figure 6-14 so that you can key the curves to the results in Table 6-5.

The final result of adding together the individual element curves using the data in Table 6-5 is shown in Figure 6-14. The final gain and phase angle curves are shown in boldface type in this figure.

Let's now try another example of graphing a complex transfer function expression.

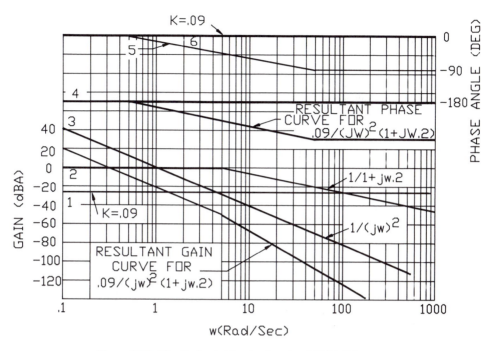

Figure 6-14   The Bode plot for the function $0.09/(j\omega)^2(1 + j\omega0.2)$.

TABLE 6-5   DETERMINING THE POINTS NEEDED FOR PLOTTING THE EXPRESSION $0.09/[j\omega)^2(1 + j\omega0.2)]$

| Curve no.: | Gain | | | | Phase angle | | | |
	1	2	3	Result	4	5	6	Result
$\omega =$	0.1	-22	$+0$	$+40 = +19$ dBA	-180	$+0$	$+0$	$= -180°$
$\omega =$	0.5	-22	$+0$	$+13 = -9$	-180	$+0$	$+0$	$= -180$
$\omega =$	1	-22	$+0$	$+0 = -21$	-180	-15	$+0$	$= -195$
$\omega =$	5	-22	$+0$	$-26 = -48$	-180	-45	$+0$	$= -225$
$\omega =$	10	-22	-6	$-40 = -68$	-180	-60	$+0$	$= -240$
$\omega =$	50	-22	-20	$-68 = -110$	-180	-90	$+0$	$= -270$
$\omega =$	100	-22	-25	$-80 = -127$	-180	-90	$+0$	$= -270$

EXAMPLE 6-4

Plot the expression:

$$\frac{100}{j\omega(1 + j\omega0.25)}$$

using the straight-line approximation method.

Solution:

Refer to Figure 6-15 for the graph of this expression. We have broken down the given expression into its elements of 100, $1/j\omega$, and $1/(1 + j\omega 0.25)$. Next, the individual elements are plotted just as we did in the previous example. We do this so that we can keep track of adding the elements together in the proper sequence to give us our final gain and phase angle summation curves. These final curves are again shown in bold on our graph.

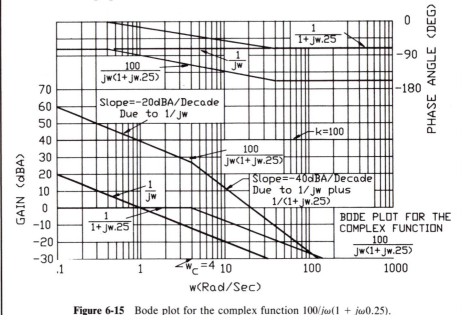

Figure 6-15   Bode plot for the complex function $100/j\omega(1 + j\omega 0.25)$.

6-6 DETERMINING A SYSTEM'S STABILITY

Up to this point, we have learned a system of mechanically plotting the frequency response curves for transfer functions of devices or systems. Little thought has been given as to what information is available from these curves concerning the behavioral characteristics of the system or of the device itself. As we find out here, we can get a surprising amount of information from our transfer function and its Bode diagram concerning the behavioral characteristics of the system itself.

As just mentioned, Bode diagrams can be useful in telling us about certain control system characteristics. Back in Chapter 2 we discussed the characteristics of positive and negative feedback circuits. In Section 2-6 we discussed how negative feedback could be used to automatically control the altitude of a hot air balloon. We described the behavior of an altitude control system that was too quick, or jittery, in its response, and we also described how that same system would behave if it were too sluggish. What we didn't mention, however, was the fact that even though we were inverting the feedback signal by 180° before adding it to the incoming input command signal, the error signal's resultant phase angle (remember, the error signal was the resultant correction or readjustment signal going into the control system) would be equal to the command signal's phase angle (whatever that happened to be) plus −180°. To clarify a point here, let's look at our hot air balloon control system once again. Assume for the moment that the output phase angle, as determined from a Bode plot made of the balloon's control system, showed a value of 0°. Remember, the Bode plot is based solely on open-loop data. Let's further assume that the input signal had a phase shift also equal to 0° and a magnitude voltage of, say, 10 volts. In other words, our open-loop system gain, as determined by the overall system's transfer function, is:

$$\frac{\text{output}}{\text{input}} = \text{system's transfer function gain, } KG = \frac{10 \angle 0°}{10 \angle 0°}$$

$$= 1 \angle 0°$$

Now, we close the loop on the balloon's system so that the 180° phase shift takes place on the output signal in the feedback circuit. Back in Chapter 2 we learned that the transfer function for a negative feedback system was $AB/(1 + AB)$. Let's assume that we are feeding back 100% of our output signal back to the input so that now $B = 1$. Since our transfer function gain is KG, we can now say that the closed-loop gain of our system can now be expressed as:

$$\frac{KG}{1 + KG} \tag{6-9}$$

We now substitute the gain $1 \angle 0°$ for KG in Eq. (6-9) to see what sort of closed-loop gain we will obtain:

$$\frac{KG}{1 + (1)KG} = \frac{1 \angle 0°}{1 + 1 \angle 0°} = \frac{1 \angle 0°}{1 + (1 + j0)} = \frac{1 \angle 0°}{2} = 0.5 \angle 0°$$

$$\hookrightarrow \text{Remember, } B = 1$$

In other words, the correction signal would be reduced by one-half and the system would undoubtedly respond very sluggishly to this reduced signal. However, in all fairness to the preceding problem, we could have selected an output signal whose magnitude was something greater than 1 volt so that the resultant closed-loop signal was larger. But, as is typically done in control system design, system behavior is usually analyzed for a system gain equal to 1, or, as it is more frequently referred to, the system is usually analyzed at *unity gain*. This means for our control system, the output signal would be exactly the same magnitude as the input signal.

Now, let's analyze the same system but with an output signal having a phase angle of −180°, as measured with our open-loop Bode plot instead of 0° just used.

Again, let's run this new *KG* figure through Eq. (6-9) to see what the resultant closed-loop gain will be:

$$\frac{KG}{1 + BKG} = \frac{1 \angle -180°}{1 + (1)1 \angle -180°} = \frac{1 \angle -180°}{1 + (-1 + j0)} = \frac{1 \angle -180°}{0} = \infty$$

Based on what we just calculated, our control system will be jolted with an infinitely high gain. What started out as an open-loop gain of unity has now turned into an infinitely large gain, and our control system has now become a wild, uncontrollable system. This is just the opposite condition of the previous experiment.

Our two control system examples have pointed out one thing to us. There must be some compromise existing between an output phase angle of 0° and −180° where we can expect a control system to behave just right, and there is. We can find this just-right point of operation either through trial-and-error, or we can use our Bode plot to obtain this vital information. But before we see how the Bode plot is used, let's try another series of experiments having to do with our balloon's control system stability. Again, keeping our open-loop system's gain magnitude ratio at unity (i.e., this is the same thing as saying a gain of 0 dB), but varying the output phase angle through the entire range of 0° through 360°, let's calculate the resultant gains. Figure 6-16 shows the results. We can see what happens at or

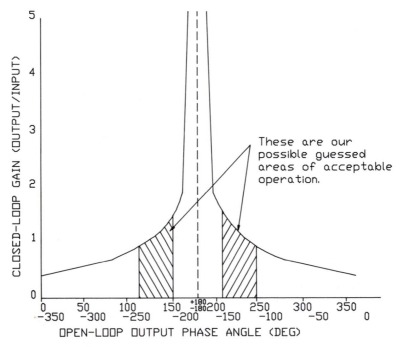

Figure 6-16   Curve showing relationship between closed-loop gain and the open-loop phase angle.

near −180°; obviously we want to stay clear of that area of phase angle values. We also want to stay away from the area where the phase angles are near zero or a little greater in absolute value. It appears that maybe a just-right condition may occur somewhere in the vicinity where the closed-loop gain is about 1, or maybe even slightly larger than 1. We don't want the gain to be too large though, otherwise our system may be too quick or jittery. Perhaps we should not go above a gain of 2 or so. According to our graph in Figure 6-16, this represents a phase angle range of around −110° to, say, −150°. (This is the same thing as saying a range of 210° to 250°.) However, this is not a very scientific method of determining what the desirable range of phase angle operation should be. As it turns out, if the phase angle is found to be within about 40° to 60° of −180° (see Figure 6-17), the

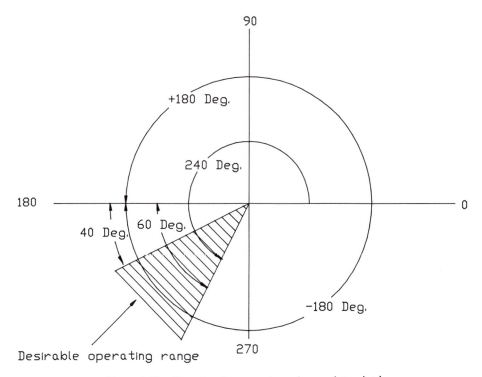

Figure 6-17. How the phase margin angles are determined.

system is going to be stable (i.e., just right) when the system's control loop is closed. In other words, a phase angle range of (−180° + 40°) = −140° to (−180° + 60°) = −120° is the desired design operating range. There is a special name given to this range; it's called the *phase margin*. You might think of it as a margin of safety. Any phase angles greater than −140° (that is, more negative than −140°) will result in a too lively or jittery of a system, while phase angles less than −120° (less negative than −120°) will result in too sluggish of a system. Remember though, the system's gain must be 1 (or to say it another way, the system's gain must be zero dBA, since 20log 1 = 0 dBA). By the way, if you are wondering how the figures of 40° to 60° were determined, their derivation is somewhat involved. At this point we have to accept their values as fact until we learn some more theory on control systems. Further explanation is given in later chapters.

Notice that in our discussion of phase margin, we referred to that portion of our curve in Figure 6-16 between 0° and −180° and not the portion from 0° to +180°. The reason for this is partially due to custom. You will find that most Bode diagrams dealing with negative feedback systems show their phase angles in the 0° to −360° range rather than in the positive angle range.

Describing the behavioral characteristics of a control system can be a little difficult. To explain what we mean when we say that a system is either too jittery or too sluggish is somewhat of a judgment call in certain instances. However, there are cases where most people would agree that a system is behaving rather poorly. As an example, think of the steering characteristics of a car. We have, at one time or another, experienced a car's steering that was probably very quick— perhaps too quick for our comfort. In other words, when turning the car's steering wheel rapidly back and forth, we developed a feeling that we were going to lose control of the car because of the too sudden response or quickness of the steering mechanism. The other extreme of poor steering is where the car's wheels seemed very slow or sluggish in responding to the turning of the steering wheel. It's almost as if there were too much slack in the steering mechanism and it took uncomfortably long for the wheels to respond to the steering command. You had the feeling that you were trying to turn the steering wheel one way while the car's body was still moving in the other direction. The steering command and the car's body were *out of phase* with each other. Obviously, this was an extreme steering condition, just as the too quick steering condition was also an extreme. We would want our steering system characteristics to lie somewhere in between. We have the similar problem in control system design. We want to be able to design the just-right system—one that is not too jittery in its response, but one that is not too sluggish either.

Bode plots are used to forecast a system's stability habits. They allow you to make a what-if determination. They allow you to see what would happen if you were to close the loop on your control system to see how it would behave from a stability standpoint. Therefore, in reality, Bode plots are open-loop diagrams based on open-loop information used to forecast closed-loop characteristics.

Based on what we have discussed so far, let's try our hand at finding out the stability characteristics of some systems.

EXAMPLE 6-5

Using the Bode diagram of Figure 6-14, find the stability characteristics of the plotted transfer function.

Solution:

Step 1. Determine the gain curve's gain crossover frequency. This is where the gain curve crosses the 0 dBA line in the Bode plot: According to Figure 6-14, the frequency at which the gain curve crosses the 0 dBA line is 0.3 rad/sec.

Step 2. At this crossover frequency, determine the phase angle from the phase angle curve: From Figure 6-14, the phase angle at $\omega = 0.3$ rad/sec is $-180°$.

Step 3. Determine the phase margin. This is done by performing the following calculation: phase margin = phase angle reading + 180°. (Note that this is the same thing as saying: phase margin = phase angle reading $-$ ($-180°$)).

$$\text{phase margin} = -180° + 180°$$
$$= 0°$$

According to our previous discussion, a phase margin of 0° produces an unstable system behavior.

EXAMPLE 6-6

Find the stability characteristics of the transfer function in Figure 6-15.

Solution:

Step 1. Determine the gain crossover frequency: From Figure 6-15 we see that the gain crossover frequency is 20 rad/sec.

Step 2. Find the corresponding phase angle at this frequency: Again from Figure 6-15, we see that the phase angle at $\omega = 20$ rad/sec is $-165°$.

Step 3. Calculate the phase margin:

$$\text{phase margin} = \text{phase angle reading} + 180°$$
$$= -165° + 180°$$
$$= 15°$$

A phase margin of 15° represents a system that is too jittery since it is outside the recommended 40° to 60° range and is closer to $-180°$.

Other examples of determining control system characteristics would be:

1. If, at the gain crossover frequency on a Bode plot the phase angle value were $-100°$, the phase margin would be $-100° + 180° = 80°$. This system would be too dead or sluggish.

2. On another Bode plot, if at the gain crossover frequency the phase angle were found to be $-135°$, the phase margin would be $-135° + 180° = 45°$. This system would be quite stable since its phase margin falls within the recommended range of 40° to 60°.

SUMMARY

In Chapter 6 we discovered that by plotting the transfer functions in a special configuration called a Bode diagram, we could obtain a much better understanding as to how the control system whose transfer function we are plotting will behave. Most transfer functions can be broken down into much simpler expressions called *elements*. Each of these elements possesses a particular Bode plot form, so that by adding these elements together during the course of making a Bode plot, more complex control system transfer functions may be plotted and analyzed.

The corner frequency on a Bode gain plot is the frequency at which a relatively sudden change in the gain curve's slope occurs. This is also called the gain curve's *break point*. Knowing the corner frequencies of a curve helps in the effort to make an estimate of the shape of that curve's Bode plot.

The behavior of a control system depends on the amount of gain in the system and on the resultant phase angle. If we represented a transfer function by its open-loop form, *KG,* then its closed-loop equivalent would be (assuming 100% feedback), $KG/(1 + KG)$.

Normally, the Bode plot is analyzed for its open-loop phase angle at unity crossover gain, and then its phase margin determined. The phase margin is the arithmetic difference between the phase angle and $-180°$. It is this figure that gives us insight into the behavioral patterns of the control system. If the phase margin is less than 40°, the system will most likely be too jittery. If the phase margin is greater than 60°, the system will most likely be sluggish in its performance. A phase margin between 40° and 60° is considered stable and desirable.

EXERCISES

The following exercises require the using of four-cycle semilog graph paper.

6-1. Draw the open-loop Bode plots of the transfer functions 1, 10, 17, and 125 all on the same graph paper.

6-2. Draw the open-loop Bode plots of $j\omega$, $(j\omega)^{-2}$, $(j\omega)^3$, and $(j\omega)^4$ all on the same graph paper.

6-3. Draw the open-loop Bode plots of $(j\omega)^{-1}$, $(j\omega)^2$, $(j\omega)^{-3}$, and $(j\omega)^{-4}$ all on the same graph paper.

6-4. Plot the expression $1/1 + j\omega 0.05$ using the straight-line approximation method.

6-5. Draw the Bode diagram for the function $15/1 + j\omega 0.15$. Determine its corner frequency (in rad/sec), the phase margin, and determine the system's stability.

6-6. Draw the Bode diagram for the expression $1 + j\omega/1 + j\omega 10$.

6-7. Draw the Bode diagram for the expression $21.5/j\omega(1 + j\omega 0.5)$.

6-8. Determine the stability of a system whose transfer function is $0.26/(j\omega)^2(1 + j\omega 0.5)$.

6-9. Do the same for the function $25.5/j\omega(1 + j\omega 0.25)$.

6-10. Do the same for the function $3(1 + j\omega 0.5)/j\omega(1 + j\omega 0.1)(1 + j\omega 5)$. Be careful with this one as far as plotting accuracy is concerned, due to the many elements involved.

REFERENCES

DeRoy, Benjamin E., *Automatic Control Theory,* New York, N.Y.: John Wiley & Sons, Inc., 1966.

McDonald, A. C. and Lowe, H., *Feedback and Control Systems,* Reston Va.: Reston Publishing Company, Inc., 1981.

Sante, Daniel P., *Automatic Control System Technology,* Englewood Cliffs, N.J.: Prentice-Hall, Inc., 1980.

Weston Components Division, *Servo Engineer's Handbook,* Weston Instruments, Inc., 1968.

Closing The Loop

7-1 CLOSING THE LOOP

We found out in Chapter 6 that the Bode diagram was a frequency response curve especially designed for interpreting the transfer function characteristics of an automatic control system. All of the data used for constructing the Bode plot was open-loop data. We used the Bode plot to make a "what if we closed the loop . . . ?" determination for us. The advantage to this method is that it is fast and fairly accurate. The disadvantage is, you don't really see the actual closed-loop gain and phase angle curves resulting from closing the loop. Perhaps you want to know what the actual gain and phase angle values are for a given frequency or range of frequencies. The open-loop Bode plots won't give you that information at all. What we need then is some sort of equivalent method that will allow us to plot a closed-loop Bode plot which will give us this data.

7-2 THE CLOSED-LOOP GAIN PLOT

Since automatic control systems are closed-loop systems, the logical question arises, are there such things as closed-loop Bode plots? The answer to this question is, yes. However, their construction can be rather difficult because of the math involved. First, it helps to review the meaning of the general closed-loop expression, $KGH/(1 + KGH)$. Look at Figure 7-1. Here we can see the difference between the open-loop versus closed-loop system. The transfer function expressions (i.e., the algebraic expressions used to describe output/input) will be

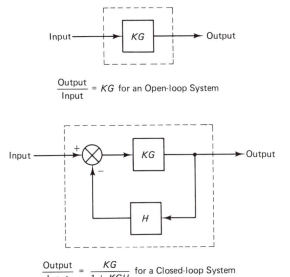

$$\frac{Output}{Input} = KG \text{ for an Open-loop System}$$

$$\frac{Output}{Input} = \frac{KG}{1 + KGH} \text{ for a Closed-loop System}$$

Figure 7-1 Open-loop and closed-loop system comparison.

radically different in each case. Let's try an example. Assume that we have an open-loop transfer function for a control system equal to $100/[j\omega(1 + j\omega 0.25)]$. We have to find its closed-loop output/input expression. The components that would generate this type of transfer function could be an amplifier (with a magnitude gain setting of 1 used as a summing point for the input control voltage or set point, and the feedback signal), another amplifier with a magnitude gain setting of 1,000, a servomotor whose transfer function is $5/[j\omega(1 + j\omega 0.25)]$ (see Section 4-11.1), and a gear train for reducing the servo's high rotary displacement output. The gear ratio for this train is 50 (remember that $K_{gear} = 1/N$; therefore, its transfer function becomes 1/50). When all of these individual component transfer functions are combined into one overall system transfer function (this is done by multiplying all the transfer functions by one another), we will obtain the foregoing overall transfer function. Actually, there are an infinite number of component transfer functions that, when properly combined, will give us our desired resultant transfer function. However, for the purpose of presenting this problem, we use the combination just given. This system, by the way, could be a system used for moving a pointer or other indicator to a specific location on a readout scale. The control signal could be sent from a remote location. The indicator will have the capability to correct its position automatically if it should be bumped or otherwise disturbed. Figure 7-2 shows the block diagram for our system. If the servomotor is forced to move away from a certain position, the servo potentiometer compensates for this shift in position by outputting a voltage of the opposite polarity, which is fed back to the differential amplifier (the summing point). A servo pot is nothing more than a precision potentiometer whose wiper is attached to a rotat-

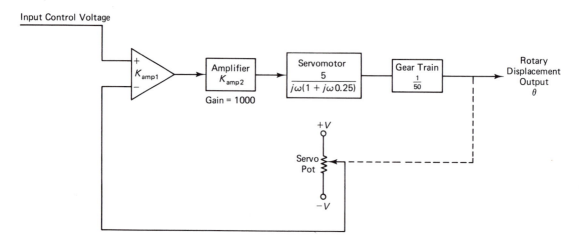

Figure 7-2 Complete automatic control positioning system.

able shaft. The shaft is usually coupled to a servomotor's shaft. As the servo's shaft rotates, the potentiometer shaft also rotates, producing a proportional variable resistance. A voltage is applied to the potentiometer so that a variable output voltage is obtained that is proportional to the amount of rotation. In our case, this output voltage's polarity will cancel a portion of the input control voltage to the amplifier, thus causing the servomotor to rotate in the opposite direction until it once again returns to its original position. For our system we will assume 100% feedback so that $H = 1$. The conversion to a closed-loop transfer function would proceed something like this:

$$\frac{KG}{1 + KG} = \frac{\dfrac{100}{j\omega(1 + j\omega 0.25)}}{1 + \dfrac{100}{j\omega(1 + j\omega 0.25)}} = \frac{\dfrac{100}{j\omega(1 + j\omega 0.25)}}{\dfrac{j\omega(1 + j\omega 0.25) + 100}{j\omega(1 + j\omega 0.25)}}$$

$$= \frac{100}{j\omega(1 + j\omega 0.25) + 100} = \frac{100}{j^2\omega^2 0.25 + j\omega + 100}$$

$$= \frac{100}{100 + j\omega - \omega^2 0.25} \tag{7-1}$$

As you can see, there is a considerable difference between the open-loop and closed-loop expressions. Notice that we didn't include a transfer function for the servo pot in our closed-loop function. The servo pot is part of the feedback loop and really doesn't enter into the system's function. It merely converts displacement into a voltage. Any device located in a feedback loop generally doesn't get included in a transfer function if all it does is convert a quantity into a voltage for the summing junctions of a system. This conversion must be 100% fed back to the summing point. Any amount less than this must then show up as a decimal portion of *H* in the equation, $KGH/(1 + KGH)$.

Unfortunately, there is no easy straight-line approximation method that we can use to plot the closed-loop curves as we had for the open-loop expressions. The only way to plot the closed-loop expression is by using the laborious point-by-point calculation method, that is, choose values for ω and calculate the resultant gain and phase angles. (Later, we learn another, much quicker method of constructing these curves.) This can be extremely time-consuming; however, Figure 7-3 is the resultant plot.

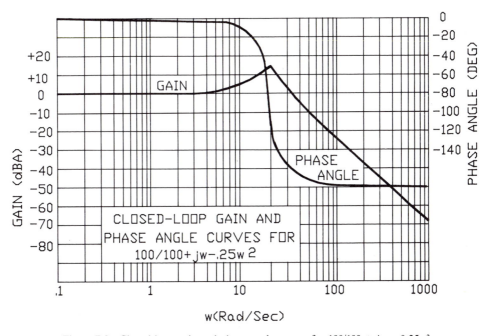

Figure 7-3   Closed-loop gain and phase angle curves for $100/100 + j\omega - 0.25\omega^2$.

After inspecting the curves in Figure 7-3, we can see immediately what one advantage there is to a closed-loop plot of this kind. We have an immediate picture of what the gain and phase angle values are for any given value of ω. We also can't help but notice the rather prominent feature of the peak occurring in the gain curve at around 20 rad/sec. This is an often-encountered feature in many closed-loop gain curves and is called the *maximum system gain,* or M_m. The frequency at which the maximum system gain occurs (in our case, 20 rad/sec) is called the *maximum gain frequency,* ω_m. And there is an additional piece of vital information that we can obtain from this peak. As long as this peak doesn't exceed a value of 3 dB, the system is probably stable or at worse, jittery, depending on the system's application. Any value higher than 3 dB describes a system that is either too jittery or unstable. This interesting piece of information ties in with a statement that was made in Section 6-6 concerning system stability. If you recall, we discussed a curve in Figure 6-16 in an attempt to explain how the phase margin was derived. Referring to this figure, if you were to determine the gain ratio using the curve at the 40° phase margin, you would obtain a value of approximately 1.4. Converting this to dBA, we would get 2.9 dBA. These two observations, the 40° phase margin gain and the closed-loop peak gain, are obviously related and are discussed later.

The gain peak in Figure 7-3 occurs at approximately 14 dBA. This tells us that the control system described by this curve is entirely too jittery. Inspection of the system's Bode plot in Figure 6-15 supports our conclusion. The phase margin at the crossover point in this diagram is only 15°. This is entirely too close to $-180°$ for proper system operation.

What would happen to our gain curve peak if we plotted a function that we knew ahead of time was sluggish rather than jittery? Let's try it. Let's plot the function, $1/(1 + j\omega - \omega^2 0.25)$. The reason why we know this particular function is sluggish, is this: Look at the Bode plot for $100/(100 + j\omega - 0.25\omega^2)$, Figure 6-15 in the previous chapter. Notice that we have reduced the constant value, 100, to only 1. In essence, what we have done here was to reduce the gain by a factor of 20log 0.01, or -40 dBA. In reality, we multiplied the transfer function by the constant, 0.01, which is the same as subtracting 40 dBA from the original gain curve. Doing this has no affect on the phase angle curve, since constants plot as 0° on all Bode plots. The resultant Bode gain curve is shown in Figure 7-4. Reducing the gain curve by -40 dBA has increased the phase margin to a value of 70°. Since this is greater than the recommended 40° to 60° phase margin, the system will have a deadened behavior. This is the same as reducing the gain of the amplifier in the system, or increasing the gear ratio in the transmission gear box connecting the servo to the servo pot. The idea is to tame the system, so to speak, from what it was originally.

The closed-loop plot of our new function is seen in Figure 7-5. We now see that the maximum gain hump is missing in the gain curve. This indicates that our system has in fact become deadened as a result of subtracting the 40 dB from the original curve. Our system is now too dead. Let's try to liven it a bit by subtracting only 20 dB from the original curve in Figure 6-15. Figure 7-6 shows the Bode plot for the transfer function, $10/[j\omega(+ j\omega 0.25)]$ that would result from subtracting this 20 dB. Notice the phase margin has now decreased to a value of only 35°. This will create a jittery control system. The closed-loop response curve in Figure 7-7 proves this out. Notice the return of the maximum gain hump at $\omega = 6$ rad/ sec. The peak of this hump occurs at little over 4 dBA. This verifies the jitteriness of our system, since the peak of our curve exceeds the recommended $+3$ dBA value mentioned earlier in our discussion.

Finally, let's adjust our original curve in Figure 6-15 so that it gives us a perfectly stable system. We do this so that we can see what the closed-loop

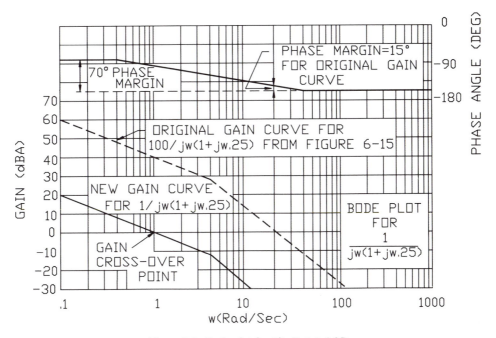

Figure 7-4   Bode plot for $1/j\omega(1 + j\omega 0.25)$.

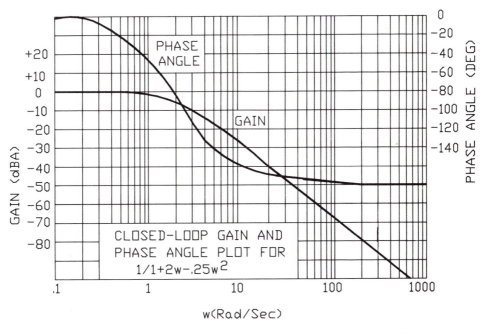

Figure 7-5   Closed-loop gain and phase angle plot for $1/1 + j\omega - 0.25\omega^2$.

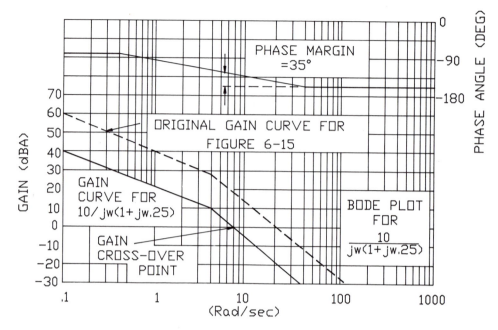

Figure 7-6   Bode plot for $10/j\omega(1 + j\omega0.25)$.

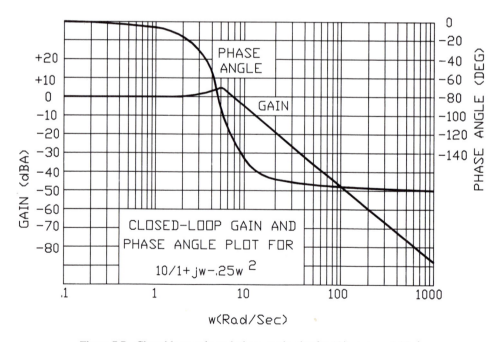

Figure 7-7   Closed-loop gain and phase angle plot for $10/1 + j\omega - 0.25\omega^2$.

frequency response curve looks like. Looking at Figure 6-15, it appears that if we were to lower our original curve by 30 dB, this would give us a phase margin of 50°. This would put us right in the middle of the recommended phase margin region for stability. Figure 7-8 shows the Bode plot and how the phase angle value was determined. The resultant closed-loop frequency response curve for this system is seen in Figure 7-9. The gain curve in this figure shows a barely perceptible rise in gain before taking the characteristic drop to the right of this rise. The maximum gain at the peak is less than 0.2 dBA. This very gradual rise in the gain is fairly characteristic of a stable system although it varies from system to system. Again, this rise may be as high as +3 dB and still produce an acceptable system.

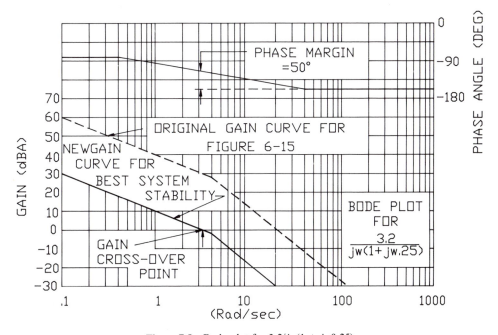

Figure 7-8   Bode plot for $3.2/j\omega(1 + j\omega0.25)$.

Another important piece of information that we can obtain from our closed-loop curves, in addition to the maximum peak gain and frequency, is the system's *bandwidth*. In Section 2-3 of Chapter 2 we defined bandwidth as the span of frequencies that is included within the 3-dB down points of a device's frequency response curve. Applying that definition to Figure 7-3, we see that the left-hand side of the gain curve is relatively flat and at approximately 0 dBA. It appears to be quite unlikely that the curve would go down to −3 dBA before reaching 0 rad/sec. However, the right-hand side of the gain curve is seen to dip to −3 dBA at 32 rad/sec. Therefore, the bandwidth of this curve is said to be equal to 32 rad/sec. System bandwidths, such as the one we have been working with here, are always

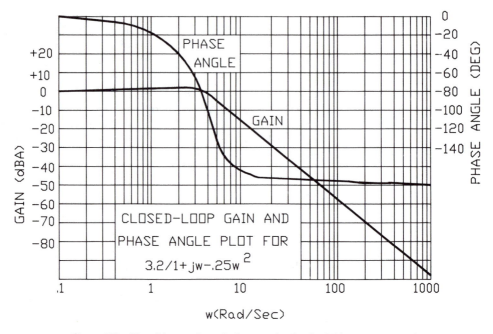

Figure 7-9 Closed-loop gain and phase angle plot for $3.2/1 + j\omega - 0.25\omega^2$.

determined from their closed-loop data, never from their open-loop Bode diagrams.

Another observation that we can make concerning the maximum system gain is its frequency value on the closed-loop response curve. Its frequency always coincides with the frequency of the gain crossover point on the Bode diagram. Be sure to verify this to yourself using the curves we have just finished discussing. This is a helpful check in comparing the accuracy of plotted data in both diagrams.

So far, we have focused our attention on the gain curves in our closed-loop response curves. Let's look now at the phase angle curves to see what information they contain for us. Earlier, we stated that the transfer function we have been working with represents the control system in Figure 7-2. Let's use the modified version of this function as in Figure 7-9. According to this curve, the bandwidth is now only 4.3 rad/sec, as opposed to the earlier bandwidth of 32 rad/sec when the amplifier gain was higher (or the gear box ratio was lower). The actual frequency at this point is equal to $\omega/2\pi$, or 0.68 Hz. In other words, if the command voltage variations on our servomotor vary any more rapidly than about 0.7 Hz, the system, in all likelihood, will not be able to keep up with these variations. And notice what the phase angle is doing when we are near the maximum gain area of the gain curve. The closed-loop phase angle curve begins changing rather rapidly. This is especially true for livelier systems such as the one described in Figure 7-7 (or such as in the extreme case seen in Figure 7-3). Obviously, it isn't desirable to have a system's output out of phase with its input. This is like the car's steering mechanism described in Chapter 6. This could cause a lot of confusion for the system's mechanisms and electronics, not to mention the amount of confusion created for the system's operator. We want to operate in that portion of the closed-loop frequency response curve where the phase angle remains very close to 0°. This is usually in the region just to the left of the maximum gain point on our gain curve.

Let's review the advantages of using closed-loop data instead of open-loop data in designing a control system. A closed-loop frequency response plot allows us to:

1. determine the bandwidth of the control system;
2. find the maximum system gain, M_m;
3. determine the frequency at which the maximum system gain occurs; and
4. find the actual phase angle occurring at a particular frequency.

In the next section we learn another method for constructing the closed-loop diagram allowing us to obtain the very same information but in a much shorter time.

7-3 USING THE NICHOLS CHART

As we have found out, constructing closed-loop frequency response curves can be a very tedious and time-consuming experience. The calculations alone account for the majority of time spent in preparing these diagrams. It helps to have a programmable calculator available for performing the numerous math steps required; however, we now look at a much easier method that's available to us. This is a method that produces fairly accurate and quick results. We will have to turn in our log paper used for making Bode plots for a special coordinate paper referred to as *Nichols chart paper*. A copy of this paper style is seen in Figure 7-10. Unfortunately, this paper is not as readily obtainable as log paper; therefore, it will be necessary to make photocopies of Figure 7-10. Doing this will allow us to work the problems that we discuss here.

Actually, we will still be using log paper for construction of the open-loop Bode plot. This information is needed so that we can use the Nichols chart to convert the open-loop data into closed-loop data. We now analyze this new method and, in the process, use an example problem to see how the conversion process works.

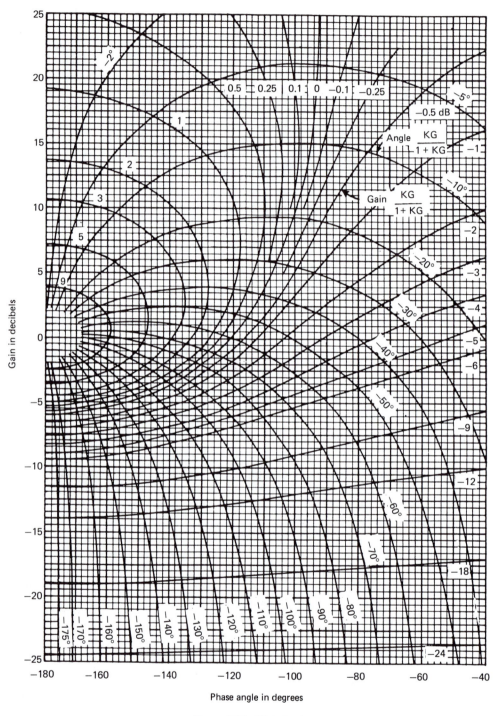

Figure 7-10 Nichols chart.

7-4 AN EXAMPLE

Let's convert the open-loop Bode diagram for the expression $100/[j\omega(1 + j\omega 0.25)]$ into its closed-loop response curve. We will have to refer back to Figure 6-15 for the open-loop Bode plot of this expression. Here is a step-by-step instruction on how to use the Nichols chart method.

Step 1:   Construct the open-loop Bode plot of your control system as usual (Figure 6-15).

Step 2:   Convert your straight-line approximation curves to smoothed curves using the conversion information given back in Section 6-4, Chapter 6. (See Table 6-3 in this section for the gain conversions and Table 6-4 for the phase angle conversions.) The results of our example conversion are shown in Figure 7-11.

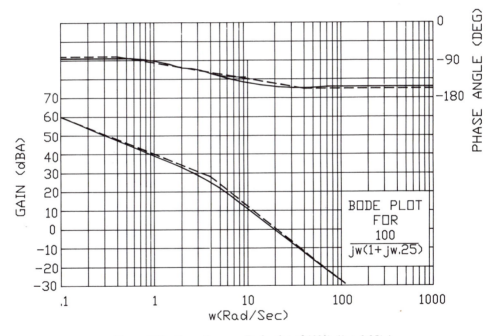

Figure 7-11   Smooth-curve Bode plot of $100/j\omega(1 + 0.25j\omega)$.

Step 3:   Choose about 10 or so ω values from your Bode plot and determine both the gain and phase angle values for each of the ω values. If necessary, make a table of these results (see Table 7-1). Obtain a sheet of Nichols chart paper and, using the chart's outer margin rectangular coordinates of gain and phase angle, transfer the preceding gain and phase angle values for each ω to this new grid.

**TABLE 7-1   OPEN-LOOP GAIN AND
PHASE ANGLE VALUES FROM
FIGURE 6-15**

ω	Gain (dBA)	Phase Angle (deg)
4.0	25	−135
6.0	20	−147
7.0	17	−150
10.0	11	−159
15.0	4	−166
20.0	−1	−170
30.0	−8	−174
40.0	−13	−174
50.0	−17	−175
70.0	−22	−178

Label each ω with its value. With a french curve, draw a smooth curve through your 10 or so points, as was done in Figure 7-12.

Step 4:   At each ω point on the gain–phase angle curve, read off the closed-loop gain values using the curved grid pattern labeled Gain $KG/(1 + KG)$. Also, read off the closed-loop phase angle values using the curved grid pattern labeled Angle $KG/(1 + KG)$. Record these new data in a separate table (see Table 7-2).

**TABLE 7-2   CONVERTED
CLOSED-LOOP DATA TAKEN FROM
NICHOLS CHART**

ω	Gain (dBA)	Phase Angle (deg)
4.0	.4	−3
6.0	.7	−4
7.0	1.1	−5
10.0	2.7	−7
15.0	8	−20
20.0	9+	−123
30.0	−3.5	−170
40.0	−11	−172
50.0	−16	−174
70.0	−22	−178

Step 5:   Plot the data obtained from the Nichols chart onto semilog paper. The results of our example conversion are shown in Figure 7-13.

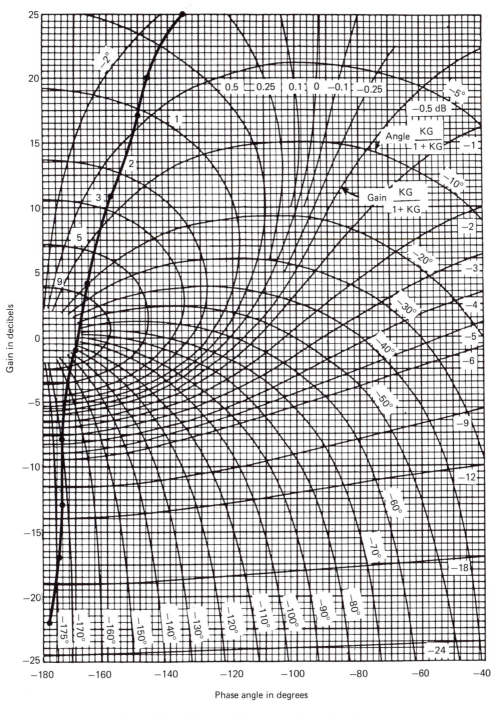

Gain in decibels

Phase angle in degrees

Figure 7-12 Using the Nichols chart for plotting Example 6-6.

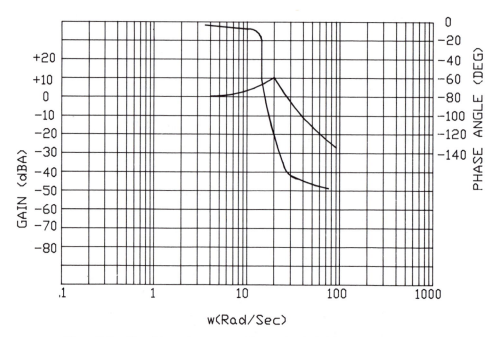

Figure 7-13 Closed-loop plot generated from using the Nichols chart method (see Figure 7-3).

As you can see, the results in Figure 7-13 compare fairly well with the hand-plotted results in Figure 7-3. There is some distortion present in the curves of Figure 7-13; however, this is not a serious problem as far as interpreting the data in the resultant curves. The maximum peak gain value obtained with the Nichols chart does not go any higher than 9 dBA, but since this is an unstable condition to begin with, there is really no need to go beyond this reading. Since our Nichols chart doesn't convert open-loop data beyond +25 dBA or below −25 dBA, the resultant closed-loop curves will be cut off at these points. Again, the trend of most curves at these points is generally established, so you can easily extend the curves, if you like, with little loss in accuracy. In our case, the curves' ends are obviously headed outward along straight lines and may be extended as straight lines if desired.

The most interesting thing about using the Nichols chart method has to do with its conversion speed. Being able to convert rapidly from the open-loop data of a Bode plot to the closed-loop data using the Nichols chart, one can easily observe the closed-loop frequency response curves as they respond to any modifications being made to the transfer function. The problem we have been working with is a good example. We know, for instance, that the transfer function, 100/

$[j\omega(1 + j\omega0.25)]$ is unstable. We ascertained this when we plotted its Bode diagram in Figure 6-15, only to discover the very small phase margin. We verified its instability when we hand-plotted its closed-loop transfer function and saw the excessively large maximum gain hump that exceeded the +3 dBA criteria. This fact was proven in the Nichols plot. However, what we could not ascertain from our frequency response curve was how to correct the unstable condition. Some trial-and-error in adjusting the values of our transfer function was necessary in order to fine-tune the correct values needed in the function. With the Nichols chart, we can now keep an eye on the gain–phase angle curve to see if we are compensating our transfer function by the correct amount necessary to create the desired results in the closed-loop plot.

Earlier, we found that by lowering the system gain by 30 dB, we could obtain a desirable phase margin of 50°. We determined this by inspecting the open-loop Bode diagram for our transfer function (Figure 6-15) and noting that 30 dB was needed in order to create a desired 50° phase margin. This 30 dB reduction created an open-loop transfer function of $3.2/[j\omega(1 + 0.25j\omega)]$. Let's plot this corrected transfer function using our Nichols chart to see if we obtain the same results as were obtained earlier plotting it by hand. We pick off our open-loop gain and phase angle values from the curves already constructed in Figure 7-8. The resultant Nichols gain–phase angle curve is seen plotted in Figure 7-14. The final closed-loop gain and phase angle curves are illustrated in Figure 7-15. Our results compare favorably with the hand-plotted results of Figure 7-9 with only minor discrepancies. Notice, however, the bandpass results. We obtained a bandpass of 5 rad/sec using our Nichols method, whereas the hand-plotted point-by-point method showed a bandpass of 4 rad/sec. This points out the possibility of discrepancies that can occur between these two methods. Only carefully laying out points on paper and carefully reading the data on the Nichols chart can help to reduce these differences.

There is yet a third method for obtaining closed-loop response curves for control systems, and it is perhaps the most accurate and quickest method of all. The method involves using a personal computer to perform all calculations and to automatically plot all data results. Software is available which can be modified to perform all open or closed frequency response plotting. The operator only has to enter the design parameters according to prompts given by a screen menu and the data is plotted in a very short period of time. Again, because of the speed of plotting by a dot-matrix printer, the operator can make modifications to the system and can see within seconds how the overall system will respond to these modifications. We should now ask ourselves this question: Of the three methods that we have now studied, the manual point-by-point method, the Nichols chart method, and the computer method, which is the most accurate? We would have to state, of course, the computer method. Not only is it unquestionably the most accurate method, but it is also the fastest method of obtaining plotted results.

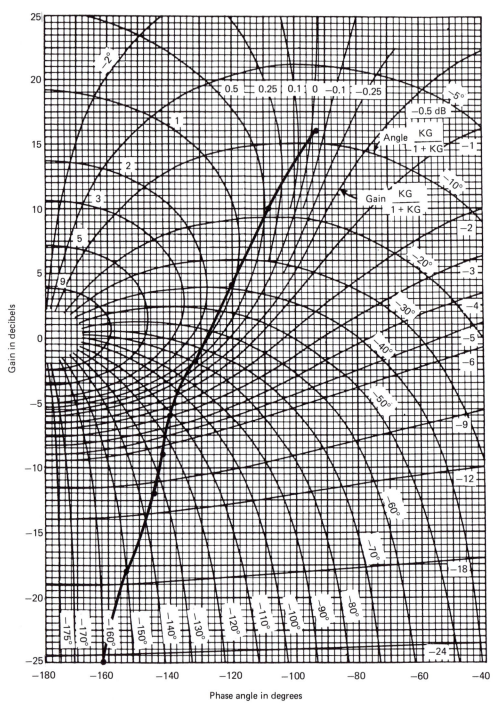

Figure 7-14 Plotting $3.2/j\omega(1 + 0.25j\omega)$ using the Nichols chart.

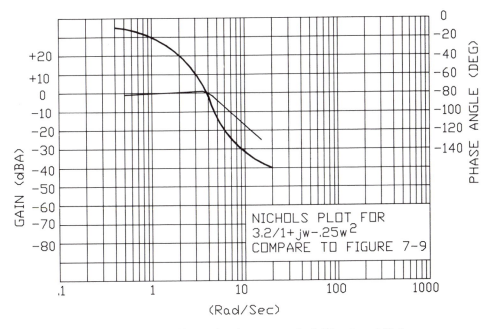

Figure 7-15 Closed-loop gain–phase curves for $3.2/1 + j\omega - 0.25\omega^2$.

SUMMARY

The Bode plot is used for predicting closed-loop behavior. The actual closed-loop data must be obtained by converting the open-loop transfer function to a closed-loop function and plotting the results. This can be a rather tedious process. On the other hand, the Nichols charting method is a fairly convenient method for generating a closed-loop plot from the open-loop transfer function. The quickest method, however, is through the usage of a personal computer and the several software programs that are now available for this sort of application.

EXERCISES

7-1. Plot the closed-loop frequency response curves (i.e., both the gain and phase angle curves) for the expression:

$$KG = \frac{25}{j\omega(1 + j\omega 0.2)}$$

7-2. Plot the closed-loop frequency response curves for the expression:

$$KG = \frac{370}{j\omega(1 + j\omega 0.25)}$$

What is the stability condition for this system?

7-3. Find the open-loop bandwidth for the system:

$$KG = \frac{25}{j\omega(1 + j\omega 0.2)}$$

Compare this figure to the closed-loop bandwidth for the same system. Do you see any correlation between the two figures?

7-4. Plot the closed-loop frequency response curves for the expression:

$$\frac{KG}{1 + KG} = \frac{1}{1 + j\omega - 0.25\omega^2}$$

What hints do you see from observing these curves that this system may be sluggish?

7-5. Construct the Bode plot for the expression:

$$KG = \frac{200}{j\omega(1 + 0.016)}$$

Find the phase margin. Determine the gain reduction needed to make this system stable. Now, using the Nichols chart method, construct the closed-loop gain–phase angle frequency response curves and determine again the amount of gain reduction needed to stabilize the system. How close are your two gain reduction results?

7-6. Construct the closed-loop frequency response curves for the expression:

$$KG = \frac{25.5}{j\omega(1 + j\omega 0.25)}$$

7-7. For the open-loop expression in Exercise 7-1, plot the closed-loop frequency response curves using the Nichols chart method.

7-8. Find the peak frequency in the gain curve and calculate the gain and phase angle at this frequency for the transfer function:

$$KG = \frac{200}{j\omega(1 + j\omega 0.333)}$$

7-9. Using the Nichols chart method, determine the reduction in gain needed, in dB, to make the transfer function in Exercise 7-3 stable. Hint: The system is considered stable when the gain curve's peak gain is less than +3 dB.

7-10. Refer to Figure 7-3. Determine the change in bandwidth resulting from reducing the peak gain in the gain curve to make this system stable.

8

Transient Analysis

8-1 INTRODUCTION

One of the most interesting aspects of automatic control systems has to do with the system's behavior when it is first turned on, or when it receives a change in command signals. Many control system designs that have been scrapped due to poor performance would have probably been otherwise acceptable if all they had to do was to operate with very slow or infrequently changing command signals. Obviously, this is an unrealistic condition for any control system. Control systems must be designed to handle rapidly changing input commands and must be capable of remaining stable during this time. In this chapter we study ways of forecasting a control system's behavior to a sudden change in command signal and how to correct an otherwise undesirable behavioral characteristic. A system's reflex to a command signal is a relatively short-lived event; consequently, we refer to this condition as a *transient response*. The term *transient* means that the system's response or behavioral reaction to a command signal is only temporary; it is a passing condition that shortly gives way to a more sustained *steady state* condition.

8-2 A SYSTEM'S RESPONSE TO A STEP INPUT

In order to analyze a control system's transient response, we must have some means of measuring the control system's output response over very short periods of time. We refer to time periods measured in milliseconds, or even in microseconds. To get a better grasp of what we are describing here, let's look first at a system described in Figure 8-1. This is an automatic chart recorder used for the automatic recording of a variable DC input voltage. This is how it works: A variable DC voltage is fed into one of the two inputs of a differential op amp (*A*) whose gain is 100. From there, the amplified output signal is fed into another

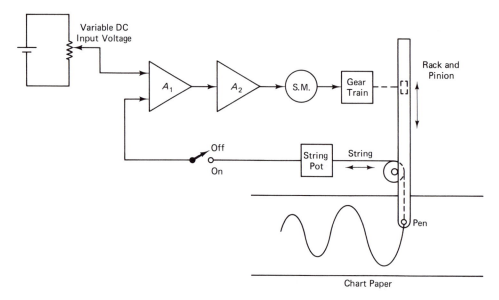

Figure 8-1   Automatic chart recorder.

amplifier (servoamplifier *B*) having an additional gain of 5. The purpose of this amplifier is to supply the required power to drive the servomotor. (The op amp by itself does not have the required power needed to do this; it only has voltage gain capabilities.) The output of the servoamplifier is then fed to the servomotor where it rotates in either direction, depending on the original polarity of the input DC voltage. As the motor rotates, a rack and pinion arm moves in and out across the moving chart paper beneath a pencil attached to the moving rack. Attached to the pencil is a wire string that stretches back to a pulley and string pot assembly. A string pot is nothing more than a potentiometer attached to a spring-loaded coil of string, called a yo-yo. As the spring-loaded string is pulled into and out of the string pot, a variable resistance, transformed internally to a variable voltage, is created at an output terminal on the pot. This variable voltage is sent to the second input of the op amp.

With the switch shown in our system in its off position, we have an open-loop system. With the switch in its on position, we have a closed-loop system. In the closed-loop configuration, the output of the string pot is connected to the op amp. This output is adjusted so that it is exactly the same amplitude as the variable input voltage being supplied to the other input of the op amp. We now forcibly place the pencil over to the edge of the moving chart paper with our hand. This position will represent 0 volts input into our system. Now we supply a 5-volt DC voltage to our system's input, causing the servomotor to jump instantly into action moving the rack and pinion with its pencil over to the other side of the chart paper. However, the servomotor keeps on moving beyond the paper, since nothing is telling it to do otherwise. The only way to stop the motor's turning is to remove the 5-volt input voltage. This is obviously not a very practical chart recorder design.

To correct this situation, we again force the pencil back over to the far side of the chart paper. This time, however, we adjust the output of our string pot, which is tracking the position of the recording pencil, so that it reads 0 volts. (Voltage adjustments are made to the string pot by means of an internal pot that varies the supply voltage to the yo-yo pot.) Next, we forcibly move the pencil over to the opposite edge of our chart paper (assuming that this is where we want a 5-volt signal to cause the recording pencil to move to, to make its 5-volt recording mark) and again adjust the string pot, but this time for a 5-volt output reading.

Now, we apply the 5-volt input signal. Again, our servomotor jumps into action, causing the pencil to move once again across the moving chart paper. This time the pencil stops at the paper's opposite edge, exactly where we had manually positioned the pencil earlier. What has happened is this: As soon as the string pot outputted 5 volts to the op amp, the op amp's output dropped to zero since its output was equal to the difference between its two input voltages (times its gain of 100, which still results in an output of 0). What would happen if we supplied, say, 2.5 volts to the op amp instead of 5 volts? Our pencil recorder would then travel only as long as there was a difference between the op amp's two input voltages, or until the string pot outputted 2.5 volts. This would occur when the pencil was mid-way across the surface of the chart paper.

Let's inspect the chart paper results from applying the 2.5-volt input signal to our voltage recording system. Refer to Figure 8-2. As you look at the chart paper, notice that time can be measured along its length. Let's assume that each division represents 100 milliseconds of time. What we are interested in is the zig-zag waveform, called a *damped sinusoidal wave,* that was produced from time = 0 to about time = 200 ms. This is our transient response condition for this particular control system. In addition, this is called a *transient response to a step input.* The term *step input* refers to the fact that a sudden 2.5 volts was applied to the system, not a gradually increasing 2.5 volts. The characteristics of these two conditions are important. If the 2.5 volts were applied gradually, we would get a distorted-looking transient response curve, one that would contain the damped sine wave feature but along a gradually increasing line. Also, the amplitudes of each individual cycle within the waveform would be reduced. We will see why in a moment.

The transient waveform is produced by a number of quantities. The cyclic homing of our recording pencil is called *hunting.* Hunting is the result of overcor-

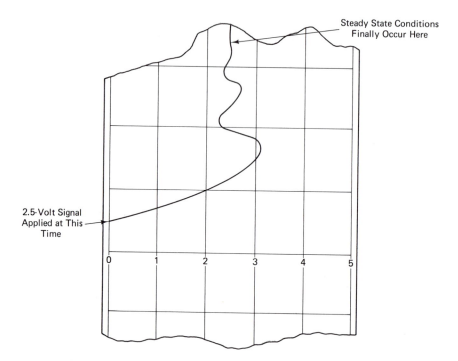

Figure 8-2   A chart recording made with the system shown in Figure 8-1.

rection and undercorrection of the system's response to a command signal. Ideally, the system's output (the pencil, in our case), would report to its exact location with no hunting. The hunting of a control system's output is caused by component inertia. In other words, in the case of the moving pencil and the rack's steel-machined arm, these components have a difficult time stopping and starting instantly because of their respective masses. Also, the frequency of oscillation is determined by several factors, as we soon discover. One such factor has to do with the system's frequency response, and another has to do with its time constants, velocity constants, and the amount of viscous damping built into the servomotor (see Section 4-11.1, Chapter 4).

There are other types of input waveforms that could be supplied to the automatic control system. Figure 8-3 shows some of these types. However, the step input is the one most frequently used for the reasons mentioned earlier. This is a square-wave input rising from zero volts to some higher voltage value, usually one volt, rising in an infinitely short period of time. More precisely, this type of input is called a *unit step input,* and is shown graphically in Figure 8-4. We use the unit step input to "shock" our control systems so that we can then study their behavioral responses. This type of input stimulus allows us to study the various characteristic outputs that will enable us to evaluate the performance of our systems.

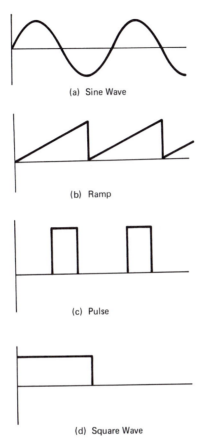

(a) Sine Wave

(b) Ramp

(c) Pulse

(d) Square Wave

Figure 8-3 Possible input waveforms for an automatic control system.

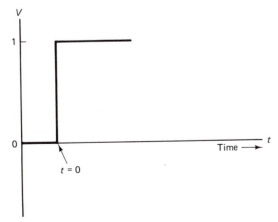

Note: Normally, $t = 0$ Begins at the Axes Intersection, but is Shown to the Right of Intersection for Clarity

Figure 8-4 A unit step input signal.

8-3 *GENERATING AND ANALYZING A TRANSIENT RESPONSE CURVE*

Before closing the switch in the feedback circuit of our recording device, our open-loop transfer function was:

$$\frac{d}{E_c} = \frac{(A_1)(A_2)(K_V)(K_{rp})}{s(1 + s\tau)(N)} \tag{8-1}$$

where d = the output displacement of the rack's arm (in.)
E_c = the control voltage supplied to the system (volts)
K_{rp} = transfer function for rotary pot (in./rad)
K_V = velocity constant of servomotor (rad/sec/volt)
τ = servomotor's time constant (sec)
N = gear ratio for rack and pinion plus gear train in Figure 8-1

Let's assume the following values for all the variables in Eq. (8-1):

$$A_1 = 100$$

$$A_2 = 10$$

$$K_V = 10 \text{ rad/sec/volt}$$

$$\tau = 0.2 \text{ sec}$$

$$N = 100$$

$$K_{rp} = 0.25 \text{ in./rad}$$

After substituting these values into Eq. (8-1), we have for our open-loop transfer function,

$$\frac{d}{E_c} = \frac{(100)(10)(10)(0.25)}{s(1 + 0.2s)(100)} = \frac{25}{s(1 + 0.2s)} \tag{8-2}$$

Notice that a transfer function for the string pot was not included in the transfer function, since it had nothing to do with the open-loop functioning of the system. Even when we close the feedback loop, we still ignore the string pot for the same reason we ignored the servo pot in our control system back in Chapter 7. The string pot merely converts the positional information coming from the servomotor into a proportional voltage that is fed back to the system's summing junction. As in our previous system, we will feed back 100% of this signal so that $H = 1$.

$$\frac{KG}{1 + KG} = \frac{\dfrac{25}{s(1 + 0.2s)}}{1 + \dfrac{25}{s(1 + 0.2s)}} = \frac{\dfrac{25}{s(1 + 0.2s)}}{\dfrac{s(1 + 0.2s) + 25}{s(1 + 0.2s)}}$$

$$= \frac{25}{s(1 + 0.2s) + 25} = \frac{25}{s + 0.2s^2 + 25} = \frac{25}{j\omega + 0.2(j\omega)^2 + 25}$$

$$= \frac{25}{25 + j\omega - 0.2\omega^2} \tag{8-3}$$

Equation (8-3) is now our closed-loop transfer function which we can now plot to see its closed-loop characteristics. We also plot the open-loop Bode plot to determine the system's phase margin. Figure 8-5 is the Bode plot and Figure 8-6 shows the closed-loop frequency response curves.

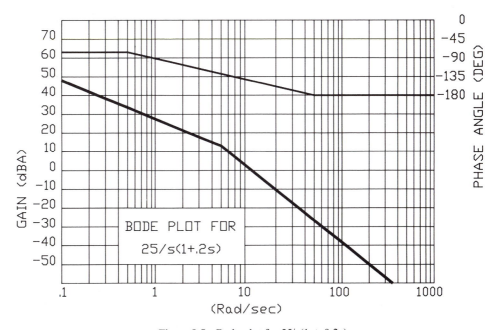

Figure 8-5   Bode plot for $25/s(1 + 0.2s)$.

We can see from the Bode plot that the phase margin for our system is only 28°. This tells us that our system is entirely too jittery to be of any use. We would find that whenever a command voltage, e_c, would be supplied to the chart recorder, the pencil would overshoot the desired voltage mark many times before eventually settling at the proper voltage mark. Of course, our closed-loop plot shows us the characteristic gain hump that exceeds the desired 3 dB limit of stability. This verifies what the Bode plot told us. While we are looking at the closed-loop plot, let's make a note of our bandwidth. We see that it is 16.5 rad/sec, or the equivalent of 2.6 Hz. We refer to Figure 8-6 later.

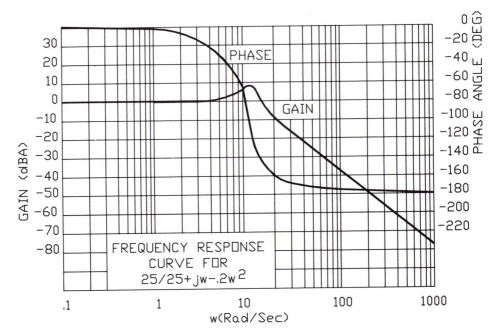

Figure 8-6 Frequency response curve for $25/25 + j\omega - 0.2\omega^2$.

8-3.1 The Natural Resonant Frequency and the Damping Factor

We want to be able to determine some additional information about our control system. It would be helpful to be able to obtain a graph of the hunting characteristics of our marking pencil as it searches for its final resting spot at the proper voltage mark on the chart paper. If we could do this, then we wouldn't have to run the chart recorder to see how it would behave. We could, instead, draw its *transient curve* out on paper ahead of time. However, in order to do this, there are two quantities that we must first calculate. We want to be able to find (a) our system's *natural resonant frequency, ω_n*, and (b) the system's *damping factor, z*. The natural frequency of any system is the frequency at which the system oscillates, rings, or vibrates naturally without any outside influence except for the initiating energy or force. The damping factor, or *damping ratio* as it is sometimes called, is the rate of "dying-out" of the oscillations. We will have a better understanding of this term in a few moments.

To find ω_n and z, we first rewrite our closed-loop function in its Laplace form: $25/(0.2s^2 + s + 25)$. The next thing we do is to supply a unit step voltage input to our system, but do it mathematically. Referring to the table of Laplace transforms in Chapter 3 (Table 3-1), we see the Laplace equivalent of a unit step input as $1/s$. In reality, we would apply a sudden one-volt shock to our system. However, the mathematical equivalent of doing this is:

$$\frac{1}{s} \times \frac{25}{0.2s^2 + s + 25} \tag{8-4}$$

Referring again to the Laplace tables, we search for a transform having the description of a closed-loop second order system (i.e., a system having terms raised to the power of 2) being excited by a unit step input of $1/s$. In Table 3-1 we find Eq. 17 that fits the description. However, we first divide the entire expression, both its numerator and its denominator, through by the term, ωn^2, to give:

$$\frac{1}{s\left[\dfrac{s^2}{\omega_n^2} + \dfrac{2zs}{\omega_n} + 1\right]} \tag{8-5}$$

The reason for doing this will become evident in a moment. Meanwhile, if we can get Eq. (8-4) to somehow confirm to Eq. (8-5), we can then find the values of ω_n and z.

To get Eq. (8-4) to look like Eq. (8-5), we first divide the numerator and denominator by 25, to get:

$$\frac{1}{s\left[\dfrac{0.008s^2}{1} + \dfrac{s}{25} + 1\right]} \tag{8-6}$$

Next, we multiply the denominator of Eq. (8-6) by 125/125 to get:

$$\frac{1}{s\left[\dfrac{s^2}{125} + \dfrac{125s}{3,125} + 1\right]} \tag{8-7}$$

Since the first term in the denominator (125) is ω_n^2, the second denominator term (3,125) must be "forced" into being ω_n, which is $\sqrt{125}$ or 11.2. So we multiply this second term by 11.2/3,125. Of course, we have to do the same thing to the $125s$ numerator term. Our expression now becomes:

$$\frac{1}{s\left[\dfrac{s^2}{125} + \dfrac{125s\,\dfrac{11.2}{3,125}}{3,125\,\dfrac{11.2}{3,125}} + 1\right]} = \frac{1}{s\left[\dfrac{s^2}{125} + \dfrac{0.448s}{11.2} + 1\right]} \tag{8-8}$$

Now that Eq. (8-4) is in the form of Eq. (8-5), we can identify ω_n as 11.2 rad/sec (i.e., a frequency of 1.78 Hz) and z as 0.224 (be sure to note that $2z = 0.448$, so you have to take $\frac{1}{2}$ the value shown in Eq. (8-8). A damping factor of 0 means that a control system will oscillate forever, each peak of oscillation having the same amplitude. A z of 1 means that there are *no* oscillations; the output, in other words, reports *directly* to its final location or value with no hunting whatsoever. However, the time it takes to get to that spot will be excessively long. This is called a *critically damped* system. Consequently, a z of 1 represents a very sluggish system and is as undesirable as a z value near 0.

At this point it is logical to ask, what is considered an acceptable value of z? From experience, it has been found that control systems exhibiting any values less than 0.4 tend to be too jittery; any values greater than 0.7 and the systems tend to be too sluggish. Therefore, any z values within the range of 0.4 to 0.7 are considered acceptable by many control system designers.

8-3.2 The Damped Frequency

Our system is obviously too jittery. And while we discuss oscillations here, we must be able to distinguish the difference between a system's *natural resonant frequency* and that system's *damped frequency*. Earlier, we implied that a system's natural frequency of oscillation was its unimpeded frequency. That is, the system would oscillate at its *natural* frequency of oscillation if there were no losses due to friction, viscous damping, and so on. In reality, since these impedances are definitely present, the system will instead oscillate at a somewhat lower frequency called its *damped* frequency, ω_d. The quantity ω_d is related to ω_n by the equation:

$$\omega_d = \omega_n \sqrt{1 - z^2} \qquad (8-9)$$

Figure 8-7 gives us a visual idea as to how the damping ratio values alter the shape of the transient curve. These are generalized curves for different z values. Note in particular how the z values produce the varying amounts of ω_d.

We are now armed with enough information to allow us to actually plot the transient waveform of our system to see what sort of settling pattern will be drawn by the pencil as it eventually settles at a final reading. However, based on our computed z value of 0.224, our recorder's pencil may take a few oscillations before settling down to its final reading.

Let's now refer to our table of Laplace transforms once again in Chapter 3. We want to find the time domain equation for Eq. (8-5) because, knowing this, we can find the amplitudes for any value of time in our transient function. According to Table 3-1, Eq. 17, the time domain equation equivalent is:

$$1 + \frac{e^{-z\omega_n t}}{\sqrt{1 - z^2}} \sin\left[\omega_n \sqrt{1 - z^2}\, t - \arctan\left(\frac{\sqrt{1 - z^2}}{-z}\right)\right] \qquad (8-10)$$

where the arctan expression returns an angle, ϕ, *that must be greater than 0 but less than 180°*.

Equating our preceding expression, Eq. (8-10), to y, and using our values for z, ω_n, and ω_d, we can now hand-plot the damped transient waveform for our control system. There is one modification, however, we must make to Eq. (8-10)

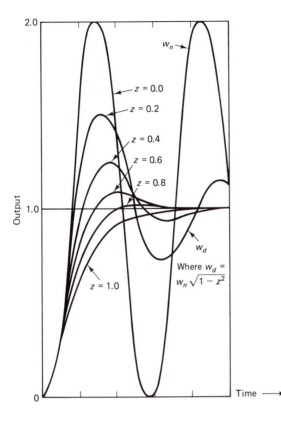

Figure 8-7 Comparison of damping factor (z) values.

in order for it to have consistent units. Notice that the expression is taking the sine of an expression containing ω_n. Since ω_n is in radians, we have to remember to convert the expression to degrees before taking the sine of it. Therefore, we modify Eq. (8-10) by multiplying the ω_n expression by 57.3°/rad to convert it to degrees. We also equate the expression to y, since this is a relationship that allows us to determine an amplitude, y, for a given value of time, t. Equation (8-10) now becomes:

$$y = 1 + \frac{e^{-z\omega_n t}}{\sqrt{1 - z^2}} \sin\left[(\omega_n \sqrt{1 - z^2} \cdot t)57.3 - \arctan\left(\frac{\sqrt{1 - z^2}}{-z}\right)\right] \qquad (8\text{-}11)$$

Using rectilinear coordinate graph paper, we can now lay out a y (amplitude) versus t (time) axis. However, it will take a considerable number of points to construct our curve, but there are some additional calculations we can do to make our plotting job easier. First, let's look at a typical transient waveform. Figure 8-8 shows such a waveform and how our recorder's waveform will probably look also. Notice the curve begins at 0, crosses the 1 axis (which is the desired final resting position for this system, and for our system too) and reaches a peak

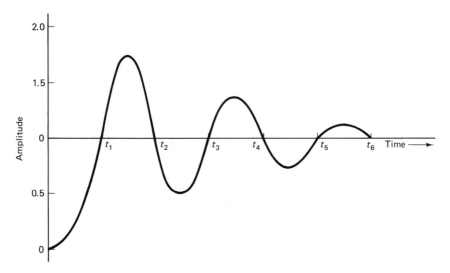

Figure 8-8   The transient response curve as a result of a unit step input quantity.

amplitude before going back down across the center axis once again. The cycle is repeated a number of times, each cycle having less of an amplitude as compared to the previous cycle. This is characteristic of a damped transient waveform. It would be helpful to us if we could calculate the crossover times, t_1, t_2, t_3, t_4, etc. Then, we could calculate the maximum and minimum amplitudes occurring midway between each crossover time and that would give us a pretty good idea what our final curve is going to look like. We can find the crossover time each time $y = 1$ in Eq. (8-11); in other words, when

$$1 + \frac{e^{-z\omega_n t}}{\sqrt{1 - z^2}} \sin \left[(\omega_n \sqrt{1 - z^2})57.3 - \arctan \left(\frac{\sqrt{1 - z^2}}{-z} \right) \right] = 1 \qquad \text{(8-11a)}$$

This event will take place whenever the sine term in Eq. (8-11) becomes 0. That is, when

$$(\omega_n \sqrt{1 - z^2} \cdot t)57.3 - \arctan \frac{\sqrt{1 - z^2}}{-z} = 0 \qquad \text{(8-11b)}$$

We solve the preceding equation for t, but first, let's substitute the expression ω_d for the expression $\omega_n \sqrt{1 - z^2}$, since they are equivalent (Eq. 8-9). We now have,

$$(\omega_d)57.3t - \arctan \frac{\sqrt{1 - z^2}}{-z} = 0 \qquad \text{(8-11c)}$$

Solving for t:

$$t = t_a = \arctan \frac{\left[\dfrac{\sqrt{1 - z^2}}{-z}\right]}{\omega_d \, 57.3} \tag{8-11d}$$

where t_a = *initial* crossover time (sec)

Since one cycle in our transient waveform is ω_d in length, and one cycle represents 2π radians, and a crossover occurs at $\omega = 2\pi/T$ intervals, in general, we can say that $\omega_d = 2\pi/T$, or

$$T = \frac{2\pi}{\omega_d} \tag{8-11e}$$

where T = the waveform's period (sec)

But, since there is a crossing every $180°$ in each cycle, that is to say, a crossing every half-cycle, we have

$$T = 2t_b = \frac{2\pi}{\omega_d} \tag{8-11f}$$

where t_b = a crossover time interval per half-cycle (sec)

or,

$$t_b = \frac{\pi}{\omega_d} \tag{8-11g}$$

These crossovers will occur for n cycles; therefore, we can modify Eq. (8-11g) further by multiplying the expression by n, and letting n become a series of integer values until the waveform we are plotting dampens out completely. Therefore, our final expression for determining all the crossover times occurring on the 1 axis of our transient plot will be the sum of t_a (Eq. 8-11) and the n modified t_b expression, in Eq. (8-11g). That is,

$$t_{x\text{-over}} = \frac{\arctan(\sqrt{1 - z^2}/-z)}{57.3\omega_d} + \frac{n\pi}{\omega_d} \tag{8-12}$$

where $n = 0, 1, 2, 3, 4$, etc.

For each crossover time determined by Eq. (8-12), we can place the mid-value between any two crossover times into Eq. (8-11) to find y, the amplitude at that particular time.

Table 8-1 lists the calculated terms resulting from using Eqs. (8-11) and (8-12) in developing the transient response curve for Eq. (8-4). The first three amplitude values listed in Table 8-1 are calculated in the following manner:

The first crossover time, t_1, for Eq. (8-4) requires using Eq. (8-12). The following constant values are obtained from Eq. (8-8):

TABLE 8-1 CALCULATED CROSSOVER TIMES AND AMPLITUDES FOR TRANSIENT CURVE RESULTING FROM APPLYING A UNIT STEP INPUT TO EQ. (8-4)

n	t	$t_{x\text{-over}}$	Mid-value	Amplitude
0	t_1	0.165		1.000
			0.309	1.473
1	t_2	0.453		0.998
			0.597	0.771
2	t_3	0.741		1.001
			0.885	1.111
3	t_4	1.029		0.999
			1.173	0.946
4	t_5	1.317		1.000
			1.461	1.026
5	t_6	1.605		1.000
			1.749	0.987
6	t_7	1.893		1.000
			2.037	1.006
7	t_8	2.181		1.000
			2.325	0.997
8	t_9	2.469		1.000

$$\omega_n = 11.2 \text{ rad/sec}$$

$$z = 0.224$$

$$\omega_d = 10.9 \text{ rad/sec, using Eq. (8-9)}$$

Since we are interested in the crossover time for $n = 0$, Eq. (8-12) now becomes:

$$t_1 = \arctan \frac{\left(\dfrac{\sqrt{1 - 0.224^2}}{-0.224} \right)}{(10.9)(57.3)} + \frac{(0)\pi}{10.9}$$

$$= \frac{\arctan(-4.35)}{624.6} + 0$$

If a calculator is used for determining the Arctan of -4.35, most likely the angle of $-77.05°$ will be obtained. However, noting that the Arctan expression must lie between 0 and 180°, we have to subtract $-77.05°$ from 180° to obtain the proper angle value for the preceding equation. Therefore, we use the angle of 102.95° as Arctan(-4.35).

$$t_1 = \frac{102.95°}{624.6}$$

$$= 0.165 \text{ sec}$$

To find t_2 in Table 8-1, we note that the Arctan expression (i.e., t_a in Eq. 8-11f) doesn't change. However, we have to calculate a new t_b using Eq. (8-11g). We must multiply that equation by an integer, as seen in the second term of Eq. (8-12), the integer being 1 in this case. Therefore,

$$t_2 = 0.165 + \frac{(1)\pi}{10.9}$$

$$= 0.453 \text{ sec}$$

The mid-value time, which would correspond to the occurrence of the first peak of the transient curve located mid-way between the times for $n = 0$ and $n = 1$, would be 0.309.

Our next task is to determine the amplitudes occurring at the three times just calculated. We do this by using Eq. (8-11). We see that the first and second crossover times will produce an amplitude of 1 by the definition and intent of Eqs. (8-11) and (8-12). However, we need to calculate the amplitude of the mid-value term, t_3. Therefore,

$$t_3 = 1 + \frac{e^{-(0.224)(11.2)(0.301)}}{\sqrt{1 - 0.224^2}} \sin[(10.9)(57.3)(0.309) - 102.95]$$

$$= 1 + \frac{0.461}{0.975} \sin(192.99° - 102.95°)$$

$$= 1 + 0.473 \sin 90°$$
$$= 1.473$$

Table 8-1 lists the other crossover time points along with mid-value points and their amplitudes.

The data listed in Table 8-1 is plotted in Figure 8-9. Notice how long it takes the pencil in our recorder to finally settle down to read one volt. According to our transient curve, it takes approximately two seconds or so before it becomes difficult to see any more maximums or minimums occurring in the curve.

8-3.3 Settling Time

What is actually done in analyzing transient curves is to select a band or range on either side of the desired set point (one volt in our case) and note the length of time that it takes the transient curve to come within this band limit. Typically, a band of ±5% within the set point is selected. In our case, we draw a line at 1.05 volts and 0.95 volts to form our settling band. Doing this, we see our settling time becoming approximately 1.2 seconds to within ±5% of the set point. This is just

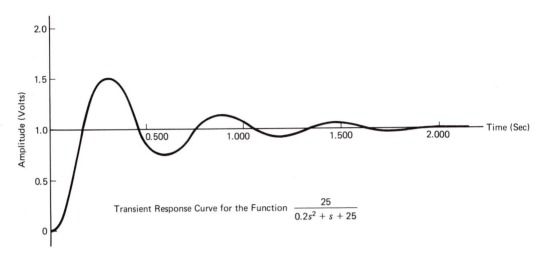

Figure 8-9 Transient response curve for the function $25/0.2s^2 + s + 25$.

an estimate, however. We can actually calculate the settling time using the following reasoning: Looking back to Figure 2-32 in Chapter 2, we see that after about three time constants, we are within 5% of being stabilized. We know that the rate of decay of our curve in Figure 2-32 (curve 2, which coincides with the behavior of our settling transient curve, as we discover later in chapter 8) is described by the term $e^{-z\omega_n t}$. Since we agreed that $\tau = 3$, then

$$e^{-z\omega_n t} = e^{-3}$$

our transient curve $\leftarrow\!\rfloor \qquad \lfloor\!\rightarrow$ from Figure 2-32

Taking the natural logs of both sides of the equation,

$$z\omega_n t = 3$$

Solving for t, which is now our settling time, we get

$$\text{settling time to within } \pm 5\% \text{ of set point } = \frac{3}{z\omega_n} \qquad (8\text{-}13)$$

8-3.4 Number of Oscillations Before Settling

We can also calculate the number of oscillations occurring before completing the $\pm 5\%$ settling time. Since Eq. (8-13) allows us to calculate the settling time to within $\pm 5\%$ of the steady state time, and since we can find the period of one damped oscillation of our transient waveform, then settling time $\div$ period will give us the number of cycles that can "fit into" this settling time span. In other words,

since $\qquad\qquad\qquad \geqslant 5\% \text{ settling time } = \frac{3}{z\omega_n}$

and the period
$$t = \frac{1}{f}$$

and since
$$\omega_d = 2\pi f, \quad \text{or} \quad f = \frac{\omega_d}{2\pi}$$

then the period for
$$\omega_d = \frac{2\pi}{\omega_d}$$

We then divide the settling time by the period of our damped waveform to get:

$$\frac{\dfrac{3}{z\omega_n}}{\dfrac{2\pi}{\omega_d}} = \frac{3\omega_d}{z\omega_n \, 2\pi}$$

$$= \frac{1.5\omega_d}{z\omega_n \pi} = \frac{1.5\omega_n\sqrt{1 - z^2}}{z\omega_n \pi}$$

Or,
$$\frac{\text{number of oscillations to}}{\pm 5\% \text{ of complete settling}} = \frac{1.5\sqrt{1 - z^2}}{z\pi} \tag{8-14}$$

For our recorder, the settling time, according to Eq. (8-13), is

$$\frac{3}{(0.224)(11.2)} = 1.20 \text{ sec}$$

This coincides with the results from Figure 8-8. The number of oscillations before settling, according to Eq. (8-14), is

$$\frac{1.5\sqrt{1 - 0.224^2}}{(0.224)(3.1416)} = 2.1$$

8-3.5 Over-shoot

Another useful equation allows us to calculate the *over-shoot* amount. The over-shoot of a transient waveform is defined as the ratio, usually expressed as a percentage, between the amount the first peak of a transient waveform exceeds its final steady state or set point value. Refer to Figure 8-10. Curve 1 in Figure 8-10 has an over-shoot ratio of 1.5/1 where the 1.5 is the amplitude of the over-shoot and the 1 is the final steady state value. Expressed as a percentage, the over-shoot amount would be 50%. In other words, the curve's peak amount exceeded the steady state value of 1 by 50%. Similarly, curve 2 has an over-shoot of 1.2/1 or 20%; curve 3 has no over-shoot, since its ratio of 1/1 gives us a percentage value of 0. To calculate the amount of over-shoot, expressed as a percentage, the following equation may be used (notice that only the curve's damping factor, z, is needed to determine the over-shoot):

$$\% \text{ over-shoot} = e^{-z\pi/\sqrt{1 - z^2}} \times 100 \tag{8-15}$$

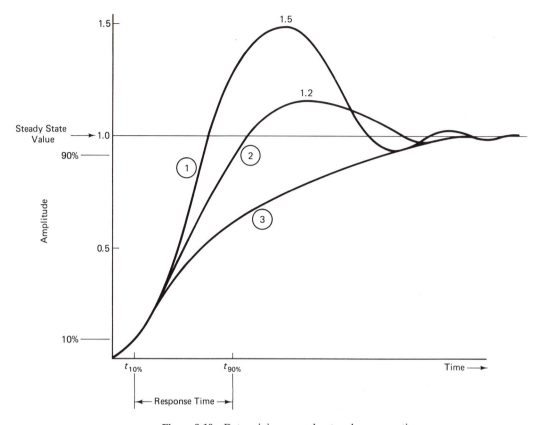

Figure 8-10   Determining over-shoot and response time.

According to Eq. (8-15), the percent over-shoot amount for our recorder is

$$e^{-(0.224)(3.1416)/\sqrt{1-0.224^2}} \times 100 = e^{-0.704/0.975} \times 100$$
$$= 48.6\%$$

Note: This result is somewhat larger in value than the result obtained in Table 8-1. The difference in values can be attributed to rounding-off errors and the method used in calculating the mid-value time of 0.309 sec and its associated amplitude of 1.473. Theoretically, this value is also the over-shoot value for this curve according to the calculation methods used.

This is equivalent to an over-shoot amplitude of 1.486 volts on our voltage recorder. Again, our transient curve plot confirms this amount.

8-3.6 Response Time

The time it takes for a system to go from 10% of its steady state value to 90% of its steady state value is called the *response time*. This term can be easily confused with a system's time constant. These are two entirely different terms. Unfortunately, many reference books and textbooks confuse these two terms by using them interchangeably. If you refer again to Figure 8-10, you see the response time term being graphically explained. Response time is used extensively by servo manufacturers. The reason for omitting the first 10% and last 10% of a performance curve is to avoid any initial or final disruptions or deviations in an otherwise linear or smooth curve, which occasionally happens in the graphing of the performance characteristics of many devices.

Now that we have defined the many important features of a transient function, we can relate some of this information to our earlier closed-loop frequency response curves. Referring back to Figure 8-6, we note that the frequency of the peak of our gain curve occurs at approximately 11 rad/sec, or 1.75 Hz. Knowing now the z and ω_n of our system, we can calculate this *closed-loop peak frequency, ω_m*.

$$\omega_m = \omega_n \sqrt{1 - 2z^2} \qquad (8\text{-}16)$$

In our case, $\omega_m = 11.2\sqrt{1 - (2)(0.224^2)} = 10.62$ Hz. Knowing this calculated value, we can now insert it into Eq. (8-4) and calculate *the closed-loop peak gain, A_m*, and the resultant phase angle.

$$A_m = \frac{25}{25 + j10.62 - (0.2)(10.62)^2} = \frac{25}{25 - 22.56 + j10.62}$$

$$= \frac{25}{2.44 + j10.62} = \frac{25}{10.9 \angle 77.06} = 2.29 \angle -77.06°$$

Converting the gain of 2.29 to decibels, 20log2.29 = 7.2 dBA. Therefore, the maximum peak is 7.2 dBA at a phase angle of −77.06°.

We can also calculate the bandwidth, ω_b, of our system instead of relying on our response curves for this information:

$$\omega_b = \omega_n \sqrt{1 - 2z^2 + \sqrt{2 - 4z^2 + 4z^4}} \qquad (8\text{-}17)$$

In our case,

$$\omega_b = 11.2\sqrt{1 - 2(0.224)^2 + \sqrt{2 - 4(0.224)^2 + 4(0.224)^4}}$$
$$= 16.78 \text{ rad/sec}$$

We now have a design problem that we must resolve with our chart recorder system. How do we correct the jitteriness of our marking pencil? Looking at the open-loop Bode plot for our system (Figure 8-5), we see that we need to increase our phase margin somewhat in order to deaden the recorder's responsiveness. One obvious approach is to reduce the gain of our system. This can be done by either reducing the amplifier gains or by increasing the gear ratio on the transmission box attached to the servomotor, or a combination of both. Or, we could increase the viscous damping on our servomotor in an attempt to dampen the oscillations. Whatever we decide to do, though, we must keep in mind that our system's bandwidth will be affected. If we want a system that is capable of responding to rapid fluctuations in voltage input, then the bandwidth should be kept as high as possible so the servomotor can respond accordingly. If we aren't interested in catching quick voltage changes, then we would probably be satisfied with the present bandwidth, or something even smaller.

To approach this problem, let's first look at Eq. (8-17) again. Notice that if you make the value of z grow larger in an effort to deaden a system's response, the value of ω_b becomes smaller. Our Bode plot bears this out. As you attempt to reduce the gain in an effort to increase the phase margin, the open-loop bandwidth decreases. Notice what happens as the gain curve in Figure 8-5 is lowered. The zero gain crossover point shifts to the left causing the phase margin to increase, but at the same time, causing the bandwidth to decrease. It's reasonable to assume that if the open-loop bandwidth decreases, so will the closed-loop bandwidth. So it seems that the only really logical thing to do, without going through a major redesign effort on our hardware, is to try to decrease the system's gain and hope that this will not affect our bandwidth all that much.

Looking at Figure 8-5, if we were to lower our gain curve by, say, 16 dBA, this would give us a phase margin of 49°. Notice too, that as the gain curve is lowered by this amount, this shifts us from the -40 dBA/decade slope of the curve and puts us on the -20 dBA/decade slope instead. This is a good place to be. A rule of thumb in controls design is, *operate on the -20 dBA/decade slope of a gain curve whenever possible.* This insures good system stability.

A decrease of 16 dBA is the equivalent of reducing our system gain by a factor of 0.158 (20log.158 $= -16$dB). The easiest way to do this is merely to reduce the gain on one of our two amplifiers on our recorder. The resulting open-loop transfer function now becomes, $3.96/s(1 + 0.2s)$. See Figure 8-11. The closed-loop transfer function becomes

$$
\begin{aligned}
\frac{KG}{1 + KG} &= \frac{\dfrac{3.96}{s(1 + 0.2s)}}{1 + \dfrac{3.96}{s(1 + 0.2s)}} = \frac{\dfrac{3.96}{s(1 + 0.2s)}}{\dfrac{s(1 + 0.2s) + 3.96}{s(1 + 0.2s)}} \\[2ex]
&= \frac{3.96}{s(1 + 0.2s) + 3.96} = \frac{3.96}{s + 0.2s^2 + 3.96} \\[2ex]
&= \frac{3.96}{j\omega + 0.2(j\omega)^2 + 3.96} = \frac{3.96}{3.96 + j\omega - 0.2\omega^2}
\end{aligned}
\tag{8-18}
$$

Our next task is to determine the value of z. This will tell us for sure whether or not our system is going to be stable. Also, from z, we can calculate what our bandwidth is going to be without having to plot a closed-loop frequency response curve. Following the procedure that we performed earlier, we use Eq. (8-5) as a pattern to help us determine our new value of z. Using Eq. (8-18),

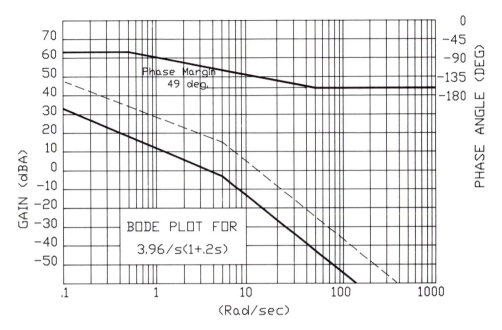

Figure 8-11   Bode plot for expression $3.96/s(1 + 0.2s)$.

$$\frac{1}{s} \times \frac{3.96}{0.2s^2 + s + 3.96}$$

$$= \frac{1}{s\left[\dfrac{0.2s^2}{3.96} + \dfrac{s}{3.96} + 1\right]}$$

$$= \frac{1}{s\left[0.051s^2 + \dfrac{s}{3.96} + 1\right]} \times \frac{\dfrac{1}{19.61}}{\dfrac{1}{19.61}}$$

$$= \frac{1}{s\left[\dfrac{s^2}{19.61} + \dfrac{19.61s}{77.66} + 1\right]}$$

$$= \frac{1}{s\left[\dfrac{s^2}{19.61} + \dfrac{(19.61)(0.057)s}{4.43} + 1\right]} = \frac{1}{s\left[\dfrac{s^2}{19.61} + \dfrac{1.118s}{4.43} + 1\right]}$$

$$\longrightarrow (77.66)(0.057) = \sqrt{19.61} = 4.43 \qquad (8\text{-}19)$$

Since $2z = 1.118$ (in Eq. 8-19), $z = 0.559$. This value of z indicates that we now have a stable system. Our new ω_n is 4.43 rad/sec. Now that we know the z and ω_n, we can calculate the new bandwidth using Eq. (8-17):

$$\omega_b = 4.43 \sqrt{1 - 2(0.559)^2 + \sqrt{2 - 4(0.559)^2 + 4(0.559)^4}}$$
$$= 5.32 \text{ rad/sec}$$

This is a substantial reduction in bandwidth compared to the earlier bandwidth result when we had the more jittery system. Converting the 5.32 rad/sec to its frequency equivalent (using Eq. 1-22 from Chapter 1) we obtain an equivalent frequency of 0.85 Hz. This means that our recorder will not be able to record voltage events at their proper magnitudes that change any more rapidly than about 0.85 Hz. We can now hand-plot the transient response curve using Eqs. (8-11) and (8-12). The resulting curve along with the calculated crossover times are shown in Figure 8-12. We can check our work by calculating the $\pm 5\%$ settling time and the oscillation numbers to settling:

$$\pm 5\% \text{ settling time} = \frac{3}{z\omega_n} \quad \text{(from Eq. 8-13)}$$

$$= \frac{3}{(0.559)(4.37)}$$

$$= 1.228 \text{ sec}$$

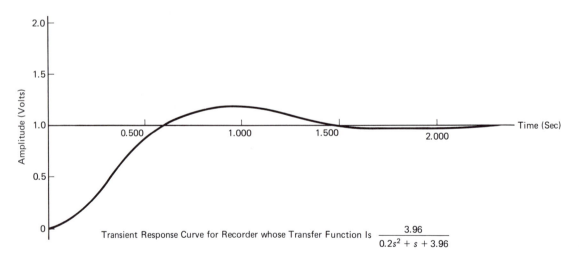

Transient Response Curve for Recorder whose Transfer Function Is $\dfrac{3.96}{0.2s^2 + s + 3.96}$

Figure 8-12   Transient response curve for recorder whose transfer function is $3.96/0.2s^2 + s + 3.96$.

To determine the number of oscillations:

$$\text{number of oscillations} = \frac{1.5\sqrt{1 - z^2}}{z\pi} \quad \text{(from Eq. 8-14)}$$

$$= \frac{1.5\sqrt{1 - 0.559^2}}{(0.559)(3.1416)}$$

$$= 0.708$$

These calculations agree with the data plotted in Figure 8-12.

As a result of decreasing our gain, we gained stability in our system but we sacrificed speed of response. It now takes over one second for our recorder to reach its final set point goal of one volt.

8-3.7 Handling Other Values of Step Inputs

Up to this point, we have dealt with a system response that has been given a step input quantity of 1. In the case of our voltage recorder, we subjected the recorder to an input of one volt to see how it would behave. But we have to ask ourselves, what about other step input values other than 1? Are they treated any differently? The answer is no. They are treated just like the unit step input. The only thing that's different are the amplitudes involved in generating the transient response curves. As an example, assume that we placed a 3.5-volt DC signal across our voltage recorder's input. Assume also that we are using the same system configuration that we just finished analyzing. How will our transient curve differ from the one in Figure 8-12? To answer this question, let's take another look at Eq. (8-10). This equation is set up for a unit step input quantity. The 1 in the front of this equation causes our plot to be generated around a horizontal axis of 1. We find out later in this chapter that the exponential expression, $e^{-z\omega_n t} + 1$ is the decaying "envelope" of our damped oscillating wave, which is our transient waveform (see Figure 8-13). We can change the magnitude of this function and shift the horizontal axis by merely multiplying the entire function of Eq. (8-10) by a constant, A. Therefore, A in our case will be the 3.5-volt step input. Our resultant waveform, as a result, will increase in magnitude by 3.5 times and become generated around a horizontal axis of 3.5 instead of the usual 1 axis. None of the times or frequencies that we calculated earlier will change. This is similar to the case of working with ordinary sinusoidal waveforms. Changing the amplitude of a sine wave doesn't affect its period or frequency.

8-4 AN EXAMPLE PROBLEM

Now that we have discussed most of the important features of a control system's transient behavior and can calculate many of the more useful parameters, let's go

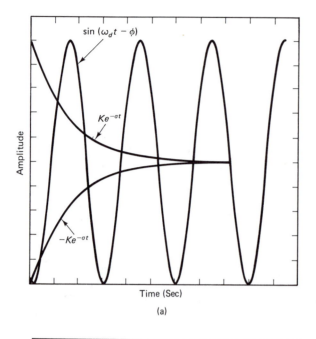

(a)

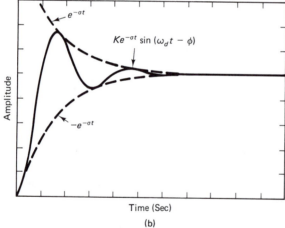

(b)

Figure 8-13 The components of the transient waveform:

$$(\omega_n\sqrt{1 - z^2}\,t - \phi), \quad \frac{1}{\sqrt{1 - z^2}}\,e^{-z\omega_n t}\sin$$

where $\sigma = \omega_n t$, $K = \frac{1}{\sqrt{1 - z^2}}$, and σ

$$= \arctan\frac{\sqrt{1 - z^2}}{-z}.$$

through some additional sample calculations. We use one of our closed-loop transfer function examples that we discussed back in Chapter 7.

EXAMPLE 8-1

Using Eq. (7-1) and assuming this is a closed-loop transfer function describing an automatic positioning system, calculate the following transient function features for this equation. Assume a step input of 2.7 volts. Find:

a. damping factor

b. the natural frequency

c. the damped frequency $\dfrac{100}{j^2\omega^2 0.25 + j\omega + 100}$ (Eq. 7-1)

d. $\pm 5\%$ settling time

e. maximum over-shoot

f. oscillation number to $\pm 5\%$ settling

g. the bandwidth

h. the closed-loop peak frequency

i. the closed-loop peak gain and phase angle

Solution:

a. To find damping factor (z):

Convert to a step input Laplace transform using Eq. (8-5):

$$\frac{1}{s\left[\dfrac{s^2}{\omega_n{}^2} + \dfrac{2zs}{\omega_n} + 1\right]} \text{ (Eq. 8-5)} = \frac{1}{s\left[\dfrac{0.25s^2}{100} + \dfrac{s}{100} + 1\right]}$$

$$= \frac{1}{s\left[\dfrac{s^2}{400} + \dfrac{s}{100} + 1\right]} = \frac{1}{s\left[\dfrac{s^2}{400} + \dfrac{0.2s}{20} + 1\right]} \qquad (8\text{-}20)$$

From Eq. (8-20), we see that $2z = 0.2$; therefore, $z = 0.1$. Also,

b. To find ω_n, we again look at Eq. (8-20), which says that $\omega_n = 20$ rad/sec.

c. To find ω_d, we use Eq. (8-9):

$$\omega_d = \omega_n\sqrt{1 - z^2} = 20\sqrt{1 - 0.1^2} = 19.9 \text{ rad/sec}$$

d. To find $\pm 5\%$ settling time, we use Eq. (8-13):

$$\pm 5\% \text{ settling time} = \frac{3}{z\omega_n} = \frac{3}{(0.1)(20)} = 1.5 \text{ sec}$$

e. To find the maximum over-shoot, we use Eq. (8-15). However, since this is a percentage over-shoot value that will be given us, we have to multiply this percentage by the step input amount to find the actual amplitude of this first over-shoot oscillation. Therefore,

$$\text{over-shoot} = e^{-z\pi/\sqrt{1 - z^2}} = e^{-(0.1)(3.1416)/\sqrt{1 - 0.1^2}}$$
$$= 0.729$$

The amplitude of the over-shoot $= 2.7 \times 0.729 = 1.97$ volts.

f. To calculate the number of oscillations, we use Eq. (8-14):

$$\text{number of oscillations} = \frac{1.5\sqrt{1 - z^2}}{z\pi} = \frac{1.5\sqrt{1 - 0.1^2}}{(0.1)(3.1416)}$$
$$= 4.75$$

g. The closed-loop bandwidth is found by using Eq. (8-17):

$$\omega_b = \sqrt{1 - 2z^2 + \sqrt{2 - 4z^2 + 4z^4}}$$
$$= \sqrt{1 - 2(0.1^2) + \sqrt{2 - 4(0.1^2) + 4(0.1^4)}}$$
$$= \sqrt{2.38}$$
$$= 1.54 \text{ rad/sec}$$

h. The closed-loop peak frequency is found by using Eq. (8-16):

$$\omega_m = \omega_n \sqrt{1 - 2z^2} = 20\sqrt{1 - 2(0.1^2)} = 19.8 \text{ rad/sec}$$

i. To find the closed-loop peak gain, A_m, and its associated phase angle, we plug the value of our just calculated ω_m into the closed-loop transfer function, Eq. (7-1).

$$A_m = \frac{100}{-(19.8)^2(0.25) + j(19.8) + 100}$$
$$= \frac{100}{-98.01 + 100 + j19.8} = 5.03 \angle -84.3°$$
$$= 14.03 \text{ dBA} \angle -84.3°$$

Based on all the calculations that were just completed, you can see that in many instances it wouldn't be necessary to plot the response curves for a control system. Much of the needed information for evaluating a system can be obtained through these kinds of calculations instead.

8-5 THE COMPONENTS OF A TRANSIENT WAVEFORM

If we look closely at Eq. (8-10), we can see that this equation is the product of two components. We see an exponential expression, $Ke^{-\sigma t}$, (where K and σ are constants) and a sine expression, $\sin(\omega_d t - \phi)$ (which looks very much like the expressions discussed in Section 1-7.3 of Chapter 1). Figure 8-13(a) shows the individual components; Figure 8-13(b) shows the combined results. Based on this approach, we could rewrite Eq. (8-10) into the following form:

$$y = Ke^{-\sigma t}\sin(\omega_d t - \phi) \tag{8-21}$$

It's easy to see that the $Ke^{-\sigma t}$ expression acts as an envelope that encapsulates the sinusoidal expression. This is the decaying function portion of the transient waveform curve that is often mentioned in transient analysis discussions. Figure 8-14 shows the relationships between the various equations used to calculate the different quantities of the transient waveform.

8-6 THE ROOT LOCUS METHOD

Another method that is frequently used among control system engineers for the analyzing of transfer functions is much more mathematical than the previous

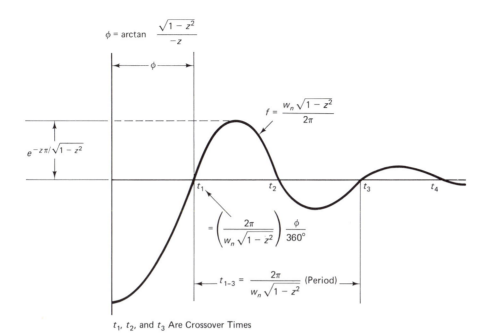

t_1, t_2, and t_3 Are Crossover Times

Figure 8-14   The components of a transient waveform.

method and requires some preliminary ground work in dealing with exponential expressions and trigonometric functions. This mathematical approach in analyzing transient behavior is called the *root locus method*.

8-6.1 Euler's Equation and the Complex S-Plane

To help us understand the root locus method, we have to go back to the two transient components of Eq. (8-21). We see a sine function mixed with an exponential function creating a rather confusing picture of the type of expression we are dealing with. It would help us if we could convert one or the other expression into the other so that we would have either all exponential terms or all trigonometric terms.

There is an identity that allows us to convert cosine expressions to powers of *e*. It is:

$$e^{i\theta} = \cos\theta + j\sin\theta \tag{8-22}$$

This equation is called *Euler's equation,* named after the Swiss mathematician, Leonhard Euler (pronounced Layonhart Oiyler). The proof for this expression is somewhat extensive. However, a proof is presented in Appendix D.

Now, let's look at another form of Euler's equation. We can also state that

$$e^{-j\theta} = \cos\theta - j\sin\theta \tag{8-23}$$

If we add Eq. (8-22) to Eq. (8-23), we will have

$$e^{j\theta} + e^{-j\theta} = 2\cos\theta$$

or
$$\frac{e^{j\theta} + e^{-j\theta}}{2} = \cos\theta \tag{8-24}$$

And since $\theta = \omega t$, then

$$\frac{e^{j\omega t} + e^{-j\omega t}}{2} = \cos\omega t \tag{8-25}$$

We now have an expression that allows us to change any cosine expression into an exponential expression. Notice that it takes *both* $+j\omega t$ and $-j\omega t$ to make one cosine curve.

What we have just proven is fine for converting cosine expressions into e^x-type expressions, but Eq. (8-21) has a sine expression in it, not a cosine expression. However, since the two expressions are related by a phase shift of 90°, we can, for all practical purposes, say they are equivalent as long as we remember to make the 90° phase adjustment. Furthermore, in Eq. (8-21), in place of the $\omega_d t - \phi$ expression, let's substitute the general expression θ. What we have now is a very generalized transient expression, $Ke^{-\sigma t}\cos\theta$ (also, let's assume that contained within our θ expression is our 90° sine-to-cosine phase shift conversion). Now, we can make a further substitution using Euler's equation so that we have

$$Ke^{-\sigma t}\frac{(e^{j\theta} + e^{-j\theta})}{2} \tag{8-26}$$

Let's leave our discussion on Euler's equation momentarily and focus our attention on this next concept: Notice the similarity in appearance between Eq. (8-22) and the complex expression $r = a + jb$. Equation (8-22) appears to be a complex expression describing the location of a point, $e^{j\theta}$, drawn on a set of real and imaginary axes. This is depicted in Figure 8-15. The similarity between this

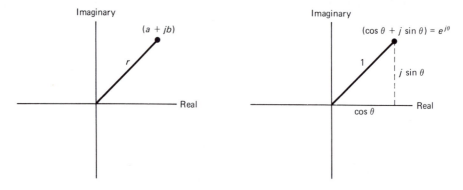

Figure 8-15 The similarities between $a + jb$ and $\cos\theta + j\sin\theta$.

and the complex number, $r = a + jb$, should now be even more apparent. Let's construct a set of real and imaginary axes and label the imaginary axes as s (or $j\omega$) and the real axes as σ (or $\omega_n t$, since $\sigma = \omega_n t$). If we were to let $\sigma = 0$ and assign a wide range of values to θ (ωt), we would create a whole series of sinusoidal waveforms whose frequencies would vary depending on the values of θ. These curves, if superimposed on our s-plane axes, would appear along the vertical axis of our s-plane plot as shown in Figure 8-16.

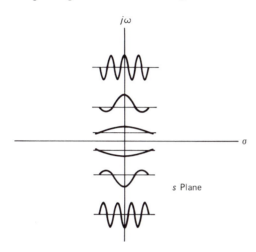

Figure 8-16 How cosine waves and their frequencies would position themselves along the $j\omega$ or s axis of the s-plane. (*Originally presented in B.E. DeRoy's Automatic Control Theory, John Wiley and Sons, publisher, 1966, and reproduced here with the publisher's permission.*)

Looking once again at Eq. (8-26), let's now see what happens when we let $\theta = 0$ and allow σ ($\omega_n t$) to vary over a wide range of values. Again, we superimpose the results onto our s-plane. Figure 8-17 shows us the outcome. Notice how the exponential slopes (representing the envelopes of our transient waveforms) increase and decrease the further you move away from the right or left of the s-axis. What we are interested in are the combined results of Figures 8-16 and 8-17 since, after all, the expressions depicted in these two figures are in fact multiplied together according to the original math expression of Eq. (8-26). Figure 8-18 shows us the final results. Notice how, as you move to the right of the $j\omega$- or s-axis, the resultant transient waveforms become increasingly unstable. It's only to the left of the s-axis where stability seems to exist.

8-6.2 The Characteristic Equation

Now that we have developed the concept of the complex s-plane and how it can help us to visualize the stability characteristics of our transient waveform, we put this concept to practical use.

Assume we have an open-loop transfer function that looks like the following:

$$\frac{\theta_{\text{out}}}{E_{\text{in}}} = \frac{(s + 3)}{(s + 1)(s + 4)} \tag{8-27}$$

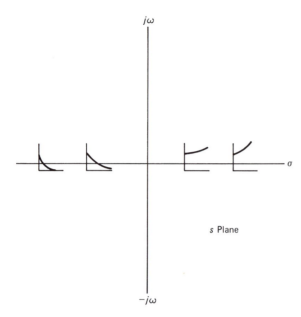

Figure 8-17 is labeled "s Plane"

Figure 8-17   The results of letting $\theta = 0$ and allowing σ to vary, resulting in curves seen along the σ-axis. (*Originally presented in B.E. DeRoy's Automatic Control Theory, John Wiley and Sons, publisher, 1966, and reproduced here with the publisher's permission.*)

The denominator of this expression, $(s + 1)(s + 4)$, is called the *characteristic equation* for Eq. (8-27). The reason for this is because the denominator contains all the important behavioral characteristics for the entire equation. We can prove this by the following method. Temporarily ignoring the numerator of our open-loop transfer function, we can break up the transfer function into two separate fractions, each having its own numerator of, say, A and B. Then we can recombine these two fractions back into a single fraction as shown:

$$\frac{(s + 3)}{(s + 1)(s + 4)} = \frac{A}{(s + 1)} + \frac{B}{(s + 4)} = \frac{A(s + 4) + B(s + 1)}{(s + 1)(s + 4)} \tag{8-28}$$

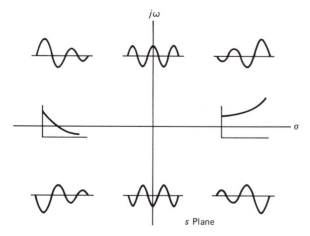

Figure 8-18 is labeled "s Plane"

Figure 8-18   Multiplying together the results of Figure 8-16 and 8-17. (*Originally presented in B.E. DeRoy's Automatic Control Theory, John Wiley and Sons, publisher, 1966, and reproduced here with the publisher's permission.*)

We now equate the numerator in Eq. (8-27) to the resulting numerator in Eq. (8-28):

$$s + 3 = A(s + 4) + B(s + 1) \qquad (8\text{-}29)$$

If we let $s = -1$ in Eq. (8-29), then

$$-1 + 3 = A(-1 + 4) + B(0)$$

and
$$3A = 2$$

or
$$A = \tfrac{2}{3}$$

If we let $s = -4$ in Eq. (8-29), then

$$-4 + 3 = A(-4 + 4) + B(-4 + 1)$$

$$3B = 1$$

and
$$B = \tfrac{1}{3}$$

Placing the values of A and B back into Eq. (8-28), we get:

$$\frac{(s + 3)}{(s + 1)(s + 4)} = \frac{A(s + 4) + B(s + 1)}{(s + 1)(s + 4)}$$

$$\frac{\tfrac{2}{3}(s + 4) + \tfrac{1}{3}(s + 1)}{(s + 1)(s + 4)} = \frac{\left(\tfrac{2}{3}\right)s + \tfrac{8}{3} + \left(\tfrac{1}{3}\right)s + \tfrac{1}{3}}{(s + 1)(s + 4)}$$

$$= \frac{s + 3}{(s + 1)(s + 4)}$$

which is what we started with.

While this is certainly not a rigorous proof that all denominators control the behavior of the entire equation, we can certainly see that this is the case for the example just given. However, this is true for all transfer function equations. In algebra and analytic geometry, the numbers -1 and -4 are called the characteristic equation's *roots*. Any numeric values that cause a fractional equation's denominator to go to zero (or, to state it another way, any numeric values that cause the entire equation to go to infinity), are called roots. The roots are referred to as being *poles* for the entire equation. On the other hand, any numeric values that cause the numerator to go to zero, and consequently, the whole fractional value to go to zero, are called *zeros*. In the preceding case, -3 would be that equation's zero.

EXAMPLE 8-2

Find the poles and zeros for the following expression:

$$\frac{\theta_{\text{out}}}{E_{\text{in}}} = \frac{23.8(1 + 0.7s)}{s(1 + 0.4s)(1 + 0.5s)}$$

Solution:

Finding the roots of the characteristic equation would result in identifying the poles for the entire equation. Consequently, solving for s in the first parenthesis of the characteristic equation: $0.4s = -1$, therefore, $s = -2.5$. Solving for s in the second parenthesis: $0.5s = -1$, therefore, $s = -2$. And don't forget the s. Letting $s = 0$ will also cause the denominator to go to zero. Therefore, s is also a pole. As a result, the poles for the given transfer function are -2.5, -2, and 0.

To find the zeros, we solve for s in the parenthesis in the numerator. Therefore, since $0.7s = -1$, then $s = -1.429$. The zero for the given transfer function is -1.429.

The reason why we want to know the poles, or roots, of a transfer function is this: Any *positive* roots occuring in that equation are a definite indication that that particular system is *absolutely unstable*. Under no circumstances can stability exist. However, we *cannot* say that because there are no positive roots present in our equation, the system is absolutely stable. On the contrary, the system could still be unstable, or jittery, because of too low of a z value. In this case, it would be mathematically proper to say that because no positive roots are present, the system is *possibly* stable.

Just one positive pole is all that it takes to make a system absolutely unstable. Recalling our discussion earlier concerning the complex s-plane, we concluded from Figure 8-18 that any plotted (σ, s) points lying to the right of the s-axis will result in an unstable system. A positive pole is one such point. While this is a simple enough observation to make, it's the determining of the poles that can sometimes be difficult. This is the case when the expression is not in its factored parenthesis form but is, instead, in the unfactored or expanded polynomial form. (A polynomial form is any algebraic expression that is a sum of terms of the form, ax^n, where the n in each term is a positive number, or power, usually arranged in descending order from one term to the next.) For instance, if Example 8-2 were stated in the form,

$$\frac{\theta_{out}}{E_{in}} = \frac{23.8(s + 0.7)}{0.2s^3 + 0.9s^2 + s}$$

instead of the given factored form, you would probably have more difficulty in determining the roots. You would first have to factor out the s term in the denominator and then, either by inspection or by using the quadratic equation, determine the other root values.

Pole determination becomes increasingly difficult as the order of the characteristic equation increases. Look at the following equation containing a third-order expression:

$$\frac{\theta_{out}}{E_{in}} = \frac{2170(1 + 0.25s)}{3s^3 + 10s^2 - 16s - 32}$$

Factoring this expression into its roots could be a time-consuming chore. Fortunately, there is an easier way.

8-6.3 The Routh–Hurwitz Stability Test

There are two cases of unfactored polynomial characteristic equations that we can automatically assume to have positive roots. As a result, these cases generate points in the right-hand portion of our s-plane, resulting in absolute instability. The first case has to do with those polynomials containing minus signs, such as $5x^3 + 4x^2 - 9$, or $16x^4 - 32$. The second case has to do with those polynomials having missing terms, such as $x^4 + 3x^3 + 7$, or $6x^3 + 5$. Factoring, as in the first case, will produce either positive real-number roots or roots containing complex numbers with positive real numbers. Either case will produce points in the right-hand portion of the s-plane and are, as a result, absolutely unstable.

If determining stability is the only thing we are interested in, and we are not interested in the root values themselves, then all we need is some method that would allow us to determine only the signs of our roots. The Routh–Hurwitz Stability Test (pronounced Rooth–Herwitz) is one such method. To use this method, we first note that our polynomial is of the general form.

$$a_0 s^n + a_1 s^{n-1} + a_2 s^{n-2} + a_3 s^{n-3} + \cdots + a_n = 0 \qquad (8\text{-}30)$$

We then arrange the coefficients associated with each term into the following pattern or matrix:

$$
\begin{array}{ccc}
a_0 & a_2 & a_4 \\
a_1 & a_3 & a_5 \\
b_1 & b_2 & b_3 \\
c_1 & c_2 & \\
d_1 & &
\end{array}
\qquad (8\text{-}31)
$$

The terms beyond a_5 must be calculated using the following equations:

$$b_1 = \frac{a_1 a_2 - a_0 a_3}{a_1} \quad b_2 = \frac{a_1 a_4 - a_0 a_5}{a_1} \quad b_3 = \frac{a_1 a_6 - a_0 a_7}{a_1}$$

$$c_1 = \frac{b_1 a_3 - a_1 b_2}{b_1} \quad c_2 = \frac{b_1 a_5 - a_1 b_3}{b_1} \quad d_1 = \frac{c_1 b_2 - b_1 c_2}{c_1} \qquad (8\text{-}32)$$

Our purpose is to fill out the terms in the far left-hand column of the array. If these terms are all positive and nonzero, there are no positive roots existing in the original equation and the system is possibly stable (i.e., it's not absolutely unstable). Let's try a couple of examples.

EXAMPLE 8-3

Determine the stability characteristics for the function:

$$\frac{25}{0.2s^2 + s + 25}$$

Solution:

This is the closed-loop transfer function whose transient waveform was plotted in Figure 8-9. The computed damping factor was 0.2; the system was not stable, but jittery. Using the Routh–Hurwitz Test: $a_0 = 0.2$, $a_1 = 1$, and $a_2 = 25$.

Calculating b_1:

$$b_1 = \frac{(1)(25) - (0.2)(0)}{1}$$

$$= 25$$

All the other factors contain zeros; consequently, all factors beyond b_1 do not exist. Finally, we arrange our factors according to the array, Eq. (8-31):

$$
\begin{array}{cc}
0.2 & 25 \\
1 & \\
25 & \\
\end{array}
$$

Inspecting the left-hand column, we see that there are no negative terms; therefore, there are no roots in the right-hand s-plane and our system is possibly stable. (Of course, we knew that it was somewhat unstable to begin with, but the Routh–Hurwitz Stability Test verifies this by implying that this equation is *possibly* stable.)

EXAMPLE 8-4

Determine the stability characteristics for $s^3 + s^2 + 2s + 8$.

Solution:

Setting up the Routh–Hurwitz array, we get: $a_0 = 1$, $a_1 = 1$, $a_2 = 2$, $a_3 = 8$, and $a_4 = 0$. Any a-terms needed beyond a_4 would also have 0 values. Calculating the other factors, we get:

$$b_2 = \frac{(1)(0) - (1)(0)}{1}$$

$$b_1 = \frac{(1)(2) - (1)(8)}{1} \qquad c_1 = \frac{(-6)(8) - (1)(0)}{-6}$$

$$= -6 \qquad\qquad\qquad = 8$$

Our array now looks like:

$$
\begin{array}{cc}
1 & 2 \\
1 & 8 \\
-6 & \\
8 & \\
\end{array}
$$

Because of the one negative term, -6, our equation is absolutely unstable. Note: You can determine the number of positive roots in the original equation by merely counting the number of sign changes that take place within the left-hand column. In the preceding case, there are two sign changes; one change is going from $+$ to $-$, and the other from $-$ back to $+$. Therefore, there are two positive roots. We can confirm this by noting the actual roots of our equation. They are $s = -2$, $s = 1/2(1 + j3.87)$, and $s = 1/2(1 - j3.87)$. The complex roots are both positive, since the real number portion (the 1 in both cases) is positive.

8-6.4 The Root Locus

The preceding discussion brings us to the concept of the *root locus*. When we talk about a *locus* of a moving object, we are speaking about a path of that object. For example, if you were to install a small light onto the rim of a wheel and then roll the wheel along a flat surface, the locus that the light would make on a time exposure photograph would be a series of looped spirals. This is the concept that we analyze here. We look at a time exposure showing the path that a series of plotted points will make on a graph as certain variables are changed within an equation.

In this section we investigate the path or locus that the roots of a given transfer function make on the *s*-plane as a quantity within the equation is changed. That quantity will be the gain, contained within the constant usually associated with a transfer function's numerator.

First-Order Functions:

Let's take a look at the expression:

$$KG = \frac{E_{\text{out}}}{E_{\text{in}}} = \frac{70(1 + 0.2s)}{1 + 0.5s} \tag{8-33}$$

Equation (8-33) is an open-loop transfer function for a phase-lead network being used in conjunction with an amplifier. The 70 term is comprised of the amplifier's gain and also the resistor ratio, $R_2/(R_1 + R_2)$, normally associated with a phase-lead network's transfer function. Equation (8-33) has a pole, $s = -2$ and a zero, $s = -5$. Converting Eq. (8-33) to its closed-loop counterpart, we get:

$$\frac{KG}{1 + KG} = \frac{70(1 + 0.2s)}{1 + 0.5s + 70(1 + 0.2s)} \tag{8-34}$$

Looking at the characteristic equation, we solve for *s* to determine its roots. We get $s = -4.90$. Figure 8-19 shows the plots for both the poles and zeros (shown as an *x* and *o* respectively) from the open-loop expression of Eq. (8-33) and the one root (shown as a ·) from the closed-loop expression, Eq. (8-34).

We now want to observe what happens when we vary the gain of our system in Eq. (8-34). In other words, instead of having a fixed value of 70 for our gain, let's make that our variable now and assign various values to it. We can then observe what happens to the root in the characteristic equation that we have plotted on the *s*-plane. Equation (8-34) now becomes:

$$\frac{KG}{1 + KG} = \frac{K_x(1 + 0.2s)}{1 + 0.5s + K_x(1 + 0.2s)} \tag{8-35}$$

Looking at the characteristic equation and solving for *s*, we get:

$$s = \frac{-K_x - 1}{0.5 + 0.2K_x} \tag{8-36}$$

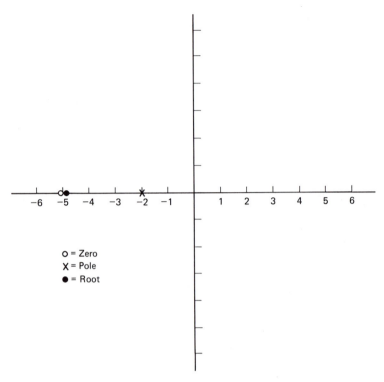

Figure 8-19   The poles, zeros, and roots for $70(1 + 0.2s)/1 + 0.5s$ and $70(1 + 0.2s)/1 + 0.5s + 70(1 + 0.2s)$.

We now pick a range of values, beginning with zero for K_x, and record the resultant values of s. Table 8-2 shows our results.

TABLE 8-2   VALUES OF THE ROOT s, RESULTING FROM ASSIGNING VARYING VALUES OF K_x TO THE CHARACTERISTIC EQUATION, $1 + 0.5s + K_x(1 + 0.2s)$

K_x	s
0	-2
1	-2.22
5	-4
10	-4.4
20	-4.67
50	-4.86
100	-4.93

Figure 8-20 shows the results of plotting the roots from Table 8-2. Notice how, beginning with a K_x of 0, the path or locus of our roots begins at -2, the zero of our open-loop function, and moves toward the final value of -5, the pole. The arrows indicate the path movement. This is typical of any *first-order characteristic equation*. Notice too, that there are no points located in the *right-hand plane* of our plot. This is the absolutely unstable region of the plane. This tells us that all first-order equations are possibly stable. There are no absolutely unstable systems when dealing with first-order systems.

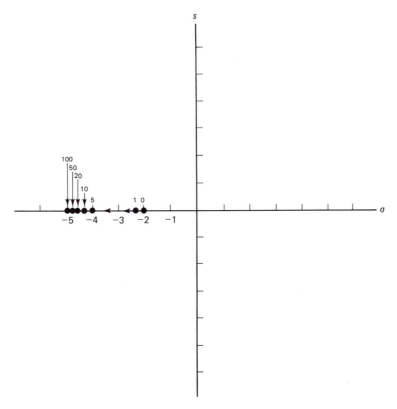

Figure 8-20   The root locus of a first order equation showing migration of roots from zero to pole.

Second-Order Functions:

Let's now analyze a second-order system. Looking back at Section 8-3 where we experimented with our voltage chart recorder, we varied the gain of our control system in an effort to stabilize it. We lowered the gain by reducing the value of the constant, 25, in the closed-loop transfer frunction, $25/(0.2s^2 + s + 25)$. This obviously had an effect on the roots of the characteristic equation, $0.2s^2 + s +$

25. The reason for this is due to the fact that as the 25 is altered in the open-loop function, $25/s(1 + 0.2s)$, the characteristic equation in the closed-loop function is also varied. That same constant, whatever value it may be, always shows up as the coefficient for the quadratic equation (the second-order expression) in the denominator that results from performing the open-loop-to-closed-loop conversion.

Let's determine the roots of $0.2s^2 + s + 25$. Using the quadratic formula, $s = [-b \pm \sqrt{b^2 - 4ac}]/2a$, we find that the roots are $s_1 = -2.5 + j10.9$ and $s_2 = -2.5 - j10.9$. Graphically, we can represent these roots as points on the σ versus s coordinate s-plane axes as seen in Figure 8-21. (Locating these points on

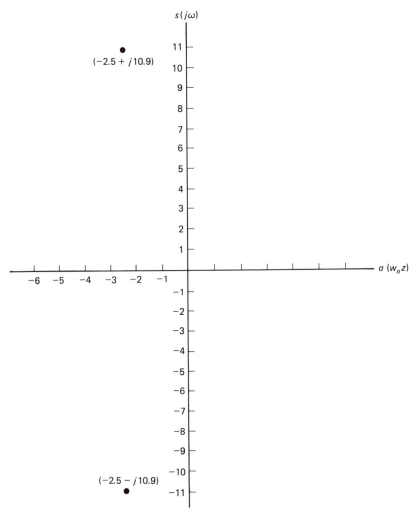

Figure 8-21 Root location for $s = -1 + j4.36$ and $s = -1 - j4.36$.

the coordinate *s*-plane as we just did is similar to locating impedances on the *R*, *j* axes of a phasor diagram. If necessary, see Appendix A to review how this is done.) As we have done before according to mathematical custom, the real numbers in our complex expressions are located along the horizontal axis; the imaginary *j* expressions are plotted along the vertical axis.

Now that we have located the roots of our characteristic equation, let's look at a more generalized expression that states:

$$KG = \frac{K_x}{s(1 + 0.2s)} \tag{8-37}$$

In place of the constant, 25, we have substituted K_x. This will result in a closed-loop characteristic equation of $0.2s^2 + s + K_x$. Next, let's plug in a range of values for K_x, just as we did for the first-order example, and for each one of these values that we choose, calculate the roots of that particular characteristic equation. Table 8-3 lists our results.

TABLE 8-3   DETERMINING THE ROOT LOCUS FOR $0.2s^2 + s + K_x$

K_x	s_1 root	s_2 root
0	−5	0
1	1.38	−3.62
2	−2.5 + j1.94	−2.5 − j1.94
3.96	−2.5 + j3.68	−2.5 − j3.68
10	−2.5 + j6.61	−2.5 − j6.61
25	−2.5 + j10.9	−2.5 − j10.9
50	−2.5 + j15.6	−2.5 − j15.6
100	−2.5 + j22.2	−2.5 − j22.2
300	−2.5 + j38.6	−2.5 − j38.6
500	−2.5 + j49.9	−2.5 − j49.9
700	−2.5 + j59.1	−2.5 − j59.1
1000	−2.5 + j70.7	−2.5 − j70.7

Now, we'll plot our results on *s*-plane coordinate axes. Figure 8-22 shows the results. Notice the arrows indicating the path of travel of these roots as the K_x value increases from 0 to a very large value. These paths are the root loci and are typical of all root loci for second-order characteristic equations. As we noticed in our first-order expression, the root locus begins at the pole of our open-loop expression when $K_x = 0$. All closed-loop root-locus plots for second-order equations *begin at the open-loop poles and progress toward their mid-point as K_x is varied from 0 to infinity.* In the case of our closed-loop expression, the mid-point happens to be the coordinates (−2.5, 0) on the *s*-plane. After the loci reach this mid-point, they split, one traveling vertically upward, the other traveling verti-

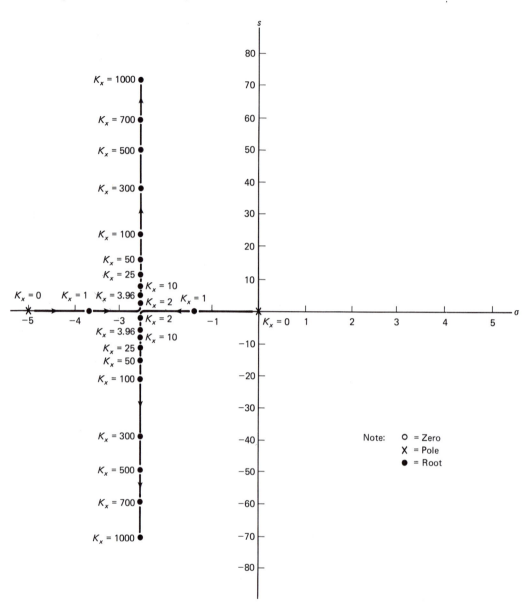

Figure 8-22 The root locus for the characteristic equation $0.2s^2 + s + K_x$.

cally downward. It's important to keep in mind what is happening here: We are varying the gain of our system's amplifier to give us the various values to our K_x. We could have just as properly, but not as easily, varied the gear ratio, K_{gear}, or the servomotor's velocity constant, K_{vel}. This is because $K_x = K_{gear} \times K_{vel} \times K_{amp}$. Varying the gain of an amplifier is so much easier to do. This can be done simply by varying a control knob or a screw-adjusted pot somewhere on the amplifier's chassis or printed circuit board.

Notice in Table 8-3 that we included the K_x values of 25 and 3.96. This was done purposely, since these values were used earlier in our analysis of the voltage recorder. Let's take a closer look at these two points, first the K_x value of 25. This point has been replotted, using the coordinates of $(-2.5, j10.9)$, on an expanded s-plane coordinate axis in Figure 8-23. We will ignore the $(-2.5, -j10.9)$ point, since it is a mirror image of the other point. However, note the numeric

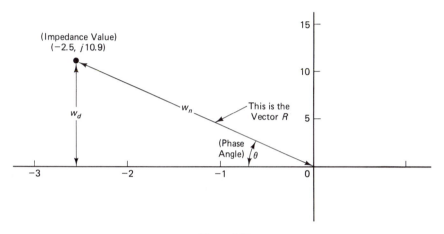

Figure 8-23

value of the j-value that we did plot, (the 10.9 value). This happens to coincide with the ω_d value that we calculated for our system when we analyzed it earlier in Section 8-3.1 for a K_x of 25. Now, if we think of our plotted point as an impedance in an electrical circuit and perform a rectangular-to-polar conversion, just as you would do in an electrical circuit, the value of the r vector is ω_n, and the resultant phase angle, θ, is the arccosine of our damping factor. According to the data that we gathered for this system in Section 8-3.1, $\omega_n = 11.2$ rad/sec and $z = 0.224$. From our root locus plot in Figure 8-23, $\omega_n = 11.2$ rad/sec and $z = \cos 77.08° = 0.224$.

We can prove to ourselves that $z = \cos\theta$ by the following relationships: If we let $x =$ the length of the horizontal base of the triangle formed by the other two sides of ω_d and ω_n, then by the Pythagorean theorem:

$$\omega_n^2 = \omega_d^2 + x^2$$

then

$$x^2 = \omega_n^2 - \omega_d^2$$

or

$$x = \sqrt{\omega_n^2 - \omega_d^2}$$

If

$$z = \cos\theta = \frac{\sqrt{\omega_n^2 - \omega_d^2}}{\omega_n}$$

then, cross-multiplying: $\quad\quad \sqrt{\omega_n^2 - \omega_d^2} = z\omega_n$

Clearing the radical: $\quad\quad\quad \omega_n^2 - \omega_d^2 = z^2\omega_n^2$

Solving for ω_d: $\quad\quad\quad\quad \omega_d = \sqrt{\omega_n^2 - z^2\omega_n^2}$

resulting in: $\quad\quad\quad\quad\quad w_d = w_n\sqrt{1 - z^2}$

This result checks with Eq. (8-9).

Hopefully by now, you have been able to see the advantage of using the root locus method. It enables a person to trace out the root locus of his or her particular characteristic equation so that the K_x value can be varied to see what value produces the most stable system. The root locus gives a good visual representation of what is happening within the control system.

Third-Order Functions:

Let's look at the situation involving third-order systems. These open-loop transfer functions are typically of the form:

$$\frac{\theta_{\text{out}}}{E_{\text{in}}} = KG = \frac{K_x(1 + s\tau_1)}{(1 + s\tau_2)(1 + s\tau_3)(1 + s\tau_4)} \quad\quad (8\text{-}38)$$

The closed-loop equivalent of this expression would be:

$$\frac{\theta_{\text{out}}}{E_{\text{in}}} = \frac{KG}{1 + KG} = \frac{K_x(1 + s\tau_1)}{(1 + s\tau_2)(1 + s\tau_3)(1 + s\tau_4) + K_x(1 + s\tau_1)} \quad (8\text{-}39)$$

The third-order characteristic equation for this expression looks somewhat menacing to handle. But there is at least one shortcut that we can look for. If any one of the time constant values in the denominator of Eq. (8-39) is reasonably close in numeric value to the time constant in the numerator, we can cancel out those close terms and wind up dealing with only a second-order equation. What we mean by reasonably close is that the zero value in the numerator is no more than about two times greater than any one of the pole values in the denominator of the open-loop expression. After cancellation, we are left with a second-order equation that can be handled in the usual way as far as determining its root locus. This will still produce reasonably accurate results. Look at the following example.

EXAMPLE 8-5

Find the damping ratio for the following expression using the root locus method:

$$KG = \frac{450(1 + 0.2s)}{(1 + 10s)(1 + 0.25s)(1 + 0.8s)}$$

Solution:

The zero for the preceding expression is -5 ($0.2s = 1$, therefore $s = -1/0.2$, or -5).

The three poles of our expression are -0.1, -4, and -1.25. Since the zero value of -5 is reasonably close to the one pole of -4, we can cancel these two expressions containing those values, resulting in a final expresion of:

$$KG = \frac{450}{(1 + 10s)(1 + 0.8s)}$$

We can now proceed with plotting the root locus of our finalized expression treating it like any other second-power system expression. In reality, the results will be quite close to the results obtained if we had actually plotted the root locus for the original third-order equation. We won't complete the solution for the damping ratio here; we only want to demonstrate the canceling procedure just discussed. Otherwise, the solution would proceed just like the example used earlier in Figures 8-22 and 8-23.

Third-order equations that can not be handled by the foregoing cancellation method must be worked out using other means. These other methods can be extremely involved and time-consuming. We discuss one approach here that will require a computer or hand calculator that has the programming capability of finding at least one real root of a cubic equation. We first describe this method, and then at the end of this discussion we go through a step-by-step set of instructions to follow. Some of the concepts presented here are presented without proof because of their complexity.

The root locus behavior for third-order characteristic equations is somewhat similar to that of second-order equations. When K_x attains values that cause imaginary roots to be generated within the characteristic equation, the root locus leaves the horizontal σ or real axis and begins traveling either up or down through the second and third quadrants (assuming, of course, that the system is not absolutely unstable). With the second-order equations, we found that the root locus left the σ-axis at a 90° or $-90°$ angle, depending on whether the locus was traveling up or down from the axis. With third-order equations, however, this angle of root locus departure cannot be depended on to always be at a right angle with the real axis. Fortunately, we can calculate this angle very easily.

$$\frac{\text{angle of root}}{\text{locus departure}} = \frac{\pm 180°}{(\text{number of poles}) - (\text{number of zeros})} \qquad (8\text{-}40)$$

Equation (8-40) works for second-order as well as third-order equations. Looking at Eq. (8-37), we see there are two poles ($s = 0$ and $s = 5$) and no zeros. According to Eq. (8-40), the angle of root locus departure $= \pm 180°/(2 - 0) = \pm 90°$. This is what we found when we constructed the root locus in Figure 8-22.

With the second-order equation, we found that the root locus departed from the real axis exactly mid-way between the poles of the open-loop equation. This is no longer true for the third-order equation. Fortunately again, we can calculate approximately where the root locus leaves the real axis using Eq. (8-41):

$$\begin{matrix} \text{asymptote} \\ \text{location} \\ \text{point, or} \\ \text{centroid} \end{matrix} = \frac{\text{sum of pole values} - \text{sum of zero values}}{\text{total number of poles} - \text{total number of zeros}} \qquad (8\text{-}41)$$

Notice that we said we can calculate *approximately*. Equation (8-41) actually locates an asymptote along which the root locus travels and gradually approaches and eventually reaches at infinity. At the departure point on the real axis, the root locus will be at its greatest distance from this line. Let's look at an example. Again, we can use Eq. (8-37) to check the preceding equation. The departure point (i.e., the mid-way point seen between the two poles in Figure 8-22 for Eq. 8-37) would be calculated as:

$$\frac{(0 + 5) - (0)}{2 - 0} = 2.5$$

This point was in fact what we found out to be true when we plotted the root locus for Eq. (8-37).

Let's now try our hand at calculating the actual departure point for a system and see how it compares with the results of Eq. (8-41). This is not a practical exercise and is done here only to show the relative comparison between an actual result and the results obtained from a much easier method that we can depend on for being fairly accurate.

EXAMPLE 8-6

Plot the root locus near the real axis for the following expression and show how its departure point compares with the asymptote or centroid location calculated with Eq. (8-41):

$$KG = \frac{K_x(1 + s)}{s(1 + 10s)(1 + 0.01s)} \qquad (8\text{-}42)$$

Solution:

First, we notice that the two poles are well outside the times two amount of the zero, -1. So we have no choice but to plot the roots for various values of K_x to see how the root locus behaves near the real axis. To do this, we rearrange the terms in Eq. (8-42) so that, when multiplied out, the characteristic equation in the closed-loop expression will have a 1 for the cubic term instead of a fractional value. This also makes the zero and pole values easier to see.

$$KG = \frac{(0.1)(100)}{(0.1)(100)} \cdot \frac{K_x(1 + s)}{s(1 + 10s)(1 + 0.01s)}$$

$\qquad \qquad \hookrightarrow \quad$ this fraction is comprised of the
$\qquad \qquad \qquad$ numbers needed to multiply the
$\qquad \qquad \qquad$ parenthetical terms by in order
$\qquad \qquad \qquad$ to convert them to the form seen
$\qquad \qquad \qquad$ in Eq. (8-43).

$$= \frac{10K_x(s + 1)}{s(s + 0.1)(s + 100)} \quad (8\text{-}43)$$

Since the K_x term represents a general coefficient, we can combine the 10 with the K_x in the numerator of Eq. (8-43) to get:

$$KG = \frac{K_x(s + 1)}{s(s + 0.1)(s + 100)} \quad (8\text{-}44)$$

Now, we proceed with determining the characteristic equation just as we did for the second-order root locus example earlier:

$$\begin{aligned}
\frac{KG}{1 + KG} &= \frac{K_x(s + 1)}{s(s + 0.1)(s + 100) + K_x(s + 1)} \\
&= \frac{K_x(s + 1)}{s(s^2 + 100.1s + 10) + sK_x + K_x} \\
&= \frac{K_x(s + 1)}{s^3 + 100.1s^2 + 10s + sK_x + K_x} \\
&= \frac{K_x(s + 1)}{s^3 + 100.1s^2 + s(K_x + 10) + K_x} \quad (8\text{-}45)
\end{aligned}$$

Looking at the characteristic equation in Eq. (8-45), we realize the work involved having to derive the roots for a cubic equation for every value we pick for K_x. However, by using a hand programmable calculator, or by using a PC, the chore becomes rather simple. We can use the simple BASIC programs for an IBM PC in Appendix C. Armed with this program or a similar program, let's construct another table (Table 8-4) to keep track of our selected K_x values and the resultant roots. The technique we use is this: We start off with a very high value for K_x, say 10,000, and then select a low value, 1,000. If our choice results in real roots only, we know our pick was too low. The root locus is still on the real axis. On the other hand, if our selection results in generating imaginary roots, our K_x value is on the root locus part that has lifted off the real axis already. We then lower our next selection of K_x. We work our selections back and forth until we "trap" the locus just as it is lifting off the real axis. We want to find the root value just at that point. The listings of our K_x values *are listed in just the order used in having the roots determined* so you can see how the final result was finally "captured."

As the table shows, the root locus leaves the real axis at a value of approximately -49.35 (splitting the difference between -49.16 obtained with $K_x = 2546$, and -49.53, the real portion of the imaginary root obtained with $K_x = 2547$). However, Eq. (8-41) says that the departure point is:

$$\frac{0.1 + 100 - 1}{3 - 1} = 49.55$$

where $0.1 + 100$ in the numerator = sum of pole values. Figure 8-24 shows the partial plot of the root locus near the real axis along with the placement of the

**TABLE 8-4   DETERMINING THE ROOT LOCUS NEAR
THE REAL AXIS FOR $s^3 + 100.1s^2 + s(K_x + 10) + K_x$**

K_x	s_1	s_2	s_3
10,000	-1.009081	$-49.55 + j86.34$	$-49.55 - j86.34$
1,000	-1.111094	-10.13	-88.86
5,000	-1.01852	$-49.54 + j49.55$	$-49.54 - j49.55$
3,000	-1.031711	$-49.53 + j21.31$	$-49.53 - j21.31$
2,000	-1.049281	-26.14	-72.91
2,500	-1.038587	-42.74	-56.33
2,575	-1.037371	$-49.53 + j5.37$	$-49.53 - j5.37$
2,550	-1.037768	$-49.53 + j1.97$	$-49.53 - j1.97$
2,545	-1.037849	-48.46	-50.60
2,547	-1.037816	$-49.53 + j.93$	$-49.53 - j.93$
2,546	-1.037832	-49.16	-49.90

asymptote line. The dimensions have been exaggerated to show the difference between the locus and the asymptote more clearly. As you can see, our use of Eq. (8-41) to approximate the point of departure is fairly accurate. Besides, since we will be dealing with points whose complex coordinates will lift them a considerable distance away from the real axis in order to maintain a stable damping ratio, this error will be reduced even more. As an example, assume that the equation we have been working with had a K_x of 5,000. The complex roots associated with this value are $s = -49.54 \pm j49.54$. The real portion of this root (-49.54) closely approximates the location of the asymptote. Figure 8-24 shows that with this value, the damping ratio, z, would have a value of 0.707.

Approximating the locus departure point, often called the *breakaway point,* using Eq. (8-41) is valid only if the asymptote is vertical to the real axis. That is, if using Eq. (8-40) to calculate the angle of root locus departure results in a $\pm 90°$ value, then we can use our approximation. Any other angle value will demand some other approach, which we do not cover in this discussion. A typical transfer function that will *not* generate a vertical asymptote is of the form:

$$KG = \frac{K_x}{s(1 + s\tau_1)(1 + s\tau_2)}$$

More advanced textbooks must be consulted for handling this type of case.[1]

There are other root loci portions associated with a cubic equation transfer function. There are loci that exist between the various pole and zero locations on the real axis; this is where you will find the real root, s_1, in Table 8-4. However, we are only interested in the vertical branch root locus, since that is the only portion of the entire root locus we use for determining system stability.

[1] Richard C. Dorf, *Modern Control Systems*, 3rd ed. (Reading, Mass.: Addison-Wesley Publishing Co., 1980), pp. 170–84.

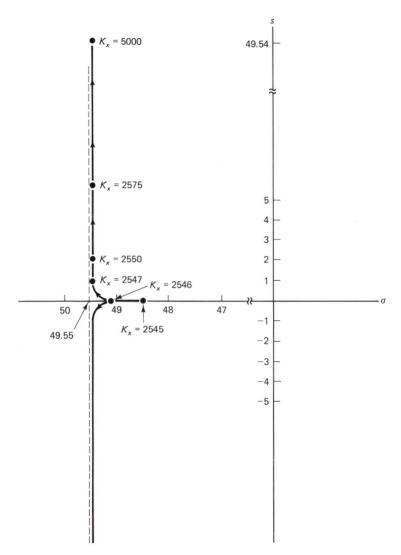

Figure 8-24   The root locus for the characteristic equation $s^3 + 100.1s^2 + s(Kx + 10) + Kx$ near the real axis.

Let's list the various steps we took in finding the root locus of a third order equation:

1. Identify the open-loop transfer function.
2. Convert to a closed-loop transfer function using $KG/(1 + KG)$.
3. Identify the characteristic equation (the third-order equation in the denominator).

4. Convert the parenthetical terms in the closed-loop expression from the $(1 + s\tau)$ form to the $(s + 1/\tau)$ form. Remember to multiply the K_x constant in the numerator by the appropriate conversion factors needed to perform the parenthetical conversions. (Refer to Eq. (8-43) to review this step if necessary.)

5. Calculate the roots of the characteristic equation using whatever means you have available for determining third-order roots. (Refer to Appendix C.)

6. Use Eq. (8-41) to determine the asymptote location on the real axis (σ-axis). The location point should be quite close to the real-number component of the complex roots obtained in step 5. Equation (8-41) is used for a check of the root results obtained in step 5.

7. We can now calculate the following information:

   a. Damped frequency: ω_d = the imaginary-number component, or j-value, of the root found in the step 5.

   b. Natural frequency: $$\omega_n = \sqrt{\omega_d^2 + x_r^2} \qquad (8\text{-}46)$$

   where x_r = the real-number component value taken from the complex root

   c. Damping ratio: $$z = \cos(\arctan \omega_n/x_r) \qquad (8\text{-}47)$$

8-7 OTHER CONTROLLER TYPES

In reality, there are three kinds of automatic control systems. These are:

1. proportional control,
2. integral control, and
3. derivative control (sometimes called differential control).

We discuss these in the next three sections.

8-7.1 The Proportional Controller

Throughout this text, we assumed that all the control systems we discussed were of the proportional type. That is, the system's input signal to the system's overall transfer function is proportional to the generated error signal. If we represented the error signal by e and the input signal to the transfer function by E, the relationship between the two could be expressed as:

$$K = \frac{E}{e} \qquad (8\text{-}48)$$

where K is a proportionality constant. Up to this point, we have been assuming a

K value of 1. In other words, *E* and *e* were the same. It's a matter of viewpoint, however. You could consider the system's amplifier gain as *K* since we are assuming the amplifier to be a proportional device. If we separated the amplifiers from the main transfer function and then let the amplifiers become our *proportional controller* as depicted in Figure 8-25, we then have a picture of a propor-

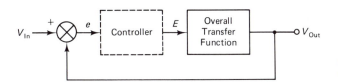

Figure 8-25   The "controller" in a feedback system.

tional controller. But what if we wanted a much quicker speed of response and a decrease in our positional error at all frequencies in response to the generated error signal? With a proportional control system, our first response would be to increase the system's gain. But increasing the system's gain would buy us a more jittery system regardless of operating frequency. We then would resort to a modification to the controller box shown in Figure 8-25 to give us the desired response.

8-7.2 The Integral Controller

The purpose of the integral controller is to increase the gain of our system at very low frequencies. We do this by introducing an op-amp integrator circuit where our controller box is located in Figure 8-25. The circuit is seen in Figure 8-26. By

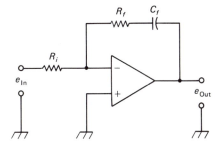

Figure 8-26   An integral control circuit.

juggling the values of resistances, capacitance, and amplifier gain in our circuit, we can actually obtain a combination proportional–integral controller whose gain curve would resemble something like that in Figure 8-27. The circuit would give us very high gain at extremely low frequencies but diminish at the middle frequencies. This is desirable because the very high gain at the very start of energizing, say, a servomotor, gives you a very fast response time. But as the transient

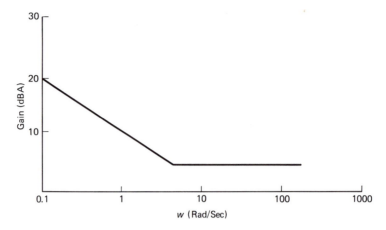

Figure 8-27   Typical gain curve for an integral control system.

response sets in, you want low gain to be present in order to decrease over-shoot as steady state conditions settle in.

8-7.3 The Derivative Controller

Ideally, once steady state conditions have been reached, we would like to have the system's gain increase once again to be prepared for the next sudden input change. This is what the derivative control does. Installing the circuit shown in Figure 8-28 in our controller box of Figure 8-25, we will have the response just

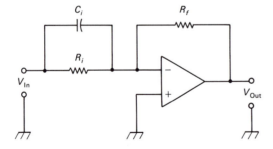

Figure 8-28   A derivative control circuit.

described. Figure 8-29 demonstrates this particular system's gain response action. Again, like the integral control system, the derivative circuit in Figure 8-28 is actually a combination proportional–derivative control system. It's possible to combine the two systems in order to obtain the increase-decrease-increase gain characteristics to give an almost ideal system response. The circuits presented here are just a few of the many circuits used for refining the desired system response.

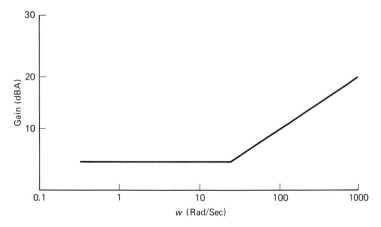

Figure 8-29   Typical gain curve for a derivative control system.

SUMMARY

The hand-plotting of the transient response curve for a closed-loop transfer function can be somewhat an involved process. However, good representations of a control system's behavior can be learned this way. There are programs, though, for the personal computer that can handle this chore nicely. One such program is found in Appendix B.

The transient response curve in an automatic control system is comprised of two subcomponent curves, one of which is a sine expression while the other is a decaying exponential function (hopefully it is decaying and not growing!). Being able to recognize these two curves allows one to determine the control system's vital response characteristics to a step input signal. Actually, there is a third component. This is a constant that represents the magnitude of the step input signal. The transient waveform "wraps" itself around this step which forms the horizontal axis of its plot.

Some of these vital statistics can be calculated from the Laplace transform of a closed-loop function "shocked" with a unit step input. The transform allows us to find the natural resonant frequency, the damped frequency, and the damping ratio of a system. Once these values are found, we can calculate values such as settling times, response times, over-shoot, etc. However, this technique is good only for second-order equations. Third-order equations require an entirely different approach, the root locus method.

Euler's equation helps us to make the leap from the trigonometric time-related plane of the transient waveform to the so-called *s*-plane. It is the *s*-plane that leads us into the discussion of the root locus concept.

When speaking about poles and zeros, reference is made to the open-loop equation. The term *roots* is used in conjunction with the closed-loop equation as is the term, *characteristic equation*. The poles and zeros of the open-loop expression determine the shape and direction of travel of the root locus plotted on an *s*-plane coordinate grid as the K_x value is changed in the closed-loop expression.

First-order characteristic equations are the simplest of all equations to deal with in automatic control systems. However, they are rarely encountered and are used only as a stepping stone to the more complex equations.

Third-order equations may have their stability characteristics determined using the Routh–Hurwitz method. The root locus method may then be applied to find ω_n, ω_d, and z. Once these quantities are found, all the other transient quantities may be calculated. The processes used here on the third-order functions may also be applied to the second-order functions if desired.

EXERCISES

8-1. Determine the poles and zeros in the following expressions and determine each equations' stability criterion based on the determined poles.

a. $\dfrac{25(1 + 0.2s)}{s(1 + 0.3)(1 - 3s)}$

d. $\dfrac{65.6(1 + 3s)}{s(1 + 0.5s)(1 + 10s)}$

b. $\dfrac{4.87(1 + 6s)}{(1 + 3s)(1 - s^2)}$

e. $\dfrac{62.4(1 + 0.25s)}{(1 + 0.99s)(1 + 25s)}$

c. $\dfrac{1044}{(1 + 0.1s)(1 + 0.5s)}$

f. $\dfrac{1,000(1 - 0.2s)}{s(1 - 0.5s)(1 + 6s)}$

8-2. Plot the transient curve of the function:

$$KG = \dfrac{12}{s(1 + 0.15s)}$$

8-3. For the expression, $KG = 100/s(1 + 0.25s)$, plot its transient response curve.

8-4. Reduce the K_x value in the expression given in Exercise 8-3 to the value, 5. Again, plot the transient response curve. What changed in this new curve? Did the transient curve become more stable or less stable?

8-5. In the revised equation of Exercise 8-4, calculate the following:

a. the natural resonant frequency,

b. ω_d, and

c. the maximum over-shoot.

8-6. Plot the root locus for the expression:

$$KG = \frac{K_x}{(1 + 8s)(1 + 1s)}$$

Determine ω_n, ω_d, and z all for a K_x value of 31.

8-7. Find the $\pm 5\%$ settling time, the number of oscillations to within $\pm 5\%$ settling, and ω_m for the expression:

$$KG = \frac{100}{s(1 + 0.16s)}$$

8-8. Find the damping factor for the expression:

$$KG = \frac{250(1 + 0.3s)}{(1 + 10s)(1 + 0.27s)(1 + 0.9s)}$$

8-9. Plot the root locus for the expression:

$$KG = \frac{K_x(1 + 11s)}{s(1 + 8s)(1 + 0.02s)}$$

Compare your results to the results obtained using Eq. 8-41 for finding the asymptote location point.

8-10. Using the Routh–Hurwitz Stability Test, find the stability characteristics for the following expressions:

a. $\dfrac{30}{0.1s^2 + s + 30}$

b. $\dfrac{100}{0.2s^2 + 3s - 100}$

c. Characteristic equation: $s^3 + s^2 + 3s + 12$

d. Characteristic equation: $s^3 - 2s^2 + 4s - 8$

e. Characteristic equation: $s^3 + 14s^2 + 6s + 22$

f. Characteristic equation: $s^3 + 2s - 7$

REFERENCES

DeRoy, Benjamin E., *Automatic Control Theory,* New York, N.Y.: John Wiley & Sons, 1966.

Sante, Daniel P., *Automatic Control System Technology,* Englewood Cliffs, N.J.: Prentice-Hall, Inc., 1980.

9
Designing A Control System

9-1 INTRODUCTION

The purpose of this chapter is to demonstrate how to apply all the facts that have been presented so far in designing and analyzing automatic control systems. In many cases, there is more than one method that can be used in this designing and analyzing process. Where possible, we discuss the other methods that can also be used. Sometimes it's a good idea to have more than one method available so that one can be used to check the results obtained with the other. And in the designing process, quite often the first design chosen turns out to be the most expensive design; therefore, alternative approaches must be used. So we discuss design examples for various automatic control systems where we can apply our newly acquired knowledge of system designing. We also want to be able to analyze these systems and make forecasts of their behavior resulting from making adjustments to them. The design approaches presented in these examples have been simplified to keep their developments shortened. However, the overall approaches to the solutions are realistic, and it's these approaches that should be studied.

9-2 EXAMPLE 1: A ROBOTIC CART STEERING SYSTEM

The first automatic control system that we design is a control system for the automatic steering of a robotic cart. Here are the details:

258

1. The cart is a self-propelled, all-electric, battery-operated cart (27 VDC, or two 13.5 VDC batteries in series) that is used for distributing and delivering interplant mail, maintenance items, and specialized inventory parts for a large manufacturing plant. You might think of this system as a robotic gofer.

2. Our cart will be designed to follow an invisible infrared reflective paint track that will be painted on the plant floor connecting the various pick-up points the cart will pass along its way.

3. A set of photocells will detect the reflection of an infrared light source mounted underneath the cart, thereby being able to detect the cart's relative position to this paint line.

4. The cart's steering must be designed to allow the cart to stay on its track despite side forces due to receiving accidental bumps or running over minor obstacles in its path. The desired damping factor for the cart's steering must be in the range of 0.5 to 0.6. This is a good compromise between a not-too-sluggish system and one that is not too jittery.

Now that we have laid down some basic ground rules for our design project, let's see how we tackle it. A good place to start is with laying out a block diagram of the system. We can fill in the detailed numerical data later. Figure 9-1(a) shows our initial concept of how we think the steering mechanism will function in a control system. Figure 9-1(a) is the schematic, while Figure 9-1(b) is the block diagram. Once we have the design outlined the way we want it, then we can refine

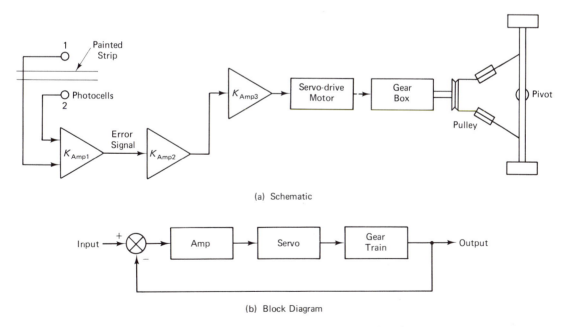

(a) Schematic

(b) Block Diagram

Figure 9-1 The initial design for the robotic gofer cart.

our drawings into detailed mechanical and electronic schematic layouts. Here is how we hope the system will work:

In our sketch we show the photocells that we plan to stride the painted strip on the floor. We will mount the cells on the cart's bottom along with the IR (infrared) light source. The IR source will provide the light energy that will reflect off the paint into either or both photo-cells. We decided to use photoresistive cells because they are inexpensive and have good IR response. The cell outputs will then go to a differential op amp whose output will be either positive or negative depending on which cell has the stronger response output. The cells' output of course is determined by how close it is to the painted strip. The error signal is amplified by a preamplifier and power amplifier. The output of the power amplifier will then drive a servomotor whose output is attached to a gear box. The gear box drives a pulley system which, in turn, rotates the steering axle toward the right or left.

Now let's see how the system will work: If the cart is moving down the center of the track, photocells 1 and 2 will be an equal distance from the reflective strip. Consequently, the signals coming from both cells will be equal. The output of the op amp, on the other hand, will be zero since its output is the difference between the two inputs. Since the output is zero (i.e., the error signal is zero), the servomotor will remain stationary and the cart will continue moving straight ahead. So far, we see no problems here. Now, let's see what happens when the cart is forced to move left due to an outside force such as an accidental bump received by a passing work person. This means that photocell 2 will now receive a stronger reflective signal than photocell 1. An error signal is now produced by the op amp because of the upsetting of the exact signal balance that existed up to this point between the two photocell outputs. The two amplifiers will now amplify this new error signal causing the servomotor to rotate, which in turn will cause the wheels to pivot right. When this happens, photocell 2 increases its output signal as the reflective strip comes closer, thereby reducing the error signal output from the op amp. The servomotor then slows down or even reverses if the error signal goes beyond zero, changing its polarity. If this happens, the steering swings back toward the left and the process starts all over again.

Obviously, we have to be careful here. We don't want a system that over-corrects or hunts. In other words, we want a system that hunts just enough to get us back on track with the minimum of oscillations in our steering. Selecting a damping factor of 0.5 or 0.6 should, theoretically, do this for us. And the very first consideration we must be aware of is that we want to make sure we have negative feedback and not positive feedback. Positive feedback would cause our steering to correct in the wrong direction. A simple oversight such as inadvertently reversing the wiring to our servomotor would do just that. Or, the same effect would be created if the wiring to the op amp's signal input would be accidentally reversed. So we want to check to make sure our prototype design is wired properly before operating.

9-2.1 The System's Dead Band

Another problem that we haven't discussed up to this point is the problem of a system's *dead band*. Dead band may be thought of as system "stickage." In our steering design, the servomotor, along with the rest of the steering control system, will exhibit a certain amount of dead band because of the low input command voltages being supplied to it. Even though servomotors are designed to operate at relatively high input voltages such as ±26 VDC or 28 VAC, most of their operating lives are spent responding to input voltages that are just a fraction of their rated operating voltage. This is because the error signals sent to the servo are usually quite small. An automatic control system spends a good portion of its time generating these small touch-up corrective signals to keep the overall system on proper course. As a result, the servomotor and the components downstream from the servo's output must be able to respond to these signals. Obviously, there is a threshold limit to any servo's ability to respond to weak signals. This is where dead band comes into play. Dead band may be defined as a range of input signals that is furnished to a device or system before that system finally responds to this input (see Figure 9-2). In Figure 9-2, rotation of this particular servomotor begins only after receiving a voltage of at least ±1.5 VDC. Dead band is usually expressed in angular displacement of the control generating the control voltage to

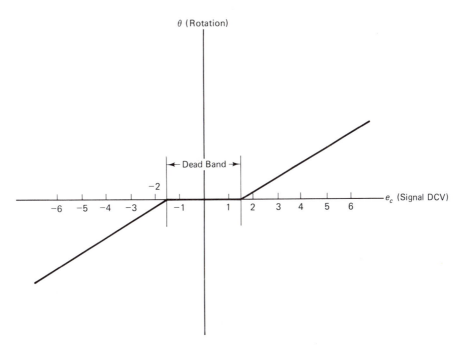

Figure 9-2   A servomotor's dead band.

the system. In the case of the servo control characteristics depicted in Figure 9-2, if a rotation of $\pm 5.7°$ of a control device is necessary to produce the ± 1.5 V for servo rotation, then the dead band is specified as $\pm 5.7°$. Dead band is created by system friction, inertia, magnetic coupling, gear train load, and backlash on the servo or any combination of these quantities. Any good servomotor should be able to respond to a voltage as low as about 5% of its rated operating voltage. Dead band may be calculated using the following equation:

$$\text{dead band } (°) = \frac{2V_{ms}}{(K_s)(K)} \tag{9-1}$$

where V_{ms} = motor starting voltage (volts)
$\quad K_s$ = control device sensitivity (volts/deg)
$\quad K$ = remaining system gain or sensitivity

Notice that only the system's gain or sensitivity constants between and including the control device (that is, the pot, synchro, or other device that, when rotated or moved, creates the activation voltage for the servo) and the servomotor are used in Eq. (9-1). The servo's K_x is *not* included in the calculation.

Dead band may also be expressed as a linear displacement rather than an angular displacement of a control input. This is explained later as we become more familiar with our steering system.

One of the easiest methods in making an otherwise unstable system stable is to reduce the system's gain. Unfortunately, the price one pays for this deceptively simple solution is an increase in that system's dead band. You can verify this by inspecting Eq. (9-1). The dead band varies in inverse proportion to the system gain. As a consequence, if we decide to lower the system's gain, we want to be sure to check out the dead band first to make sure its increase can be tolerated.

9-2.2 Selecting the Components and Generating the Open-loop Transfer Function

The heart of our cart's steering system will undoubtedly be the servomotor. We want to be sure that it not only has the proper time constants and velocity constant, but that it also has enough torque to do the steering for us. It's unlikely that we will be able to find a servo with a built-in gear train that can generate the needed torque. We may have to purchase an additional gear box to operate off of the servo's already geared-down output in order to create the torque that we need. We'll have to keep this in mind when we size the servomotor for the steering requirements.

Determining the steering force for our cart is relatively easy. We need to be able to measure the amount of force required to rotate the front axle wheels. This can be measured by using a calibrated fish scale attached to one end of the steering cable that slips over the pulley attached to the gear box. The amount of pulling on the scale that causes the proper steerage will be the force we will use in calculating the needed servo torque. Let's assume this has been done and we found that a force of 3.2 lb. was sufficient to give us the proper steering action. To calculate the torque, we measure the gear box's pulley over which the steering cable is wound, and then multiply its radius times 3.2 lb. Let's assume a radius of 3 inches; the torque will be 3.2×3 or 9.6 lb.–inches. Since this torque amount is not continuous but only intermittent, we should keep this in mind when we size our servo. This will help keep the cart's cost down by allowing us to use a smaller servo.

At this point, we can start looking for DC servomotors with 9.6 lb.–in. capabilities, and if we're successful, we can eliminate the gear box. If we can't find a servo with that kind of torque output, then we will have to shop around for a gear box to augment the gear train on the servo that we do decide eventually to use.

Along with searching for the proper torque output, we want to make sure that our torque is outputted at an acceptable speed. Obviously, we can generate as much torque as we want with the smallest servomotor available by simply using the highest gear ratio available for a gear train. But the output shaft's rotation would be barely detectable. We must decide on an acceptable speed for our 9.6 lb.–in. torque. If we moved the steering cable through a 6-inch travel in one second, that would give us an acceptable turning velocity. This means that our drive pulley (the 3-inch radius pulley) would have to rotate with an angular velocity of 6 in./sec $\div$ 3 in. = 2 rad/sec. Since one radian = 57.3°, then the number of revolutions per second would be

$$2 \text{ rad/sec} \times 57.3°/360° = 0.318 \text{ rev/sec, or } 19.1 \text{ rpm}$$

We should mention one additional item here concerning the physical size of servomotors: The body sizes or diameters of the motor housings are cataloged in whole-number sizes. The values of these numbers are the housings' diameters in tenths of inches. For example, an 08-sized servomotor is one having a motor frame diameter of approximately 0.8 inch. A size 18 housing is one that is approximately 1.8 inches in diameter.

A likely candidate for our servomotor is described in the specification sheet in Figure 9.3. Here we see a DC servo that has a peak torque capacity of 163 oz.–in. or 10.2 lb.–in.[1] At first glance, these specifications seem to meet our requirements nicely. Since this torque has been specified as peak, we can assume that this is the servo's stall torque (i.e., its torque output at 0 rpm), since the stall torque is a servo's maximum developed torque. However, notice that the rpm appears to be quite high, 573 rpm. Since there are no torque-speed curves accompanying these specs, let's draw a simple one for our own use from the information that's given. Refer to Figure 9-4. We patterned the shape of our curve after the

---

[1] This catalog data is now obsolete, according to the Singer Co., and is presented here only as an example.

SIZE 38 CONTINUOUS ROTATION DC TORQUERS

GENERAL CHARACTERISTICS
CM3 6631 003, 004, 005, & 006

Peak Torque (oz•in)	163
Power @ Peak Torque @ 25°C (watts)	77
Continuous Torque @ 25°C (oz•in)	100
Power @ Continuous Torque @ 25°C (watts)	29.5
No Load Speed (r/min)	573
Friction Torque (oz•in)	2.5
Ripple Torque (avg. to peak) (%)	5
Maximum Winding Temperature (°C)	165
Moment of Inertia (oz•in•s²)	42×10^{-3}
Damping Factor (oz•in/rad/s)	2.5
Electrical Time Constant (ms)	1.6
Mechanical Time Constant (ms)	17
Motor Constant @ 25°C (oz•in/$\sqrt{watts}$)	18.6
Theoretical Acceleration (rad/s²)	3.7×10^3
Weight (oz)	21

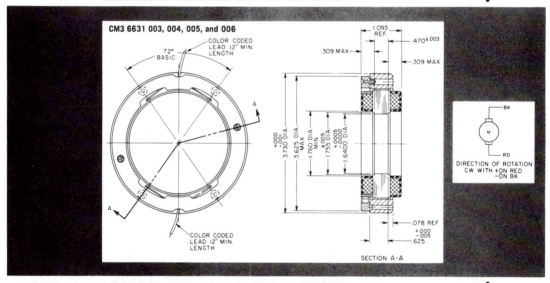

SPECIFIC WINDING CONSTANT CHARACTERISTICS

PART NUMBER	CM3 6631			
	003	004	005	006
Sensitivity (oz•in/amp)	59.5	26.2	119	238
Resistance @ 25°C (ohms)	10.2	1.98	40.8	163
Voltage @ Peak Torque @ 25°C (volts)	28	12.4	56	112
Current @ Peak Torque (amp)	2.74	6.23	1.37	0.685
Back EMF (mV/rad/s)	420	185	840	1680
Inductance (mH)	16.3	3.2	65	261

Figure 9-3 Catalog data sheet. (*Courtesy of The Singer Company, Kearfott Guidance and Navigation Division, Little Falls, N.J.*)

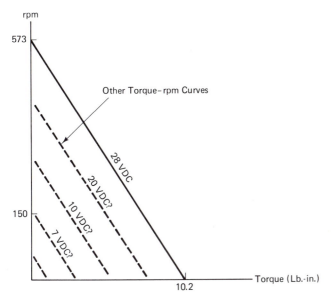

Figure 9-4   An estimated torque versus rpm curve.

one in Figure 4-30. Unfortunately, we have no data available for the lower control voltages as we see in Figure 4-30. The problem we have here for the lower control voltage curves is, we don't know how to space them from the 28 VDC curve. We only know that they will be roughly parallel with our known curve. We also know that for much lower rpm speeds, which is where we will be operating most of the time, the output torque will be substantially reduced. We have to keep in mind that whenever corrective signals are sent to the servo, these signals will be quite low in voltage since they are momentary in nature and won't have the time to build up to full operating voltage. This means that we will be operating quite low on whatever torque versus rpm curve we have to work from. In addition, we will most likely have to size a gear box for the servo in order to build up the torque capabilities once again.

Since it wasn't specified what kind of servo it is we are to use here, we will select a viscous-damped motor. Therefore, its transfer function is of the form, $\theta_{out}/E_{in} = K_v/s(1 + \tau s)$. (See Table 4-2.) Our next job is to determine a value for K_v, the servomotor's velocity constant. According to Eq. (4-20) in Chapter 4,

$$K_v! = \frac{\omega_f}{E_{cv}}$$

where $K_v!$ = catalog velocity constant (rad/volt)
ω_f = No-load or freewheeling speed at 100% of rated control voltage (rad/sec)
E_{cv} = 100% of rated control voltage (volts)

We could guess at the value of ω_f at 25% of rated control voltage using our rather crude torque versus rpm curves in Figure 9-4. But that is what it would be—a crude guess. (A call to the manufacturer will usually get the answers you need, but not always.) Since this 25% data is entirely missing, our data would be considerably more accurate if we used the 100% rated voltage data instead and then derated our results using the methods outlined in Section 4-11 of Chapter 4. Therefore,

$$K_v! = \frac{(573)(2\pi)}{(28)(60)} \frac{\text{rad}}{\text{volt}}$$

$$\quad\quad\quad \hookrightarrow 2\pi/60 \text{ converts the}$$
$$\quad\quad\quad\quad\quad \text{rpm of 573 to rad/sec}$$

$$= 2.143 \text{ rad/volt}$$

Since this is data for $K_v!$ (i.e., catalog data), we use Eq. (4-24) to convert $K_v!$ to "real" data. That is, $K_v = 2 \times Kv!$, or 2.143 × 2 = 4.286 rad/volt.

According to the catalog data, the mechanical time constant for our servo is $\tau! = 0.017$ sec. Therefore, according to Eq. (4-23), $\tau = 2 \times \tau!$, or $\tau = 2 \times 0.017 = 0.034$ sec. Notice that an electrical time constant of 0.0016 sec is also given, but because of its very small value, it can be neglected. The reason for this is because of the location of the corner frequency for this particular time constant. The corner frequency would be located at 625 rad/sec on a Bode plot, far to the right on the plot compared to the ones that we have done so far. Therefore, the 0.0016 sec time constant would really not contribute significantly to the overall gain or phase curve. It would contribute if this system were a high-gain, extremely fast response type system, but this is not the case with our steering system. The transfer function for our servo now becomes:

$$K_{\text{VDSM}} = \frac{\theta_{\text{out}}}{E_{\text{in}}} = \frac{4.286}{s(1 + 0.034s)} \quad\quad (9\text{-}2)$$

Now, let's look at the other components of our system. The two photocells will have a transfer function of $K_{\text{photo}}/(1 + s\tau)$. (See item 10 in Table 4-2.) In order to keep the response time to a minimum for our light detection system, we pick a phototransistor for our photocell. These are known for their very fast response times, typically less than a microsecond or two. We want to make sure that our phototransistor has a good spectral response in the infrared region, so we want to be certain this is specified clearly in the photocell's catalog data. Another reason for choosing the phototransistor is that it gives us the additional advantage of possessing a fairly high sensitivity of gain factor. For our transfer function, we assume τ to be negligible and will not be concerned with a value for K_{photo} at this time. We could determine its exact value now, but it will be easier to compensate for whatever value it turns out being by simply adjusting, instead, the gains of the various amplifiers within our system. Or, if necessary, if the gain is too high, we can install a pot. In most cases, this will be true for many transducers. And to a certain extent, this is also true in the selection of a gear train. These components are multiplied together, along with the servo's K_v, to form the product, K_x, in the numerator of the system's final transfer function. Of course, if we knew these constant values at this point, we would use them. In the case of the phototransistor and our system design, we can go ahead and order the devices knowing that we can make system adjustments later with whatever units we have.

This brings us to the next stage of developing our system. We are ready now to determine the K_x value in our system. First, though, let's look at a block diagram of our steering system (Figure 9-5) and its transfer function. The feed-

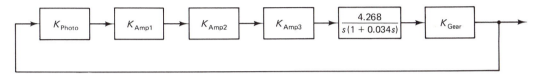

Figure 9-5   Block diagram for robotic gofer cart steering system.

back path is through the painted strip and its relative positive with the photocell system. This is shown as simply a line connecting the output of the steering system back to its input. The transfer function for our system is now:

$$KG = \frac{(K_{photo})(K_{amp1})(K_{amp2})(K_{amp3})(K_{gear})(4.286)}{s(1 + 0.034s)} \tag{9-3}$$

$$KG = \frac{K_x}{s(1 + 0.034s)} \tag{9-4}$$

9-2.3 Generating the Closed-loop Data

We leave our transfer function in this form so that we can go now to a root locus analysis. We use this analysis to determine the value of K_x needed to produce the desired damping factor.

Figure 9-6 is a plot of our root locus using the process outlined in Section 8-6 of Chapter 8 for second-order equations. We draw our K_x on a vertical line midway between the two poles of 0 and −29.41 on the real axis. Since we want a damping factor of 0.5 to 0.6 (let's split the difference and say 0.55), we can show angle θ as arccosine 0.55 = 56.63°. We can now determine the j-value produced by K_x, which is the same as ω_d. This will be 14.71 × tan 56.63 = 22.33 rad/sec. To find ω_n, we let $\omega_n = \sqrt{22.33^2 + 14.71^2}$, which is 26.74 rad/sec. To find K_x, we must first go back to Eq. (8-5) in Chapter 8. Notice that before the calculating of ω_n, we manipulated our closed-loop expression so that it looked like Eq. (8-5). We did this by performing the following steps:

1. Convert the open-loop expression into a closed-loop expression, writing that expression in the s form rather than the $j\omega$ form.

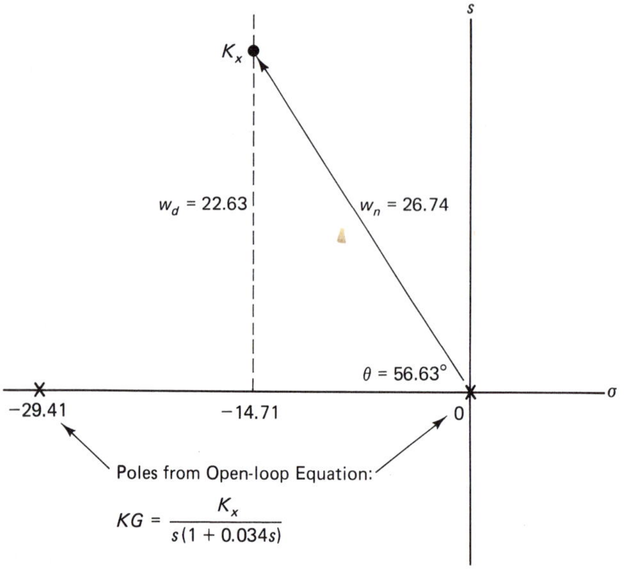

Figure 9-6 Root locus plot for $KG = K_x/s(1 + 0.034s)$.

2. Then we divided both the numerator and denominator by K_x and multiplied the denominator by a factor/factor to reduce the s^2 term to 1. (The factor/factor equals 1).

3. We then found ω_n by taking the square root of the term under the s^2 term, according to Eq. (8-5).

Let's do the same thing now for our expression, except in our case we know ω_n already. What we have to find is K_x instead. Therefore:

1. Our closed-loop expression becomes:

$$\frac{K_x}{s(1 + 0.034s) + K_x} = \frac{K_x}{0.034s^2 + s + K_x}$$

2.

$$= \frac{1}{\dfrac{0.034s^2}{K_x} + \dfrac{s}{K_x} + \dfrac{K_x}{K_x}}$$

$$= \frac{1}{\dfrac{0.034s^2}{K_x} \times \dfrac{29.41}{29.41} + \dfrac{s}{K_x} + \dfrac{1}{1}} \quad (9\text{-}5a)$$

$$= \frac{1}{\dfrac{s^2}{29.41K_x} + \dfrac{s}{K_x} + 1}$$

3.

$$\omega_n = \sqrt{(29.41)(K_x)} \quad (9\text{-}5b)$$

Or,

$$26.74 = \sqrt{(29.41)(K_x)}$$

Solving for K_x:

$$26.74^2 = 29.41 \, K_x$$

$$K_x = \frac{26.74^2}{29.41}$$

$$= 24.31$$

Looking at Eq. (9-5b), we see that the value, 29.41, under the radical sign, is the reciprocal of the time constant, τ, which is 0.034 sec. We can now make the general statement that:

$$\omega_n = \sqrt{\frac{K_x}{\tau}} \tag{9-6}$$

or,

$$K_x = \omega_n^2 \tau \tag{9-7}$$

Now that we have found a value for K_x, we have to determine our K_{gear}. But first, let's check our damping factor in Eq. (9-5) to make sure nothing has changed. This will also check the validity of our K_x value:
From Eqs. (9-5) and (8-7):

$$\frac{s}{(24.31)} \times \frac{(0.041)}{(0.041)} \times \frac{(\omega_n)}{(\omega_n)} = 2zs$$

$\hookrightarrow$ This fraction removes the 24.31 allowing the forcing-in of ω_n according to Eq. (8-5)

Solving now for z, we get

$$2z = (0.041)(26.74)$$

or,

$$z = 0.55$$

We see now that our K_x value checks out and z has not changed.

Now we are ready to proceed with finding K_{gear}. Since we estimated the operating speed of our servo to be approximately 150 *rpm*, and we determined earlier that we needed an input *rpm* to our drive pulley of 19.1, we will need a gear reduction box having a gear speed ratio of $rpm_{\text{in}}/rpm_{\text{out}}$, or 150/19.1, or 7.85. (Since 1 *rpm* is equivalent to 2π radians per minute, *rpm* is nothing more than an equivalent form of angular velocity. Consequently, since we already know that a gear transmission's speed ratio, N, is $\omega_{\text{in}}/\omega_{\text{out}}$ from inverting Eq. (4-30), we can also state that a gear transmission's speed ratio $= rpm_{\text{in}}/rpm_{\text{out}}$. Since $K_{\text{gear}} =$

$1/N$, our resultant transfer function will be 0.127 (see Section 4-12 in Chapter 4). If we can achieve this particular reduction, then we will increase our torque output by the same factor, that is, 7.85 times. With this information, our K_x can now be calculated as:

$$K_x = 24.31 = (4.286)(K_{photo})(K_{amp1})(K_{amp2})(K_{amp3})(0.127)$$

then $$(K_{photo})(K_{amp1})(K_{amp2})(K_{amp3}) = \frac{24.31}{(0.127)(4.286)} = 44.7 \qquad (9\text{-}8)$$

What this figure of 44.7 represents is the amount of combined amplifier gain needed, times K_{photo}, to stabilize our steering system.

Theoretically, we wouldn't need to perform any response curves for our system, since we can calculate virtually every feature of these curves. However, there are certain things we wish to avoid that are mathematically difficult to predict. Figure 9-7 is a good example. Figure 9-7 is the open-loop Bode plot of

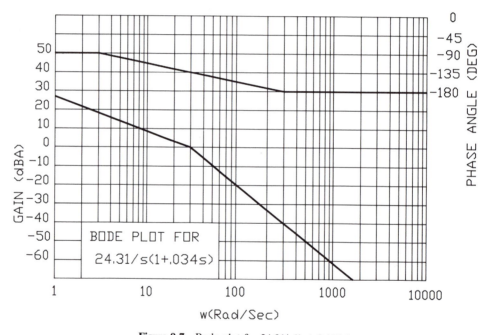

Figure 9-7   Bode plot for $24.31/s(1 + 0.034s)$.

our steering control system up to this point. Everything looks normal; the phase margin is right where we would like it to be. But notice where the zero gain crossover point occurs—right at the corner frequency or bend in our gain curve. This is coincidental, and not too desirable. The reason is this: You always have better system stability operating within the −20 dBA portion of a gain curve rather than the −40 dBA portion. The phase margin changes more slowly for a change in operating frequency in the -20 dBA area of the curve. It would be nice if we could move the zero gain crossover point to the left of its present position without reducing our phase margin too much. Let's investigate this possibility. If we were to lower the gain of our system, our gain curve would lower, causing the gain crossover point to shift toward the left. According to our Bode plot, we could lower our gain by 6 or 8 dB and still maintain a desirable phase margin. At the same time, we have moved our gain crossover point far enough away from the corner frequency to assure us stability with our steering control system. But what about our dead band? In the case of our steering control system, the dead band would be measured at the photocell pick-up. It would become a linear measurement of how far the reflective track would be from the photocell to finally cause the servomotor to start. Our dead band equation would be:

$$\text{dead band (in.)} = \frac{2 \times \text{motor starting voltage}}{(K_{\text{amp}}) \text{ (photocell sensitivity)}}$$

$\llcorner\!\!\rightarrow$ units are

volts/in.   (9-9)

To actually determine this dead band distance without making measurements would be difficult. We will have to be satisfied with just making relative comparisons. In other words, if we were to actually lower our gain by 8 dB as was just suggested, our dead band would increase by 2.51 times as compared to the original dead band.

Earlier, we stated that whenever the gain was decreased in a system, the dead band would increase. Rather than have that happen, where we would wind up with a system having little or no response for critical small error signal corrections, let's look into another alternative. Let's consider what would happen if we were to add a rate generator to our system. An important function of a rate generator is to furnish damping to a servomotor by shifting the gain curve toward the left in the Bode plot without affecting the system's dead band. If the system becomes too deadened as a result of the increased viscous damping created by the rate generator, one could simply increase the system's gain to compensate for this and obtain a decrease in the system's dead band at the same time.

Figure 9-8 shows a page of specifications for various sized tachometers or rate generators (either name is used extensively in the industry). The problem is choosing the right one for the job. Since the rate generator's output is dependent on the rpm of the servomotor that it is attached to, we will want to choose a model that generates enough voltage with the available rpm to equal at least one-half the servomotor's control voltage. This should give us a fair degree of control over the servo's operating speed with the full output of the rate generator being fed back to the servo. Most likely what we will do, though, is to feed back only a portion of the rate generator's output, just enough to give us the desired damping. This can be done through a potentiometer. The tachometer with the 20.8 $\pm3\%$ volts/1,000 r/min. output looks like a good choice to begin with. It's transfer function, stated

DC TACHOMETERS

TYPICAL ELECTRICAL AND MECHANICAL DATA

Part Number	CMO 9608		R9608		CRO 9610		CUO 9611			CUO 9612 001	CVO 9612 001
	001	024	002	003	001	005	001	002	003		
Size	8	8	11	11	11	11	12	12	12	12	18
*Output (volts/1000 r/min)	3±1%	3±5%	2±3%	7±3%	7±1%	19±3%	7±2%	7±2%	20.8±3%	7±5%	100±5%
Linearity (% to 3600 r/min)	0.05	1	0.05	0.05	0.05	0.05	0.05	0.05	0.05	0.05	0.2 (0—1000 rpm)
Output Impedance (ohms)	150	150	1500	1700	275	1000	275	275	1000	300	3000
Ripple Voltage (above 100 r/min) (%) (min.)	2	3	3	3	3	3	3	3	3	1.5	1.5
External (Test) Load Resistor (ohms)	100k	100k	200k	100k	100k	100k	100k	100k	100k	100k	100k
Directional Error (max.) (%)	.2	1	.2	.2	.2	.2	.25	.25	.25	.25	.3
Temperature Sensitivity† — Over −54°C to +100°C (%/°C) (max.)	±.01	—	—	—	±.01	—	±.01	±.01	±.01	±.01	.04
Temperature Sensitivity† — Over −15°C to +71°C (% Variation from 25°C value) (max.)	—	±.06††	±.15	±.15	—	±1.5**	—	—	—	—	—
Max. Speed (r/min)	12,000	3,600	12,000	12,000	12,000	8,000	12,000	12,000	8,000	12,000	1,200
Friction Torque (in-oz) (max.)	0.20	0.30	0.25	0.25	0.25	0.25	0.25	0.25	0.25	0.25	0.5
Rotor Moment of Inertia (gm-cm²)	2	2	7	7	7	16.5	7	7	16.5	7	175
Weight (oz)	2.0	2.0	3.0	3.0	2.8	5.0	3.0	3.1	4.5	3.0	15.0

*Better than 3% can be supplied
**Over temperature range of −54°C to +100°C
†Speed-sensitive output voltage variation
††Over +10°C to +60°C range

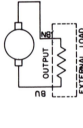

CRO 9610 001, 005, CUO 9611 SERIES, & CUO 9612 001
Polarity is positive when unit is driven in a counterclockwise direction.

R9608-002, R9608-003
Polarity when driven counterclockwise is BN (positive) and BU (negative). The external load resistor is applied across output during all tests.

CMO 9608 001, 024
Polarity when driven counterclockwise is Terminal 1 (positive) and Terminal 2 (negative).

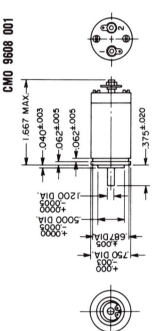

CMO 9608 001

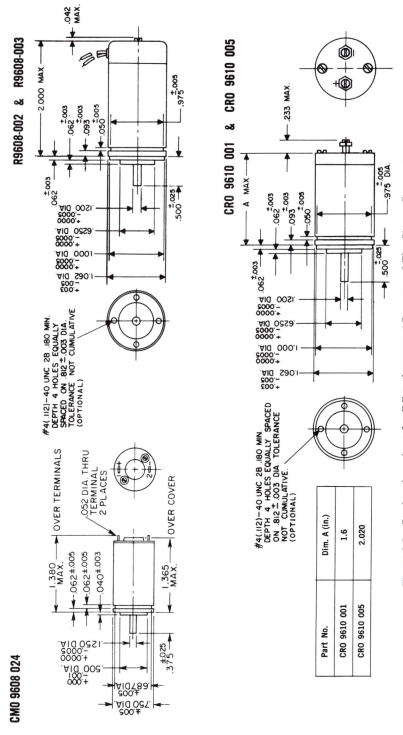

Figure 9-8 Catalog data sheet for DC tachometer. (*Courtesy of The Singer Company, Kearfott Guidance and Navigation Division, Little Falls, N.J.*)

in units of radians and seconds, would be 20.8 volts/$(1{,}000 \times 2\pi)/60 = 0.199$ volts/rad/sec. These are the units needed to make this transfer function compatible with the others in our system.

Figure 9-9 is a revised block diagram of Figure 9-1 showing our steering control system. The tacho is shown in the feedback path. We now have to deter-

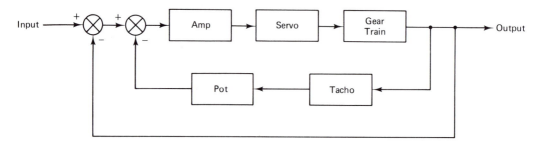

Figure 9-9   Block diagram of steering system using a tacho.

mine how this is going to alter the transfer function of our system in the way that we want. The tacho becomes the H in the general closed-loop expression, $KG/(1 + KGH)$. When we first investigated this equation, we agreed to allow $H = 1$ for all our closed-loop problems. However, in this case, since the tacho is in the closed-loop circuit, it now becomes equal to H. In other words, $E_{out}/\theta_{in} = K_{rg}s = 0.199s$ (see reference 15 in Table 4-3 in Summary of Chapter 4). And, the overall transfer function now becomes:

$$= \frac{24.31}{0.034s^2 + s + (0.199s)(24.31)} = \frac{24.31}{0.034s^2 + s + 4.84s}$$

$$= \frac{24.31}{0.034s^2 + 5.84s} = \frac{24.31}{s(0.034s + 5.84)}$$

Dividing both the numerator and denominator in the preceding expression by 5.84, we get:

$$\frac{4.163}{s(0.006s + 1)} \tag{9-10}$$

Figure 9-10 shows the Bode plot of Eq.(9-10). Notice how the corner frequency has been shifted to the right of its original location. Also, the gain crossover point appears to have been shifted to the left, resulting in a larger phase margin. We can easily adjust our phase margin now either by adjusting the output pot of our rate generator or by adjusting the system's amplifiers. Adjusting the pot will not affect the system dead band as will adjusting the amplifier gain.

We now have a very reasonable chance of producing a workable robotic cart steering system. Because of our efforts in calculating system parameters and plotting the anticipated response curves, we can be reasonably assured of being

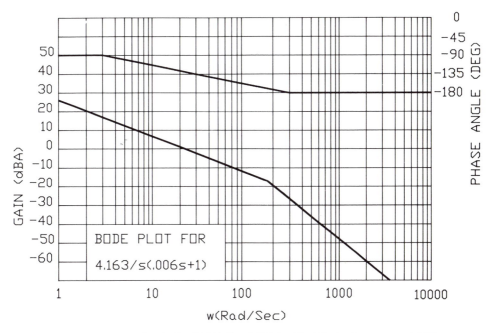

Figure 9-10   Bode plot for $24.31/s(1 + 0.034s)$ with shifted gain curve, or $4.163/s(0.006s + 1)$.

able to produce a system with desirable performance characteristics. Realistically, there will most likely have to be additional fine-tuning of the various components being used in our steering system once it is built, but at least we now have a usable system to work with.

9-3 EXAMPLE 2: A COMPASS-DIRECTED AUTO-PILOT CONTROL SYSTEM FOR A BOAT

In this second example, we want to design an automatic rudder control system for a small boat. This system will enable the ship's operator to set a particular coarse at the wheel and then have the rudder automatically report to that position while being corrected by a compass for any deviations from that position. Figure 9-11 is a schematic diagram of the proposed system. Again, this will be a DC control system. It is assumed that the boat has available a semiregulated high-voltage, high-current DC supply for operating a large DC servomotor.

We begin our design once again with choosing a servomotor. We pick a motor having a large torque output capability. The one described in Figure 9-12 will do nicely. We assume that torque measurements have already been determined as to what is required for rotating the rudder under the highest boat speeds encoun-

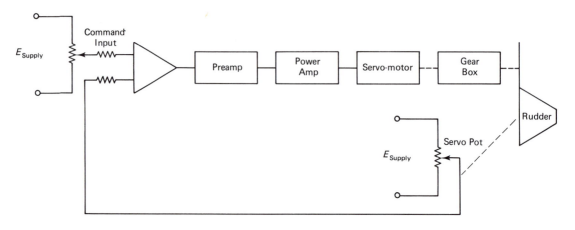

Figure 9-11 Initial design for an autopilot control for a boat.

tered. Additional torque will be available, since we will have to gear down the high rpm output of this motor. Notice that the motor's manufacture has already saved us the trouble of deriving its transfer function. It is, $1.6/[s(1 + 0.022s)]$. However, this is based on catalog data; therefore, the real world transfer function becomes:

$$\frac{\theta_{\text{out}}}{E_{\text{in}}} = \frac{1.6 \times 2}{s(1 + 0.022s \times 2)} = \frac{3.2}{s(1 + 0.044s)} \tag{9-11}$$

We can check the catalog's K_v! by noting that it lists the no-load speed as 3,600 rpm, or $3,600 \times 2\pi \div 60$, or 377 rad/sec. The listed rated control voltage is 230 volts; therefore, its $K_v! = 377/230 = 1.64$ rad/sec/volt. This checks with the transfer function's K_v value. This also gives us assurance that the method used in our first example for assuming a value for the K_v in Section 9-2 was probably correct.

The real world expression of our transfer function is based on operating at approximately 25% of the rated control voltage which, in our case, would be 57.5 VDC. Our problem now is to decide what the rpm would most likely be at this operating voltage, since the catalog data gives no clue as to what we could expect. We can only guess that the rpm will vary proportionally to the supplied voltage (assuming constant supply voltage to the motor's shunt field winding). Therefore, at 25% of rated control voltage, the operating rpm will be in the neighborhood of 25% of 2,700 rpm (we use a rated rpm figure at 0.33 horsepower (hp) from the catalog's chart) or 675 rpm. This figure will help us size the proper gear train needed to drive the boat's rudder.

Speaking of the boat's rudder, we now have to determine the amount of torque required to turn the rudder under worst conditions. Worst conditions would be experienced while the boat is traveling at maximum speed, say 15 mph,

Characteristics* — FD84-51-1

Rated Power Output (hp)	0.6
Duty	Intermittent
Armature Rated Voltage (Vdc)	80
Armature Rated Current (amp)	9.4
Armature Resistance (ohms)	1.2
Armature Compensation	
Field Resistance (ohms)	1.1
Armature Total Resistance (ohms)	2.3
Armature Total Inductance (mh)	3
Shunt Field Voltage (Vdc)	100
Shunt Field Current (amp)	0.3
Shunt Field Resistance	
at 25°C (ohms)	300
No Load Speed (r/min)	7,800
Locked Torque (in·oz)	350
Torque Constant (in·oz/amp)	18
Starting Voltage (Vdc)	4
Armature Back EMF Constant	
(v/rad/s)	0.127
Inertia (in·oz²)	15
Theoretical Acceleration	
at Stall (rad/s²)	9,000
Friction Coefficient	
(in·oz/rad/s)	0.43
Time Constant (s)	0.09
Frequency Response (Hz)	2

*Data valid when motor operates from Kearfott
FDF144-2211-1 generator with If = 0.15 amp.

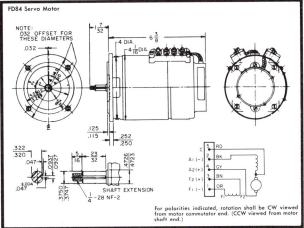

FD84 Servo Motor

NOTE:
.032 OFFSET FOR
THESE DIAMETERS

For polarities indicated, rotation shall be CW viewed
from motor commutator end. (CCW viewed from motor
shaft end.)

SIZE 45 WOUND FIELD DC SERVO MOTOR   0.33 HP
Type FD87

This 1/3 horsepower, 3400 r/min, 230 volt direct current motor is an open, fan cooled unit. Construction features include flange mounting, ball bearings, die cast aluminum housing, lubrication per MIL-G-23827, and number 18 gauge leads. This unit can be operated continuously at ambients of −55°C to +55°C wherein the power loss (input/output) does not exceed 350 watts or the total winding temperature does not exceed 105°C.

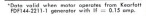

FD87-12-1

FLD AMPS: 470
LOCKED AMPS: 13.7
LOCKED TORQUE: 96

Characteristics — FD87-12-1

	ARMATURE	FIELD
Voltage (Vdc)	230	70
Current (amp)*	13.7	0.47
Resistance (ohms)**	15.4	145
Inductance (mh)*	1.5	—
Torque (ft·oz)*	96	
No Load Speed (r/min)*	3600	
Inertia (in·oz²)	26	
Theoretical Acceleration		
at Stall (rad/s²)	17,100	
Friction (in·oz/rad/s)	3.1	
Time Constant (s)	0.022	
	Θm	1.6
Motor Transfer Function		
	Vm	$S(.022S + 1)$
Weight (lb)	13.5	

*At Stall
**Includes armature resistance of approximately 9
ohms and compensating winding of approximately
6 ohms.

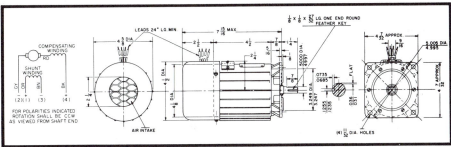

Figure 9-12   Data sheet for a DC servomotor. (*Courtesy of The Singer Company, Kearfott Guidance and Navigation Division, Little Falls, N.J.*)

and turning the rudder suddenly to 90° to the direction of travel. The force created on the rudder can be approximated by the following equation:

$$F_{\text{drag}} = 1.16v^2A\sin\alpha \qquad\qquad (9\text{-}12)$$

where F_{drag} = the drag force on rudder (lb.)
 $\quad\quad v^2$ = velocity of boat (ft/sec)
 $\quad\quad A$ = area of rudder (ft²)
 $\quad\quad \alpha$ = angle of rudder to travel line (°)

Assuming a rudder area of 4 square feet, the resultant force would be (1.2)(15 × 5,280/3,600)² (4) = 2,246 lb., obviously much too great a force to deal with at that speed. On the other hand, if the boat's forward speed were, say 5 mph, and the rudder suddenly turned to 90°, the F_{drag} would be only 250 lb. This means that it will take a 250 lb.–ft. torque to rotate the rudder under this condition. (We assume that the 250-lb. force acts through the center of the 2 × 2 ft. rudder area, thereby creating a 250-lb. torque through the rudder's steering post.) We size our steering gear transmission for that condition instead. Let's assume that we can allow 4 seconds for the rudder to rotate 90°. This means that it will take 4 times 4 seconds, or 16 seconds to rotate 360°. This is a speed of 0.063 rps or 3.75 rpm. Our gearing ratio will not be 675/3.75 or 180 : 1. We now have to determine our *torque constant, K_τ*. This ratio will enable us to determine the output torque of our servomotor at any operating voltage. It is calculated using the following method:

$$\text{torque constant, } K_\tau = \frac{\text{stall torque}}{100\% \text{ control voltage}} \qquad (9\text{-}13)$$

In our case, the torque constant would be 6 lb.–ft. ÷ 230v = 0.026 lb.–ft./volt. This ratio changes little regardless of the motor's operating voltage. In our case, since we estimated our rpm at 57.7 VDC to be 675 rpm, our torque generated at that speed would be 0.026 lb.–ft. × 57.5 volts = 1.5 lb.–ft. of available torque. Operating through our 180 : 1 gear train, the available output torque at the rudder will now be 180 × 1.5, or 270 lb.–ft. This means that there is enough torque to turn the rudder a full 90° in four seconds at 5 mph. Our concern now is, will this slow turning rate create too sluggish of a system for our steering? Also, we will have to be careful that no attempt is ever made to turn the rudder suddenly to 90° at boat speeds higher than 5 mph; otherwise the rudder will stall out and not move. However, there is one consolation in our design: Regardless of how fast the boat is traveling, the relatively small touch-up course corrections due to wind and water currents will most likely involve only small rudder movements. This means that the rudder drag will be kept small, resulting in much faster response times for our control system. Figure 9-13 shows our control design results so far. What isn't shown in the block diagram is the servopot attached to the rudder to indicate the rotation amount. The pot sends back a voltage proportional to its

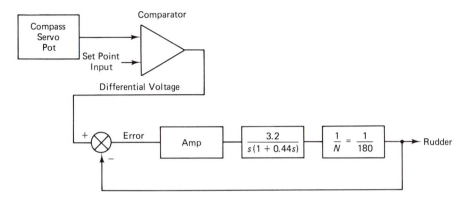

Figure 9-13   Autopilot control system after sizing gear train.

rotation and is compared to the difference voltage coming from a voltage comparator and servopot attached to the boat's compass.

Our first step in analyzing our initial design results is to plot the root locus (Figure 9-14). Our system transfer function is:

$$\frac{\theta_{\text{out}}}{E_{\text{in}}} = \frac{K_{\text{amp}}\,3.2}{180s(1 + 0.044s)} \tag{9-14}$$

where

$$K_x = \frac{(K_{\text{amp}})(3.2)}{180} \tag{9-15}$$

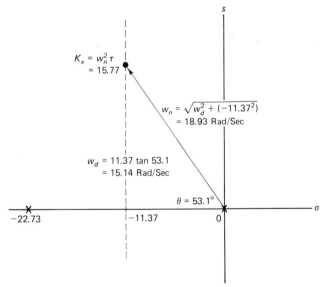

Figure 9-14   Root locus for $K_{\text{amp}}\,3.2/180s(1 + 0.044s)$.

We identify the open-loop poles and place a locus mid-way through the distance between them, knowing that whatever K_x value we select, it will lie on this vertical line. We pick θ to be the arccosine of 0.6 (assuming we will want a z of 0.6) which is 53.1°. We can now calculate ω_d, which is (11.37) (tan 53.1°), or 15.14 rad/sec. Using Pythagorean's equation, we find ω_n to be 18.93 rad/sec. Using Eq. (9-7), we can now determine K_x, which we find to be 15.77. And we can now find our amplifier gain by using Eq. (9-15) and solving for K_x:

$$15.77 = \frac{(K_{amp})(3.2)}{180}$$

Solving for K_{amp}:
$$K_{amp} = \frac{(15.77)(180)}{3.2}$$
$$= 887$$

Since rudder response time to a 90° rotation command is a concern, let's determine the settling time for the rudder. In Chapter 8, we found the $\pm 5\%$ to settling time to be $3/z\omega_n$. In our case, that figures out to be 0.26 sec. This is small compared to the four seconds needed for the rudder to arrive at a 90° position. It appears that this is the only concern that could cause a problem with this particular design, aside from the fact that servomotors of the size that we are using in this design are quite expensive. Now that we have completed an autopilot control system design for our boat, we study another problem, a design problem for a remote valve positioner.

9-4 EXAMPLE 3: A REMOTE VALVE POSITIONER

Our third example of an automatic control system uses AC circuitry instead of DC circuitry just to get you familiar with AC control hardware. This particular project is a remote controlled valve stem positioner. The valve could be part of a fluid control system in a chemical plant, or perhaps part of a flow control system for gases or coolants in a nuclear facility. This system is more of an instrument control system. That is, the hardware is physically small and is capable of low torque generation only. Supply voltages for this system will be 115 VAC, 60 Hz. Figure 9-15 shows our design layout. The block diagram for this system is shown in Figure 9-16. Our set point input to our system is shown as θ_{in} in Figure 9-15. The control scheme will work like this: The control transmitter's rotor is the setpoint input. The amount of rotation, θ_{in}, relative to the angle of the control transformer's rotor, determines the phase and magnitude of the AC signal going into the amplifier. The phase angle is measured relative to the AC supply voltage. The servomotor's rotor direction and amount of rotation is determined by the amplifier's output voltage phase and magnitude. The gear train's rotational output is mechanically coupled to the valve's stem. It is also coupled to the rotor of the control transformer. If the two rotors are at the same angle of rotation, the

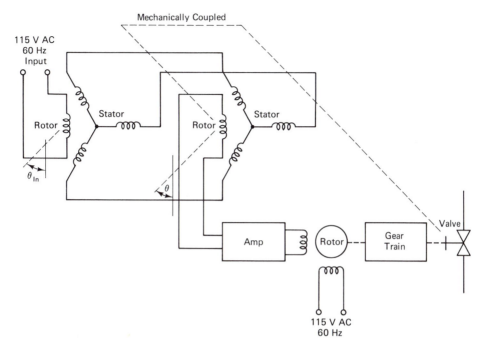

Figure 9-15   Schematic of remote valve positioner.

error signal will be zero. Any difference in these two angles will create an error signal in the form of an AC voltage, which will cause the servo to turn in the proper direction to reduce this difference to zero.

Once again, we begin our design project by searching the design catalogs for the proper servomotor. Because of the relatively light torque requirements required to operate the valve, we can probably get by with using about a number 8 or 10 servo. Looking through the specifications, we come across a number 11 servomotor, our only choice according to the data sheet, that operates at 115 VAC 60 Hz. See Figures 9-17(a) and 9-17(b). (Notice that many of the other servos operate at 400 Hz according to this sheet. This is because 400 Hz is the line frequency most commonly encountered in aircraft. This higher frequency is used to reduce the sizes, and therefore the weights, of power transformers needed for voltage rectification on board the aircraft.) We can now assemble the transfer constant

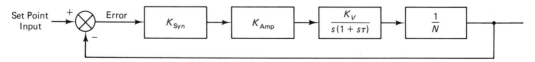

Figure 9-16   Block diagram of remote valve positioner.

SERVOMOTORS

ELECTRICAL AND MECHANICAL CHARACTERISTICS FOR SIZE 10 AND SIZE 11 60 AND 400 HERTZ SERVO MOTORS

SIZE	PART NUMBER	VOLTAGE (volts) PHASE 1	PHASE 2 SERIES	STARTING (max.)	STALL TORQUE (in. oz.) (min.)	NO LOAD SPEED (rpm.) (min.)	POWER INPUT PER PHASE (watts) (max.)	POWER FACTOR	*CURRENT PHASE 1	PHASE 2 SERIES	*R (ohms) PHASE 1	PHASE 2 SERIES	*X (ohms) PHASE 1	PHASE 2 SERIES	*Z (ohms) PHASE 1	PHASE 2 SERIES
10	R118-1	26	26	.9	.30	6500	3.1	.63	166	166	98.5	98.5	123	123	157	157
	R118-2	26	26	1.5	.34	3500	3.75	.75	180	180	116	116	94	94	148	148
	R124-1	26	26	.9	.30	6500	3.1	.63	166	166	98.5	98.5	123	123	157	157
	R124-4	26	80	3.0	.28	3500	3.75	.75	180	52	116	1160	94	1015	148	1540
	P124-06	26	36	1.3	.28	6500	3.1	.63	166	120	98.5	189	123	236	157	301
	P121-02	115	115	3.0	.28	10,000	2.8	.72	33	33	2520	2520	2430	2430	3500	3500
	CL9 0118 017	26	36	1.25	.30	6500	3.1	.63	166	120	98.5	189	123	236	157	301
	CL9 0121 003	115	40	1.0	.28	10,000	2.8	.65	36	105	2080	248	2430	290	3200	381
	CL9 0121 009	26	36	.9	.27	10,000	3.1	.65	161	116.5	105	201	122.5	235	161	309
11	R119-2†	115	115	3	.60	6200	3.9	.575	53	53	1250	1250	1780	1780	2175	2175
	R119-36	115	36	.95	.60	6200	3.9	.575	53	169	1250	123	1780	174	2175	213
	CR0 0130 570	115	40	1.0	.35	10,000	3.9	.54	50	150	1240	144	1940	224	2300	267
	CR0 0132 670	115	36	.9	.70	6500	3.9	.514	60	185	995	99.4	1647	166.2	1933	194
	CR4 0132 012	115	36	1.0	1.5	6500	9.5	.50	147	471	391	38.3	676	66.2	784	76.5
	CR0 0133 670	115	36	.9	.75	4500	3.9	.58	52	169	1270	124	1800	174	2210	214
	CR4 0164 002**	115	36	1.0	.60	3000	3.9	.83	37	117	2580	255	1733	172	3108	308

* At Stall
** 60 hertz units. All others are 400 hertz.
† R119-2A is functionally equivalent to the MK 14 MOD O Bu/Weps motor.
Minimum temperature for all units is −54°C.
Maximum operating temperature (ambient plus motor temperature rise) for all units is +165°C.
All motors operate continuously at stall.

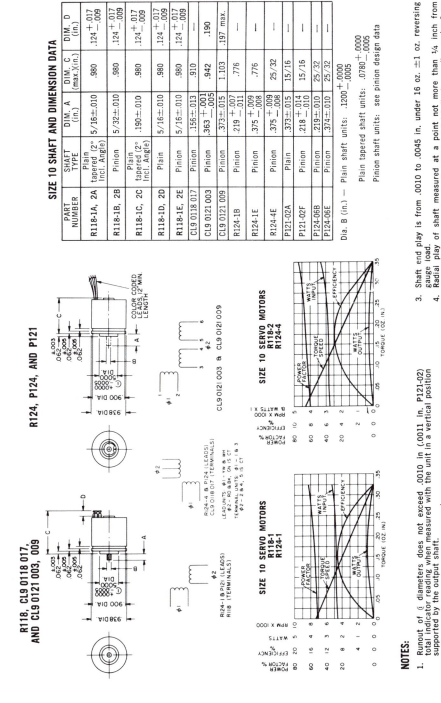

SIZE 10 SHAFT AND DIMENSION DATA

PART NUMBER	SHAFT TYPE	DIM. A (in.)	DIM. C (max.)(in.)	DIM. D (in.)
R118-1A, 2A	Plain tapered (2° Incl. Angle)	$5/16 \pm .010$	.980	$.124 \, {}^{+.017}_{-.009}$
R118-1B, 2B	Pinion	$5/32 \pm .010$	.980	$.124 \, {}^{+.017}_{-.009}$
R118-1C, 2C	Plain tapered (2° Incl. Angle)	$.190 \pm .010$	.980	$.124 \, {}^{+.017}_{-.009}$
R118-1D, 2D	Plain	$5/16 \pm .010$	.980	$.124 \, {}^{+.017}_{-.009}$
R118-1E, 2E	Pinion	$5/16 \pm .010$	.980	$.124 \, {}^{+.017}_{-.009}$
CL9 0118 017	Pinion	$.156 \, {}^{+.013}_{-}$	.910	—
CL9 0121 003	Pinion	$.363 \, {}^{+.001}_{-.005}$	.942	.190
CL9 0121 009	Pinion	$.373 \pm .015$	1.103	.197 max.
R124-1B	Pinion	$.219 \, {}^{+.007}_{-.011}$	.776	—
R124-1E	Pinion	$.375 \, {}^{+.009}_{-.008}$	.776	—
R124-4E	Pinion	$.375 \, {}^{+.009}_{-.008}$	25/32	—
P121-02A	Plain	$.373 \pm .015$	15/16	—
P121-02F	Pinion	$.218 \, {}^{+.014}_{-.010}$	15/16	—
P124-06B	Pinion	$.219 \pm .010$	25/32	—
P124-06E	Pinion	$.374 \pm .010$	25/32	—

Dia. B (in.) — Plain shaft units: $.1200 \, {}^{+.0000}_{-.0005}$

Plain tapered shaft units: $.0780 \, {}^{+.0000}_{-.0005}$

Pinion shaft units: see pinion design data

R124, P124, AND P121

R118, CL9 0118 017, AND CL9 0121 003, 009

SIZE 10 SERVO MOTORS
R118-2
R124-4

SIZE 10 SERVO MOTORS
R118-1
R124-1

NOTES:

1. Runout of ⓒ diameters does not exceed .0010 in (.0011 in. P121-02) total indicator reading when measured with the unit in a vertical position supported by the output shaft.

2. Composite error of pinion assembled in unit does not exceed .0014 in. total indicator reading. Runout of plain shaft diameter does not exceed .0008 in. total indicator reading for Size 10 units. Runout of ⓒ for R119 plain shaft units does not exceed .001 in. total indicator reading.

3. Shaft end play is from .0010 to .0045 in. under 16 oz. ± 1 oz. reversing gauge load.

4. Radial play of shaft measured at a point not more than ¼ inch from bearing face does not exceed .0008 in. total indicator reading under 8 oz. $\pm \frac{1}{2}$ oz. reversing gauge load applied not more than ¼ inch from bearing face.

5. R119-☐J shaft is standard for combination with Kearfott Size 11 gearheads.

Figure 9-17(a) Servomotor data sheet. (*Courtesy of The Singer Company, Kearfott Guidance and Navigation Division, Little Falls, N.J.*)

SIZE	PART NUMBER	*EFFECTIVE R (ohms)		DC RESISTANCE (ohms)		*PARALLEL TUNING CAPACITOR FOR UNITY POWER FACTOR (µfd)		ROTOR MOMENT OF INERTIA (gm cm²)	THEORETICAL ACCELERATION (rad/sec²)	TIME CONSTANT (sec.)	WEIGHT (oz.)
		PHASE 1	PHASE 2 SERIES	PHASE 1	PHASE 2 SERIES	PHASE 1	PHASE 2 SERIES				
10	R118-1	250	250	53	53	2	2	.46	46,000	.0148	1.45
	R118-2	190	190	53	53	1.7	1.7	.46	52,000	.0087	1.45
	R124-1	250	250	53	53	2	2	.46	46,000	.0148	1.45
	R124-4	190	2000	53	584	1.8	.16	.46	44,000	.0116	1.45
	P124-06	250	480	53	91	2	.95	.46	46,000	.0148	1.45
	P121-02	5280	5280	740	740	.09	.09	.46	44,000	.024	1.7
	CL9 0118 017	250	479	53	91	2	1.04	.46	46,000	.0148	1.45
	CL9 0121 003	4920	559	740	41	.08	.37	.46	43,000	.0244	1.7
	CL9 0121 009	248	475	166	37	1.3	.98	.46	43,000	.0240	1.7
11	R119-2	3800	3800	438	438	.16	.16	1.07	39,450	.0168	4.5
	R119-36	3800	372	438	42	.16	1.5	1.07	39,450	.0168	4.5
	CRO 0130 570	4250	476	356	42	.15	1.25	.76	32,600	.032	3
	CRO 0132 670	3770	376.2	255	26.5	.175	1.75	.76	69,700	.0113	3
	CR4 0132 012	1570	153	105	11.2	.44	4.5	1.2	139,000	.0097	3
	CRO 0133 670	3790	372	340	33	.15	1.51	.76	74,000	.0086	3
	CR4 0164 002	3745	371	1600	156	.48	4.8	1.07	39,450	.0085	4.5

When phase 2 windings are connected in parallel, the following relationships exist:
Current = 2 x phase 2 series values;
R, X, Z, **Effective R**, and **DC Resistance** = ¼ phase 2 series values;
Voltage = ½ phase 2 series values;
Parallel Tuning Capacitance for unity power factor = 4 x phase 2 series values.

TERMINAL UNITS
ϕ1—1 & 3, ϕ2—2 & 5, 6 & 4
LEAD UNITS
ϕ1 — YW & WH
ϕ2 — BK & GN, RD-BK & RD

CR4 0164 002, CRO 0130 570, CRO 0132 670, CR4 0132 012, AND CRO 0133 670

R119

*CR4 0164 002 (which has tapped holes on mounting face) is 1.391 in. (max.) long.
CR4 0132 012 (which has radial leads) is 1.187 in. (max.) long.

SIZE 10 & 11 PINION DESIGN DATA

No. of Teeth	13
Diametral Pitch	120
Pressure Angle	20°
Standard Pitch Diameter (in.)	.1083
Tooth Form	Full Depth Involute
AGMA Quality Class	Precision I
Outside Diameter (in.)	.1247 +.0000 −.0010
Testing Radius (in.)	.0546 +.0000 −.0012
Generating Pitch Radius (in.)	.0538 minimum
Material	AISI 416
Hardness	Rockwell C28 to 38

SIZE 11 SERVO MOTOR CRO 0132 670

SIZE 11 SERVO MOTORS R119-2 R119-36

SIZE 11 SHAFT DIMENSION DATA

PART NUMBER	SHAFT TYPE	DIM. A (in.)
R119-☐ A	Pinion	.494 +.009 −.006
R119-☐ D	Plain	.494 +.009 −.006
R119-☐ E	Pinion	.408±.010
R119-☐ J	Pinion	.384±.009
CRO 0130 570 CRO 0132 670 CRO 0133 670	Pinion	.437±.015
CR4 0164 002	Pinion	.494±.015
CR4 0132 012	Plain	.437±.015

TYPICAL KEARFOTT DRIVING AMPLIFIERS FOR SIZE 10 AND SIZE 11 SERVO MOTORS

MOTOR PART NO.	AMPLIFIER PART NO.
CL9 0118 017, CL9 0121 009	C70 3148 001
R124-4☐	C70 3148 001 (transformer required)
CL9 0121 003 P124-06☐	C70 3189 001
R119-36☐, CRO 0132 670	C70 3148 001
R119-2☐	C70 3189 001 (transformer required)
CR4 0164 002	C70 3302 001
CR4 0132 012	C70 3516 001

Literature detailing other driving amplifiers for these units is available on request.

Figure 9-17(b) Servomotor data sheet. (*Courtesy of The Singer Company, Kearfott Guidance and Navigation Division, Little Falls, N.J.*)

from the given data. The time constant, $\tau! = 0.0085$ sec. Therefore, $\tau = 2 \times 0.0085$ sec, or 0.017 sec. The $K_v!$ value, according to Eq. (4-20), is $(3,000)(2\pi)/60 \div 115 = 2.73$ rad/sec/volt. $K_v = 2 \times 2.73 = 5.46$ rad/sec/volt. Our transfer function now becomes:

$$KG = \frac{5.46}{s(1 + 0.017s)} \tag{9-16}$$

The next piece of hardware that we must select is the control transmitter and control transformer pair. Since we are using 60 Hz for our supply frequency, we must also select 60 Hz synchro units. The units that we choose are marked in Figure 9-18. But notice the sensitivity data given for them. The control transmitter's sensitivity is listed as 44 mV/deg, whereas the control transformer's sensitivity is given as 70 mV/deg. Since these units are designed to work with each other, we should have a mV/deg figure that is given for the pair. Unfortunately, manufacturers rarely do that, so consequently our only choice in this matter is to ignore the figure given for the control transmitter and use the one given for the control transformer. We assume that this figure is valid for the two units operating as a pair. In other words,

$$
\begin{aligned}
K_{CX-CT} &= 70 \text{ mV/deg} \\
&= \frac{0.070}{0.0175} \frac{\text{V/deg}}{\text{rad/deg}} \\
&= 4 \text{ V/rad}
\end{aligned}
$$

Now that we have established transfer functions for our synchro pair and servomotor, we have to determine the system's overall transfer function. Since we don't know K_{amp} and $1/N$, we must go to a root locus plot to determine our K_x as we have done in the other design examples. Finding that value, we can then determine the combined value of K_{amp} and $1/N$. Once that is determined, we can then make some reasonable guesses as to what our gear train ratio should be.

Going to an s-plane plot, we lay out our poles as shown in Figure 9-19. We have a pole occurring at $\sigma = 0$ and at $\sigma = -58.82$. Our root locus will be located mid-way between these poles, or at $\sigma = -29.41$. Only the upper vertical locus is drawn. Again, selecting a z of 0.6, angle θ becomes the arccosine of 0.6, or 53.1°. Then,

$$
\begin{aligned}
\omega_d &= (29.41)(\tan 53.1°) \\
&= 39.17 \text{ rad/sec}
\end{aligned}
$$

and,

$$
\begin{aligned}
\omega_n &= \sqrt{\omega d^2 + 29.41^2} \\
&= \sqrt{39.17^2 + 29.41^2} \\
&= 48.98 \text{ rad/sec}
\end{aligned}
$$

SIZE 11 - 60 AND 400 Hz SYNCHROS

TYPE	PART NUMBER	PRIMARY	INPUT VOLTAGE (volts)	INPUT CURRENT (mA) (max.)	INPUT POWER (watts) (nom.)	INPUT IMPEDANCE (ohms) (output open circuit)	OUTPUT IMPEDANCE (ohms) (input open circuit)	DC ROTOR RESISTANCE ±15% (ohms)	DC STATOR RESISTANCE ±15% (ohms)	OUTPUT VOLTAGE (volts)	TRANSFORMATION RATIO	SENSITIVITY (mV/degree)	PHASE SHIFT (°)	TOTAL NULL VOLTAGE (mV) (max.)	FUNDAMENTAL NULL VOLTAGE (mV) (max.)	MAX. ERROR FROM E.Z. (minutes) (max.)	FRICTION @ 25°C (gm cm)	ROTOR MOMENT OF INERTIA (gm cm²)	WEIGHT (oz.)
CX	RS911-1	R	26	280	.95	107∠81.7°	18.1∠79.5°	9	3	11.8	.454±4%	206	5	26	17	*	4	2	4
	RS911-4	R	26	90	.30	359∠81.3°	60∠78.5°	28.5	10.5	11.8	.454±4%	206	4.7	26	17	*	4	2	4
	RS911-7	R	26	136	.45	236∠81°	40∠78.5°	19	5.75	11.8	.454±4%	206	4.7	26	17	*	4	2	4
	RS911-2	R	115	60	.80	2210∠82.3°	1130∠81.3°	159	137	90	.783±4%	1570	4	94	59	*	4	2	4
	RS911-3	R	115	50	.97	2598∠81.8°	60∠78.5°	218	10.5	18.2	.158±4%	318	5	42	26	*	4	2	4
	RS911-5	R	115	30	.44	4670∠81.1°	2510∠79.3°	320	318	90	.783±4%	1570	5	94	59	*	4	2	4
	RS911-6	R	115	70	1.03	2060∠80.8°	18.1∠79.5°	160	3	11.8	.103±4%	206	5.6	26	17	*	4	2	4
	26V 11 CX 4c	R	26	130	.37	244∠82.3°	41.4∠82°	—	—	11.8	.454±.009	206	4.25	19	12	7	5	2	4.7
	11 CX 4e	R	115	31	.49	4210∠82°	2170∠78.2°	—	—	90	.783±2%	1571	4.5	75	45	7	5	2	4.7
	R911-03**	R	6.3	195	.58	22+j33	4.8+j5.5	—	—	2.5	.398±4%	44	31	18	12	*	4	2	4
TX	26V 11 TX 4c	R	26	280	1.0	106.1∠83.1°	18.5∠79.5°	—	—	11.8	.454±.009	206	4	—		7	5	2	4.7
CDX	RS941-1	S	11.8	165	.25	74.3∠79.7°	86∠77.1°	17	10.5	11.8	1.154±4%	206	6	26	17	*	4	2	4
	RS941-4	S	11.8	65	.097	195∠80.8°	231.2∠76.7°	49	21	11.8	1.154±4%	206	7.4	26	17	*	4	2	4
	RS941-2	S	90	60	.6	1640∠80.7°	1990∠77.3°	385	195	90	1.154±4%	1570	4.7	94	59	*	4	2	4
	26V 11 CDX 4c	S	11.8	150	.25	76.5∠79.5°	88.3∠77.3°	—	—	11.8	1.154±.023	206	6	26	17	7	5	2	4.7
	11 CDX 4b	S	90	49	.53	1820∠82°	2165∠78°	—	—	90	1.154±2%	1571	4	90	60	7	5	2	4.7
	R941-03**	S	2.4	80	.075	17+j30	37+j35.2	—	—	2	.980±4%	35	31	18	12	10	4	2	4
CT	RS901-1	S	11.8	165	.25	74.3∠79.7°	418∠78.3°	54	10.5	22.5	2.203±4%	393	6	53	34	*	4	2	4
	RS901-3	S	11.8	21	.03	577∠80.7°	3340∠79.2°	385	60	22.5	2.203±4%	393	4.2	40	30	*	4	2	4
	RS901-2	S	90	20	.18	5470∠80.8°	3340∠79.2°	385	555	57.3	.735±4%	1000	4.5	94	59	*	4	2	4
	26V 11 CT 4d	S	11.8	86	.142	131∠79.7°	704∠79.8°	—	—	22.5	2.203±.044	393	4.7	18	15	7	5	2	4.7
	11 CT 4e	S	90	18	.20	5025∠80.4°	3370∠80.36°	—	—	57.3	.735±.015	1000	5	60	32	7	5	2	4.7
	R901-03**	S	2.4	25	.02	49+j90	400+j500	—	—	4	1.90±4%	70	26	25	15	*	4	2	4.7

* Available with 5, 7, or 10 minute accuracies. When ordering, preface basic part number with accuracy required e.g., for a 7 minute RS911-1, order 7RS911-1. 10 minute components require no numerical prefix.

** 60 hertz units. All others have a frequency of 400 Hz.

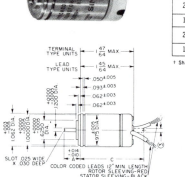

TERMINAL TYPE UNITS 1 47/64 MAX
LEAD TYPE UNITS 1 45/64 MAX

Outline dimensions for Bu/Weps units are in accordance with MIL-S-20708.

BuWeps Type	Military Part No.	Pinion Shaft Data (All Units)†	
26V 11CX4c	M20708/8C-001	No. of Teeth	21
11CX4e	M20708/2C-001	Diametral Pitch	120
26V 11TX4c	M20708/6D-001	Pressure Angle (°)	20
26V 11CDX4c	M20708/9C-001	Std. Pitch Dia. (in.)	.175 +.000/−.002
11CDX4b	M20708/81C-001	Max. Root Dia. (in.)	.155
26V 11CT4d	M20708/7C-001	Outside Dia. (in.)	.1872 +.0000/−.0005
11CT4e	M20708/1D-001	Tooth Form	Full Depth Involute

† Shaft length on all units is .555 in. to stop on shaft

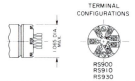

TERMINAL CONFIGURATIONS

RS900
RS910
RS930

RS901, RS911, and RS941 units are supplied with leads. Identical units equipped with terminals are identified by basic part numbers RS900, RS910, and RS940 respectively.

Figure 9-18 Synchro data sheet. (*Courtesy of The Singer Company, Kearfott Guidance and Navigation Division, Little Falls, N.J.*)

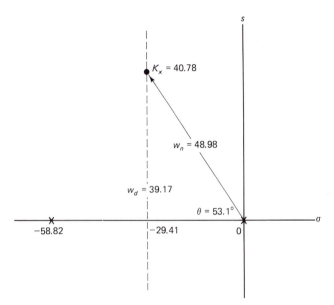

Figure 9-19   The root locus for the valve controller's transfer function.

We can now calculate K_x using Eq. (9-7):

$$K_x = \omega n^2 \tau$$
$$= (48.98^2)(0.017)$$
$$= 40.78$$

And since:
$$K_x = (K_{CX-CT})(K_v)(K_{amp})(1/N)$$
$$40.78 = (4 \text{ V/rad})(5.46 \text{ rad/V})(K_{amp})(1/N)$$
$$(K_{amp})(1/N) = 1.87 \tag{9-17}$$

Our next step is to select a reasonable gear ratio for our gear train. This decision is based on how quickly we want our valve to open and close and how much torque is available. Assuming that very little torque is required for this particular valve application, let's assume further that we wish to have the valve's stem rotate at a speed of 10 rpm. Assuming an error-correcting rpm for the servomotor as being 25% of 3,000 rpm, or 750 rpm, we can now calculate our gear ratio:

$$\text{Gear ratio, } N = \frac{\theta_{in}}{\theta_{out}} = \frac{750}{10}$$
$$= 75$$

Then, from Eq. (9-17):

$$K_{amp} = (1.87)(75)$$
$$= 140.3$$

Often, it is difficult to order the exact gear ratio needed from the manufacturer. You must choose the closest ratio available and then compensate for the change in the system's gain by adjusting the amplifier. If compactness is required in the automatic control system, you can often specify that the gear train be directly attached to the servomotor's casing for the purpose of creating one unified system.

Since we are dealing with a control knob that is to be rotated to a desired setpoint (the control knob being attached to the rotor of our CX) for the purpose of positioning our valve stem, let's calculate the dead band within the control system using the recommended gain calculated from Eq. (9-17). We do this using Eq. (9-1). However, we also need to know the servo's starting voltage. The catalog information in Figure 9-17 lists the starting voltage as 1.0 volt. For the K_s value in Eq. (9-1), we substitute our earlier calculated value of 4 V/rad for K_{CX-CT}. Therefore,

$$\text{dead band} = \frac{(2)(1.0)}{(4)(140.3)}$$
$$= 0.0036 \text{ rad, or } 0.21 \text{ deg}$$

The figure of 0.21° means that the input control rotor on the CX has to be rotated a minimum of 0.21° before causing movement in the servomotor. However, the system's hesitation to react to an input command doesn't stop at the servomotor. There is also the gear train's *backlash* to contend with. Backlash is the dead band of gear trains. It is the amount of rotation needed at the gear train's input to just cause rotation in its output. Backlash is caused by imprecise machining of gears, and certain amounts cannot be prevented. Most likely, there is also slack in the flexible mechanical coupling used in attaching the gear train's output shaft to the valve stem. All of these factors add to the overall "slop" within the control system. This causes an increase in the system's response time that is added to the dead band of the synchro and servomotor combination. However, these additional factors are difficult to obtain in data sheets and must be measured later after the entire system is assembled. For the present, we neglect these other factors beyond the servo's dead band.

The complete system transfer function is:

$$KG = \frac{40.78}{s(1 + 0.017s)} \tag{9-18}$$

Figure 9-20 is the open-loop Bode plot of our valve positioner system. Our phase margin is within the recommended 40° to 60° range.

In the examples that we have covered so far, we have assumed that we could easily obtain the components needed to make the system function the way we wanted it to function. As any engineer or technician will verify, this is not a very realistic situation. Often, the component you need the most is the component that is the most difficult or impossible to obtain (no doubt a corollary of Murphy's

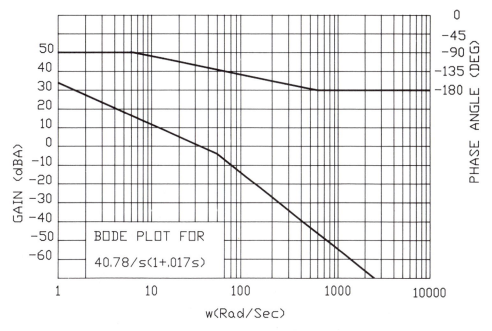

Figure 9-20   Bode plot for $40.78/s(1 + 0.017s)$.

law). The next example demonstrates how to make the best of a design situation where only certain hardware components are available.

9-5 *EXAMPLE 4: AN OSCILLOGRAPH*

This example is somewhat similar to the one we discussed in Chapter 8 concerning the voltage-reading chart recorder. Oscillographs are recording oscilloscopes. That is, they produce a hard copy output of the waveform that is being analyzed. Again, as in the previous design example, we deal with instrument-type servos in our design because of the relatively light-duty applications involved.

In our design, we assume that we must work with an amplifier whose gain is fixed at a value of 11. Because of cost restraints, we aren't allowed to add any additional amplifiers to the system. We assume, too, that the servomotor we have to work with has a fairly high K_v. Its transfer function is:

$$\frac{\theta_{\text{out}}}{E_{\text{in}}} = \frac{35}{s(1 + 0.018s)}$$

We also must work with a gear train having a gear ratio of 50:1. Therefore, the system's overall transfer function that we have to work with is:

$$KG = \frac{7.7}{s(1 + 0.018s)} \qquad (9\text{-}19)$$

Figure 9-21 is the block diagram of our system.

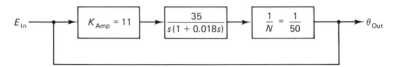

Figure 9-21 Block diagram of oscillograph for Example 4.

Our first task is to look at the system's Bode plot to see what it is we have to work with. Because of the amplifier's relatively low gain, we can, without looking at the Bode plot, guess that we will probably be working with a fairly sluggish system. Figure 9-22 shows us that our guess was correct. We see an extremely sluggish system. The phase margin is greater than 80°. For the record, we can calculate the damping ratio. Using the methods outlined earlier, we calculate a factor of $z = 1.343$. Notice that this value exceeds 1. We really haven't discussed damping factors having values this large. We can guess, however, (and guess correctly) that values this great do represent extremely sluggish systems. This

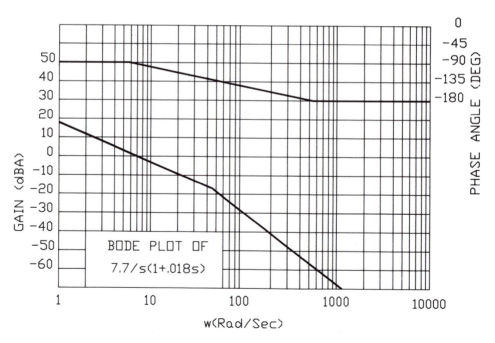

Figure 9-22 Bode plot for $7.7/s(1 + 0.018s)$.

extremely high value for z is undoubtedly due to the rather low K_x factor in our transfer function. Since we can't increase the gain of our system as we would normally do (that was a stipulation given us in our problem), we instead have to look at another alternative. We can install a phase-lag network. The phase-lag network will increase the phase angle of the output (that is, make it more negative) as compared to the system's input, thereby decreasing the phase margin making the system less sluggish. Looking back to Section 4-16.4 in Chapter 4, we see how to construct this network (Figure 4-47). Following the general rule of thumb for these networks, we make $\tau_2 = 10\tau_1$. Also, we can guess at a reasonable value for the capacitor, C. Let's make $C = 1{,}000 \ \mu F$. (For this high capacitance, we would want to use a tantalum electrolytic capacitor to take advantage of its inherent consistency with aging and good temperature stability. Bear in mind, though, that we could have picked virtually any size capacitor for this application, since it becomes a matter of developing the proper combinations of capacitance and resistances for creating the desired time constants. There are an infinite number of these combinations.) We also pick a value of 100 ohms for R_1. To determine the value of R_2, we set up the following equations:

Since

$$\tau_2 = 10\tau_1$$

and

$$\tau_1 = R_2 C$$

and

$$\tau_2 = (R_1 + R_2)C$$

then

$$\tau_2 = 10\tau_1 = 10R_2(1{,}000 \times 10^{-6})$$
$$= (100 + R_2)(1{,}000 \times 10^{-6})$$

Therefore,

$$R_2 = \frac{100}{9}$$
$$= 11.1 \ \Omega$$

Since this is an odd-size resistor, we can approximate the value of R_2 with a 10-ohm resistor instead.

With the installation of our lag network, our transfer function becomes:

$$KG = \frac{(1 + 0.01s)7.7}{s(1 + 0.11s)(1 + 0.018s)} \qquad (9\text{-}20)$$

Because of the complexity of this function, a straight-line Bode plot is not practical. A programmable calculator that has been programmed for calculating the gain and phase angle of our transfer function is recommended. The resultant open-loop plot is seen in Figure 9-23. It's interesting to note here that an approximation to the actual Bode plot could have been made by merely cancelling the $(1 + 0.01s)$ expression with the $(1 + 0.018s)$ expression, since they are approximately equal. The simplified expression of $KG = 7.7/[s(1 + 0.11s)]$ could then be plotted. Figure 9-23 shows this approximation superimposed over the actual curves. You can see that considerable inaccuracy is introduced at the high end of

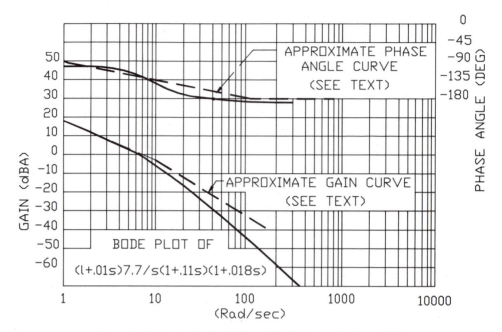

Figure 9-23 Bode plot for $7.7(1 + 0.01s)/s(1 + 0.11s)(1 + 0.018s)$.

the gain curve, but near the zero gain crossover point there is a fair correlation between the two curves. However, we are interested in the phase margin. We see now that the phase margin has decreased to a respectable 54°. Since this is within the desirable range, let's now calculate our new damping ratio. Unfortunately, our transfer function is a cubic equation in its closed-loop form. However, we can use the cancelled form, $KG = 7.7/[s(1 + 0.11s)]$, for a fair approximation (see Example 8-5 in Section 8-6.4, Chapter 8). The damping ratio is found by converting to the closed-loop form and using Eq. (8-5):

$$\frac{KG}{1 + KG} = \frac{7.7}{s(1 + 0.11s) + 7.7}$$

$$= \frac{1}{s\left(\dfrac{0.11s^2}{7.7} + \dfrac{s}{7.7} + 1\right)}$$

$$= \frac{1}{s\left(\dfrac{s^2}{8.367^2} + \dfrac{1.087s}{8.367} + 1\right)} \qquad (9\text{-}21)$$

$$z = 0.543$$

This is an obvious improvement over the original damping ratio.

We have neglected the bandwidth up to this point in our design of the oscillograph. Since we are concerned with the oscillograph's ability to respond to a wide range of input frequencies, let's see what sort of bandwidth we have with the newly installed phase-lag network design. Using Eq. (8-17),

$$\omega_b = \omega_n \sqrt{1 - 2(0.543^2) + \sqrt{2 - 4(0.543^2) + 4(0.543^4)}}$$
$$= 8.367 \sqrt{1 - 0.590 + \sqrt{2 - 1.179 + 0.348}}$$
$$= 8.367 \sqrt{1 - 0.590 + 1.081}$$
$$= 8.367 \sqrt{1.491}$$
$$= 10.22 \text{ rad/sec}$$

This is the equivalent of 1.63 Hz. Judging from this very low bandwidth, our oscillograph could only be used for the plotting of very slow cyclic waveforms, approaching DC. Figure 9-24 is our revised block diagram showing the installed phase-lag compensating network.

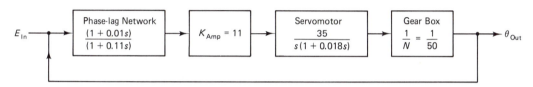

E_{In} — Phase-lag Network $\dfrac{(1 + 0.01s)}{(1 + 0.11s)}$ — $K_{\text{Amp}} = 11$ — Servomotor $\dfrac{35}{s(1 + 0.018s)}$ — Gear Box $\dfrac{1}{N} = \dfrac{1}{50}$ — θ_{Out}

Figure 9-24   Block diagram of oscillograph with newly installed phase-lag network.

In summarizing our use of the phase-lag network, you have to be careful in its application to a system. You can't always depend on a phase-lag network, for instance, to delay the system's phase angle to make a sluggish system into a desirably stable system. The system may be stable at the zero-gain crossover point, but that same system may become unstable at some lower frequency. This type of system is called a *conditionally stable system*. The next design example demonstrates this problem.

9-6 EXAMPLE 5: A CONDITIONALLY STABLE SYSTEM

We now take our design example from Example 4 and modify it somewhat. Instead of using an amplifier having a fixed gain of 11, we will change that gain to, say, 300. This will give us a K_x of $\frac{300 \times 35}{50}$, or 210. Our open-loop transfer function now becomes,

$$KG = \frac{210}{s(1 + 0.018s)} \tag{9-22}$$

The Bode plot for this revised system is shown in Figure 9-25. The phase margin of 32° indicates an unstable system as a result of the higher gain. We now add the compensating network, $(1 + s)/(1 + 10s)$. Earlier, we mentioned that there are an

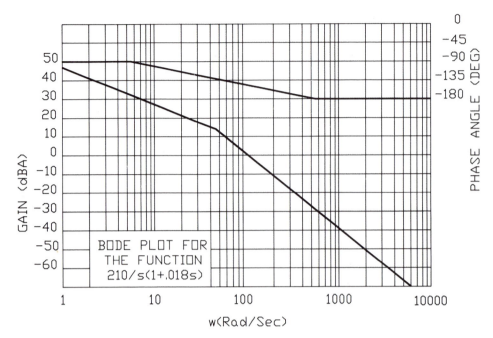

Figure 9-25 Bode plot for the function $210/s(1 + 0.018s)$.

infinite combination of capacitances and resistances that could be used to satisfy the time constant requirements for a compensating network. There are also an infinite combination of $\tau_2 = 10\tau_1$ to choose from. We picked the foregoing combination merely to prove a point in this discussion. We want to see what happens when we use this choice in combination with our transfer function, Eq. (9-22). Adding the new phase-lag network makes our function become:

$$KG = \frac{(1 + s)210}{s(1 + 10s)(s + 0.18s)} \tag{9-23}$$

Again, in order to plot this rather complex open-loop function, we want to use a programmable calculator to make a point-by-point curve plot. Figure 9-26 shows the results. At the zero gain crossover point, we find a phase margin of 68°, indicating a somewhat sluggish system. But notice what happens if the gain were to be reduced 43 dB. The phase margin becomes 35°, resulting in an unstable system. We experience a situation that is somewhat similar to a spinning top. As long as the rotational velocity is high the top remains stable; reduce its velocity and it immediately becomes unstable. We have, as a result, a conditionally stable system in the case of our top. In our control system, if the gain of a system is to be varied for any reason during its operation and a compensating network is present, be sure that you don't have this type of existing situation just described. If you do, you will want to avoid those particular gain settings. It's interesting to note

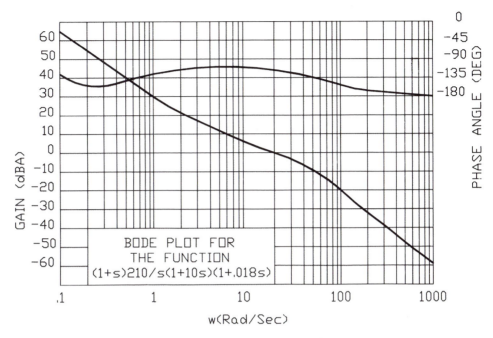

Figure 9-26   Bode plot for the function $210(1 + 1s)/s(1 + 10s)(1 + 0.018s)$.

that the inertially damped servomotor (the IDSM) has characteristics very similar to the phase-lag network and is sometimes referred to as a phase-delaying device. It is possible to have conditionally stable systems using phase-lead networks. In any case, constructing the system's Bode plot will uncover any unique characteristics that may require special design or operating attention.

9-7 EXAMPLE 6: A DIGITAL CONTROLLER FOR A HYDRAULIC PRESS

We haven't talked about digital systems up to this point. They can become rather complex, electronically. However, because of certain desirable characteristics (which we discuss later), digital automatic control systems are becoming very popular in industry. In this next example, we won't delve into the electronic or digital theory behind this particular system. It's more important at this point to simply understand the overall basics as to how the system operates. The digital circuitry can be studied in other books (see the references at the end of this chapter).

Figure 9-27 shows the schematic of an hydraulic press with an automatic load control system. The load control is a keyboard and display where an operator can type in the desired tonnage to be applied by the hydraulic press. The

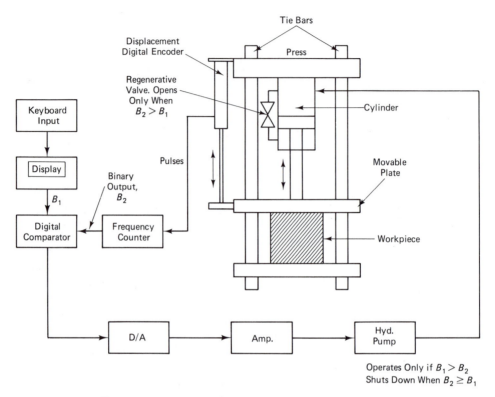

Figure 9-27 Schematic of digitally controlled hydraulic press.

workpiece to be pressed is placed between the one movable plate attached to an hydraulic ram or cylinder, and the lower stationary plate. When the hydraulic pump is energized, the blind end (the end opposite the piston's rod end) of the cylinder becomes pressurized. This causes a clamping force to be applied to any object placed between the two large plates. If the pressure becomes excessive for any reason, we can open the regeneration valve. This will equalize the cylinder's pressure, thereby relieving the pressure between two large steel plates.

The press is held together, so to speak, by large circular bars or pillars of steel, called *tie bars*. The larger presses usually have four of these arranged in a square. The lower movable plate can then ride up and down on these bars.

We must come up with a method for determining the clamping pressure created by the press. Assuming that the cylinder is pressurized, the movable plate squeezes the object between its plates. While this is occurring, the clamping cylinder exerts a force in the opposite direction by pushing against the top plate of the press. You can envision this better if you picture yourself in place of the cylinder, standing on the lower movable plate and pushing with your hands up against the top plate. You must do this in order to create any squeezing action between the two lower plates. The cylinder has to do the very same thing in this instance. As a matter of fact, the tie bars forming the frame of the press will become stretched during this squeezing action. As the cylinder expands between the upper and lower plates, the tie bars experience a stretching action, which, if it becomes excessive, could cause the bars to break or snap.

The amount of stretch or length change in any common metal, such as steel, can be easily equated to the amount of force or pressure causing that stretch. That relationship is:

$$\text{length change} = \frac{\text{(applied force)} \times \text{(length of metal)}}{\text{(cross-sectional area)} \times \text{(modulus)}} \qquad (9\text{-}24)$$

The modulus, otherwise known as the *modulus of elasticity,* for steel is 3×10^7 lb./in.^{2} Therefore, if the applied force to a steel bar is known in pounds, the cross-sectional area is known in square inches, and the length of the bar is known in inches, then the bar's change in length can be easily calculated using Eq. (9-24). It now becomes a matter of determining how much the bars will stretch in our press whenever the hydraulic cylinder is energized forcing the movable plate down onto the bottom fixed plate.

The method we use for determining the amount of stretch taking place in the tie bars is through an *optical linear encoder* (see Figure 9-28). This is a device that can generate an electronic pulse for every 0.0001 in. of deflection of its plunger rod. By counting these pulses we will know, within a tolerance of ±.0001 in., how much movement of the plunger rod has taken place and, therefore, how much deflection has taken place within each tie bar. This is assuming that the encoder is attached to the press as shown in Figure 9-27.

The control portion of the system will work like this: With the output of the optical encoder feeding into a frequency counter, we can count the encoder's output pulses. The count is then converted to a binary number, B_2, and sent to a digital comparator. This circuit is similar in principle to our differential amplifier. It has the ability to subtract one input from another and then present the difference at its output. However, whereas the op amp is an analog device, the digital comparator performs its subtracting digitally. And there is no amplification involved either, as is generally the case with any digital circuit. The other input, B_1, to this comparator comes from a keyboard entry made by the press operator.

The output of the comparator circuit, which is in the form of a binary signal resulting from the arithmetic difference between B_1 and B_2, is sent to a digital-to-analog converter. The resulting analog signal is then amplified to drive a hydraulic pump which, in turn, supplies hydraulic fluid to the press's hydraulic ram. An important feature in the hydraulic circuitry of this system is the regenerative valve shown attached to the side of the ram. This valve was mentioned earlier. When the valve is closed, the cylinder operates normally. However, when the valve is opened the pressure on the blind end of the piston is allowed to equalize the pressure on the piston's rod end; the downward force will then stop.

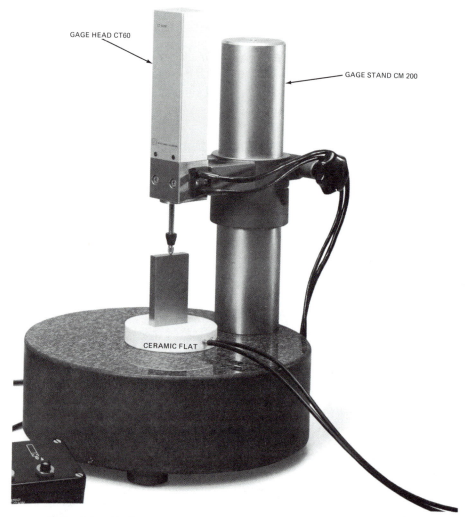

GAGE HEAD CT60

GAGE STAND CM 200

CERAMIC FLAT

Figure 9-28 The linear optical encoder. (*Courtesy of BEI Motion Systems Company, San Marcos, Calif.*)

We now need additional logic circuitry to operate the regenerative valve and hydraulic pump. This circuitry will cause the regenerative valve to open only when $B_2 > B_1$. Also, the hydraulic pump will not operate when $B_2 \geq B_1$, or it will operate when $B_1 > B_2$. Logic statements such as these can become confusing at first glance but are very common in digital circuits. Let's study them a little more closely to see how they make the circuit function. Let's assume the operator wishes to create a force of, say, 50 tons on a die that is being tested. The 50-ton figure is entered on the keyboard and is automatically displayed on an alphanumeric readout screen. At this same time, the decimal figure is converted to a binary number (B_1) and sent to the digital comparator. Since at this time the press is not operating, the pressure in the main cylinder is zero; therefore, B_2 is zero. Since B_1 is greater than B_2, the hydraulic pump turns on. As the pressure increases inside the cylinder, the magnitude of the binary number B_2 also increases. Theoretically, when B_2 equals B_1, the pump will shut down. However, because of the existence of electrical and mechanical time constants within the system, it is quite likely that the pressure within the cylinder will overshoot the desired set point of 50 tons. If this happens, B_2 will become greater than B_1 and the hydraulic pump will shut down. At this same time, the regenerative valve will open, causing the pressure to equalize inside the cylinder. B_2 will then become less than B_1, causing the regenerative valve to open again and the pump to turn back on starting the pressure building process all over again. With every cycle that the system would make, the set point value of 50 tons would be approached more closely until eventually the system will settle at that point.

Obviously, one drawback to this system design is that the system behaves like one having a damping factor near zero. One step that could be taken to reduce the number of oscillations (i.e., to increase the damping factor) is to place a restriction in series with the regenerative valve to reduce the pressure reduction rate inside the cylinder. Or, the speed of the hydraulic pump could be reduced to reduce the pressure build-up rate within the cylinder. Another method, which may be the most effective, is to build in a dead band area in the greater-than logic of the regenerative value such that if $B_2 > B_1$ by, say, 2 tons, only then will the regenerative valve open.

When dealing with digital circuits such as the one just discussed, the rules for determining the system control characteristics change somewhat. This example was presented to introduce this point. There are methods for analyzing digital systems that resemble the methods we have discussed for our analog systems, but because of their complexity we don't discuss them here. For one thing, determining the transfer functions for large mechanical devices such as hydraulic components is a fairly complicated procedure. For cases like these, it is sometimes quicker to build the system based on a preliminary design, and then through experimentation determine the proper system components for the desired system behavior. We discuss additional systems similar to this one in Chapter 10.

9-8 EXAMPLE 7: BANG-BANG SERVO SYSTEMS

Of the automatic control system types that presently exist, *the bang-bang servo* is probably the simplest to design, operate, and understand. Because of this, this type of system is probably the most commonly used today for obtaining automatic control. The bang-bang servo is a system that reacts to an input signal in a step function manner. That is, the system is either fully on or fully off. There is usually some sort of sensing device that monitors the output of the system. When a particular level of output quantity is reached, usually determined by the system operator's set point input, the system is turned off. As the output diminishes as a result of being turned off, a minimum output quantity is reached causing the system to turn back on. The cycle is then repeated over and over. Probably the best example of a bang-bang servo is the thermostatically-controlled heating system for a home. Figure 9-29 is a diagram of just such a system. It shows how the

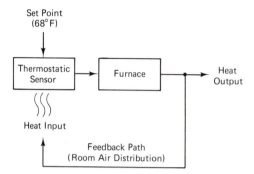

Figure 9-29   The thermostat.

furnace's output is fed back to the thermostat that has been given a set point input of 68°. In reality, we know the thermostat doesn't actually sense the furnace's heat directly. If it did, the furnace would oscillate rapidly on and off, and as a result the house would not become heated. Instead, the thermostat is usually located away from a direct heater discharge and in an area where the room air has had time to mix with the furnace's output before actually becoming sensed. The basis of operation of this system is the basis of all bang-bang servo systems. The term *servo* is applied loosely in the case of our furnace example, since there are really no servomotors present in the system.

Let's take a closer look at the system just described and attempt to analyze its behavior. Figure 9-30 shows a timing diagram of the furnace's input and its output as it arrives, so to speak, back at the thermostat. This then becomes the

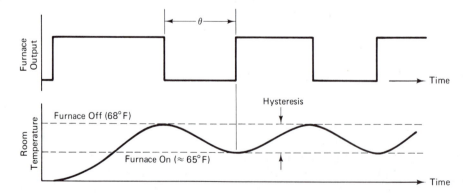

Figure 9-30   Timing diagrams for a typical furnace and thermostat system.

furnace's *input* once again. The first diagram, labeled *System Output,* shows the step function characteristics of the furnace operation itself. The second diagram, labeled *System Input,* shows the temperature fluctuations at the thermostat along with the set point input (i.e., the thermostat setting). The first thing we should note about these two diagrams is that they both have the same frequency. The second important characteristic of these diagrams is the phase shift. Just as we had a phase shift to contend with in our other servo systems, we have a phase shift to contend with here. This out-of-phase relationship between the furnace's output and its input is shown as the phase angle θ in Figure 9-30. This phase angle amount is determined by several quantities:

1. the thermostat's response time,
2. the thermostat's *hysteresis,*
3. the house's heat loss rate, and
4. the furnace's ability to come up to operating temperature.

Perhaps as you read through the preceding list, you can see striking similarities between this system and the others we discussed earlier. The frequency and phase angle are obvious, but what about time constants? Can you see how we could derive time constants for the house, thermostat, and furnace? Usually, you don't see time constant terms published for these items, but nevertheless, they do exist and can be calculated if desired.

In item 2 of the preceding list, we mentioned the term *hysteresis.* Hysteresis, in our application here, is the characteristic of a device or instrument that has two different switching points for the same sensing quantity. These two points are determined by whether the sensing quantity is increasing in magnitude or decreasing in magnitude. In the case of our thermostat, the thermostat may decide to turn the furnace on at one temperature but may turn off the furnace at a different temperature, even though the thermostat was given only one temperature setting. This discrepancy in switching temperatures is caused by such things as static friction versus motion friction (two entirely different values), bearing clearances, direction of ''play'' in mechanical parts, and changes in physical dimensions due to temperature changes. These are just a few of the factors that figure into the occurrence of hysteresis. Figure 9-30 shows where the hysteresis quantity occurs in our furnace–thermostat system.

In some cases, hysteresis is a desirable quantity. Take a look at Figure 9-30 again. If hysteresis were mostly eliminated from the thermostat's design and construction, how would this affect the cycling time of our furnace? Figure 9-31 shows us the results. The room temperature would certainly be far more consistent, with very little fluctuation in magnitude. However, the price paid for this is a furnace that cycles far more frequently and in shorter intervals. Most present-day furnaces are not constructed to withstand that kind of high-duty rate or service. As a result, many thermostats are now designed to allow the user to program in

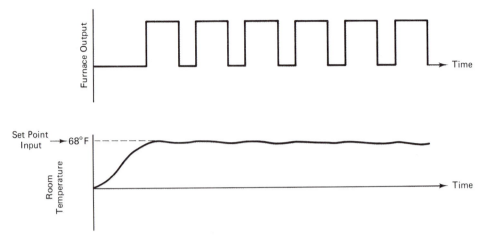

Figure 9-31 Timing diagram for a furnace–thermostat system in which most of the hysteresis has been removed from the thermostat.

just the right amount of hysteresis needed to produce the best heating effect for the house in which the system is installed.

Another common application of the bang-bang servo system in use today is the relay-operated motor positioning control shown in Figure 9-32. The relay coil is a three-position relay. In the first position, the relay energizes the motor to

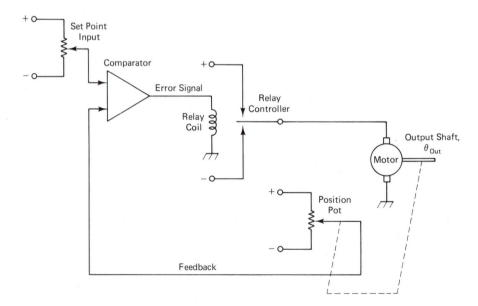

Figure 9-32 A relay-operated bang-bang servo system.

rotate in one direction, while the second position of the relay causes the motor to rotate in the opposite direction. The third position, or middle position, is the off position in which the motor is not energized. Notice that the motor is either fully on in one direction or another or fully off, a characteristic of the bang-bang servo. The advantage of the relay-operated system is that it is relatively inexpensive and that for a small power input to the system, one can control large amounts of power at the output. The disadvantage with this system is the lack of accuracy in positioning the motor. Since it is either completely on or completely off, there is no modulating of the error or control signal. It is this finely tempered signal that allows the motor to reach an exact position in the analog or digital type of servo control.

The characteristics of the relay used in a bang-bang servo are very important. The off position in the three-position relay just discussed is the relay's dead band. Also, this relay will have a certain amount of hysteresis. That is, the voltage needed to pull in its armature will be different from the voltage needed to release the armature. This is typical of this kind of electromechanical device. Figure 9-33 explains the characteristics of dead band and hysteresis and compares them to a so-called ideal relay that has no dead band or hysteresis.

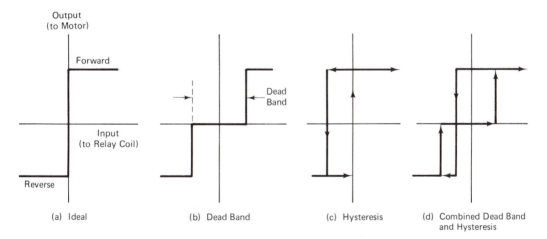

Figure 9-33 Relay characteristics used in a bang-bang servo system.

Now that we have discussed both the conventional (analog and digital) types of automatic control system and the bang-bang servo automatic control system, we should compare their output characteristics. A characteristic of the bang-bang servo is its habit of excessive hunting as compared to the conventional systems. Figure 9-34 compares the two general systems. Notice that in addition to the excessive hunting of the bang-bang system, the final settling point is somewhere within the dead band of the switching relay. The greater the dead band of a switching relay, the greater the inaccuracy of the system's positioning ability.

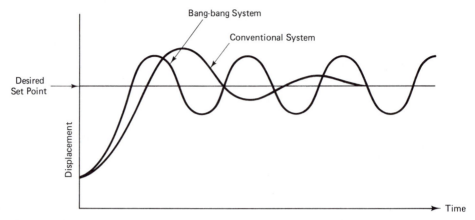

Figure 9-34 Comparing response curves of a bang-bang servo system versus a conventional control system.

There is also dead band within the conventional system, but not to the degree that you have with the bang-bang system.

One way to reduce the amount of hunting and the frequency of oscillation in a bang-bang servo is to introduce a certain amount of viscous damping to the motor. Unfortunately, when this is done the settling time is increased. The overall speed of the system is slowed.

SUMMARY

The designing of an automatic control system is heavily dependent on good engineering catalog data obtained from the components' manufacturer. Without this information, the design work can become very frustrating. Designing these systems not only requires a good knowledge of automatic control theory; it also requires a fair knowledge of physics. Without physics, designing a control system to operate and control a mechanical device that has to obey certain physical laws becomes an impossibility.

Digital controllers are becoming more and more popular as automatic controlling devices. The reliability of digital components along with their miniaturization is making them a very attractive alternative to the analog type of control system. Example 6 in this chapter was a good representation of a typical digital control system.

Bang-bang servo systems are yet another alternative to the analog and digital control systems. They have the advantage of being simple in design and inexpensive to manufacture. Their greatest disadvantage is their inaccuracy of operation as compared to the other systems.

EXERCISES

Because of the complexity of automatic control design problems, the following is a list of suggested design problems. You will need to obtain catalog data from the various servo components manufacturers in order to undertake these suggested projects. The final submitted design should show all data calculations, block diagrams, curve plots, and design catalog data from which all component performance data was obtained.

9-1. Design an x–y curve plotter that can be used with a computer for plotting graphs and curves.

9-2. Design a positioning system for locating a tool on an NC (numerical control) machine.

9-3. Design a speed control system for controlling the rotational velocity of a rotating tool for an NC machine.

9-4. Design a cruise control system for an automobile taking into account variable factors such as road grade, windage, and in-town driving.

9-5. Design a flow control system for a beverage bottling company's filling system.

9-6. Design a radar tracking system that locks onto a high-altitude flying object such as an airplane.

9-7. Design a positioning system that controls the rotating or swivelling of a robot's main body or torso as it reports to the various positions for performing manufacturing functions.

9-8. Design a bi-level temperature controller (i.e., one that controls both heating and cooling) for an environmental chamber that must have its internal temperature maintained at a specific value. Temperature range requirements are between 0°F and 200°F.

9-9. Design a remote positional controller for a robot arm located in a hazardous environment and used while the arm's operator is in a safe location several miles away.

9-10. Design a camera positioner that can lock onto a faint light source at night for the purpose of tracking and recording the movements of that light source.

REFERENCES

BAECK, HENRY S., *Practical Servomechanism Design,* New York, N.Y.: McGraw-Hill, Inc., 1968.

DEEM, BILL R., KENNETH MUCHOW AND ANTHONY ZEPPA, *Digital Computer Circuits and Concepts,* Reston, Va.: Reston Publishing Co., 1980.

DORF, RICHARD C., *Modern Control Systems,* Reading, Mass.: Addison–Wesley Publishing Co., 1980.

MALVINO, ALBERT PAUL AND DONALD P. LEACH, *Digital Principles and Applications,* New York, N.Y.: McGraw-Hill, Inc., 1975.

SANDIGE, RICHARD S., *Digital Concepts Using Standard Integrated Circuits,* New York, N.Y.: McGraw-Hill, Inc., 1978.

Process Control Systems

10-1 INTRODUCTION

Chapter 9 showed us how to go about designing some fairly basic automatic control circuits. These designs were just complicated enough to give you a flavor as to what designing is all about. In this chapter, we discuss how various process control systems work. A process control system is comprised usually of one or more automatic control systems dedicated to *regulating* the manufacturing of a particular product. Enough variety of processes are presented here to give you an idea of how other automatic control engineers and technicians apply the principles discussed in earlier chapters of this book. In many cases, the application of automatic control theory to making a particular process work is not obvious. You have to remove machine shrouds, open control box doors, and dismantle parts in general to see how the controls theory is applied by the processes' system designer. While we can't do that here, the next best thing is to describe as earnestly as possible how the systems work and how the principles behind them operate.

10-2 ROBOTIC CONTROL SYSTEM

Figure 10-1 shows the type of robot that we discuss here. This is a *jointed-arm spherical robot*. It is used in the automotive industry for welding car frames on an assembly line. Because of its ability to reach into tight areas and corners, this type of robot is used to reach those areas within the frame that are difficult for human welders to reach. This robotic design has what is commonly referred to as

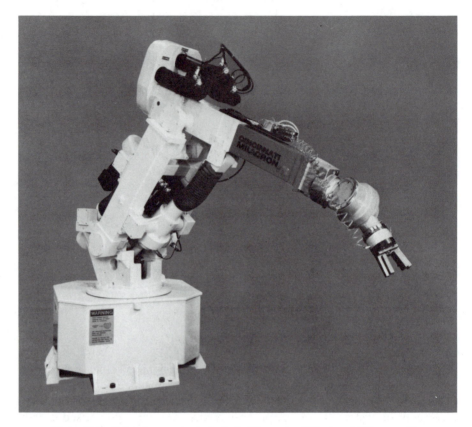

Figure 10-1 Jointed-arm spherical robot. (*Courtesy of Cincinnati Milacron, Cincinnati, Ohio.*)

six degrees of freedom (Figure 10-2). With these freedoms of movement, this particular robotic design has access virtually to every point or location within a certain sphere or volume surrounding the robot. But how does the robot know how far out it is supposed to reach or how far should it rotate, and how does it know when it is finally there?

Basically, the control mechanisms for all six motions within the robot could be designed to be roughly the same. We look at one method for controlling the positioning of the robot's arm. Figure 10-3 shows the mechanism for controlling the amount of swiveling of the robot's waist or base; this amount is the θ indicated in Figure 10-2. This system is similar to the hydraulic digital controller example described in Section 9-7. The operator merely keys in the coordinates for the robot to swivel to. This information may also be supplied to the robot by means of a computer program or software associated with an attached computer or *programmable controller*. (We talk about programmable controllers in Chapter 12.)

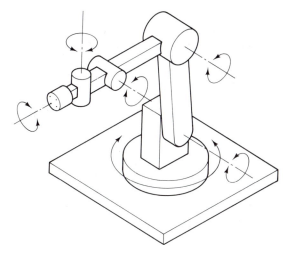

Figure 10-2   The six degrees of freedom for a robot.

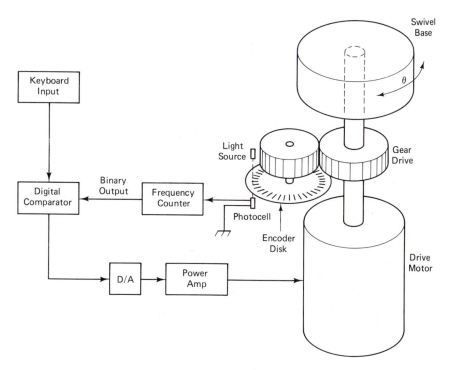

Figure 10-3   Diagram for a robotic keyboard input control system.

The digital comparator then calculates the binary arithmetic difference (the error) between the desired position (the set point) and the swivel's actual location as read by the optical encoder attached to the swivel's drive shaft. The magnitude and sign of this arithmetic difference determines the direction of rotation of the swivel motor.

In order for this system to function properly, there has to be a preselected "home" position for the robot so that the encoder doesn't put out ambiguous or conflicting information.

Let's look at an example. Using Figure 10-2, imagine taking the robot's arm and rotating the arm 180° so that it extends in the opposite direction. This would place the robot's arm to the rear. Let's assume that this is our homing position and it is facing, say, north. At this location we adjust our encoder so that the frequency (pulse) counter is outputting a zero count. That is, the binary number output is 00000000000000, or fourteen zeros. (The number of zeros indicates in this case that the maximum binary number the counter can read is 11111111111111, or fourteen ones. The decimal equivalent of this binary number is 2^{14}, or 16,384.) Now, let's swing the robot's arm in the opposite direction from the home point and swing it a full 360° until we are once again at home but arrived there from the opposite direction. Let's say that our frequency counter now shows a binary reading of:

10010011010001

The decimal equivalent of this number is 9,425. An encoder was selected to generate this number by choosing one having an encoder disk 3 inches in diameter and having 1,000 lines per inch. That is, this encoder will generate 1,000 pulses for every inch of movement around its circumference. When installed on the robot, it can be swiveled, in theory, to within ±0.001 inch of the desired location. Some sort of mechanical stop would have to be installed at home so that the robot's rotation can't go beyond this point, thereby causing the encoder reading to "roll over" to zero. If this happened, the robot's comparator output would become confused, causing uncontrollable rotation.

Let's make a further assumption that a positive error signal coming from the digital comparator will drive the swivel motor clockwise, while a negative error signal will cause the motor to drive the robot's base counterclockwise. Having established the full range of swiveling motion and binary output numbering of our encoder, let's swing the robot to some intermediate location to where the encoder output, according to the frequency counter, is now reading binary 01010011010001. If a binary number of, say 00001011011000 is entered into Figure 10-3's comparator by means of software or by keyboard, the arithmetic difference between this number and the binary number representing the robot's location becomes:

01010011010001
− 00001011011000
————————
01000111111001

Since the remainder is positive, the swivel motor will rotate clockwise and will continue to rotate clockwise until the remainder reduces to zero. In reality, because of the inertia of the robot's body, what is most likely to happen is that the motor will overshoot the zero–difference mark somewhat, therefore causing a negative error difference to be generated. The motor will then reverse itself in an attempt to reduce the error to zero once again, but again overshooting the zero mark somewhat. The error, however, should be smaller. The motor's polarity will once again become reversed, therefore starting the homing or hunting all over again, each time coming closer to the zero mark. This is similar to what takes place in an analog control system. However, because of the definite on or off state of each pulse being counted, there is a possibility that the motor could continue correcting itself either side of zero, oscillating back and forth perhaps as little as ± 0.001 inch. In other words, because of the very fine resolution of the optical encoder, and because of the possibility of the robot's swivel movement landing very near the edge of one of the optical lines on the encoder, the encoder could easily become confused as to whether or not to generate a pulse. It therefore oscillates between the on state and the off state indefinitely or until it receives a new command.

To prevent the hunting action just described from taking place, an artificial dead band can be created. If the resolution of 1,000 pulses per inch isn't needed, merely going to a lower resolution on the encoder will lessen the problem. If the resolution can't be sacrificed, viscous damping can be introduced to the swivel base. One method is to use a hydraulic pump that the swivel must continuously work against. Unfortunately, this is a waste of energy, but at least it is an effective cure without sacrificing accuracy. Another method is electronically to introduce a dead band. This is done by lifting the restrictions somewhat as to what constitutes a zero at the comparator's output. The comparator can be duped into thinking that if it gets to within binary ± 00000000000011 (± 3 pulses, or 0.003 in.), the error is as good as zero. Obviously, the price you pay for this artificial dead band is accuracy. Nevertheless, this is another effective way to prevent hunting at the zero error point.

Because of the similarities between the digital and analog forms of automatic control, let's take a look at the analog form to see if there are any advantages or disadvantages. The analog form by its very nature does not deal with discrete numbers or levels. In other words, there is no such thing as being close to the edge of an adjacent value or number as in a digital system. The voltage values are continuous. Consequently, there is no digital hunting due to the lack of dead band as in the case of very high-resolution digital systems.

Digital systems do have significant advantages over the analog systems. It is relatively easy to filter out extraneous electrical noise from digital systems. For the most part, electrical noise is analog in nature; therefore, building filters to separate this interference from the desired digital signals is relatively easy. Trying to separate analog noise from desired analog signals, on the other hand, can be difficult and frustrating.

Another advantage of digital control over analog control is that digital controls interface naturally to the computer. Since the computer is a digital device, it uses the same language as the controlling circuitry, therefore eliminating any need for analog-to-digital or digital-to-analog conversions.

10-2.1 Problems with Using Digital Encoding

With the use of digital encoding, such as the digital optical encoder, new problems are created concerning generating the output signals. In the aforementioned robotic control, we assumed that the swiveling of the robot's waist was an uninterrupted movement, the movement being registered by the number of pulses generated by the attached optical encoder. What happens if the robot accidentally strikes an object causing it to stop momentarily, reverse from the rebound, and then start up again in the original direction? The momentary rebounding or reversing adds additional pulses to the output count of the encoder. The robot could easily become lost as a result. It will have lost track of its command by reporting to the set point too early. This is where *absolute encoding* is used. Absolute encoding has the capability of keeping track of any reversals that may take place by subtracting any attempts to add additional counts due to these reversals. As a result, the robot can keep track of its location and know absolutely where its position is.

Another form of absolute encoding is where the encoder outputs a binary number instead of a single pulse. The binary number itself acts as an address, therefore telling the robot exactly what position it is in. However, there are problems with this system too. Look at Figure 10-4. Here we see a simple schematic of a typical encoder of the binary number-producing type. Looking at the left-hand illustration, notice what happens as the encoder wheel rotates counterclockwise. Theoretically, all three photocells should switch to whatever the new binary number is. According to the illustration, photocell 1 should remain low, photocell 2 should switch high, and photocell 3 should switch low, forming the binary number 010, or decimal 2. In reality, however, the three photocells will not switch to their new states simultaneously. There will be intermediate or spurious values of binary numbers generated as the division line on the encoder separating the two binary numbers 001 and 010 passes underneath the photocells. Other possible random numbers may be 000 and 011.

To get around this problem, a different sort of counting code is used. It's called the *gray code*. The gray code is so arranged that when counting up or down from one decimal equivalent to the next, there is only one binary digit or bit that changes to form that new number. Figure 10-5 shows a simple encoder wheel layout using the gray code. As a result, as the wheel is spun beneath the sensors, the only possible random numbers that could be generated during the numeral change are equal either to the present number or the number that is just about to be generated. This is a very simple scheme and a very effective one. There are

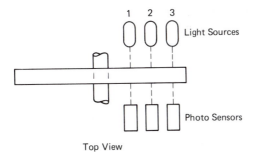

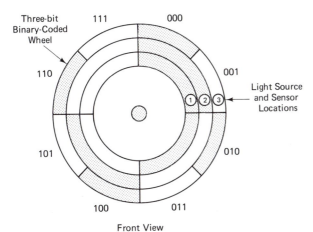

Figure 10-4   A three-bit binary encoder system.

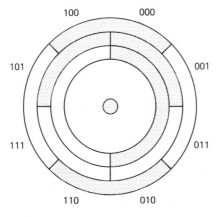

Figure 10-5   A three-bit Gray code wheel.

other methods used in addition to using the gray code, but this one is probably the most popular at the present. Whatever method is chosen, however, will require a means for converting the encoder output to a binary number to remain compatible with the rest of the operating system.

Virtually all robots used in industry are part of some larger automatic control system. The robots themselves, of course, are comprised of many automatic control systems working in concert to create the proper movements within the robot. There are basically three types of power sources that a robotic designer has to work with: hydraulic, pneumatic, and electric power sources. Each of these sources has its own unique hardware design problems as far as designing the needed control systems. Convincing arguments can be made for using any of these systems, and consequently, you will find all three sources used extensively in industry.

10-3 VISION SYSTEMS

Using vision in automatic control systems is perhaps one of the most exciting aspects of controls design today. It is a field that is developing rapidly and presents many technological challenges to the engineer. A vision system is a system that uses an image of an object to extract data and to cause another system or other systems to react. Stated more simply, a vision system is like a TV system that has been programmed to do some sort of function when it sees something it recognizes. Much of the vision system development has been performed by the military for military applications. Specifically, low-flying missiles have been given programmable vision systems to recognize certain flight paths and target areas. What the camera sees determines the reaction of the automatic control system controlling the missile's flight direction.

To understand how vision control systems operate, we must first understand how the vision or camera's optical system works. The best way to approach this topic is to first understand how a photographic camera works. Figure 10-6 shows a simple schematic of a still photographic camera showing the lens system that projects the image of an object onto the camera's film. In place of the film, however, let's place a specially made-up plate comprised of many tiny photo-cells. Each photocell will be either turned on or turned off depending on the object's light intensity at the location of that particular cell. If the cell is on, we can call its particular signal a 1. If the cell is off, that is, there is little or no light falling on the cell, the cell's output is a 0. (Remember, the 1 can represent any voltage level, not necessarily 1 volt, while the 0 represents some lower voltage, not necessarily 0 volts.) We then scan each row of the photocells using a special electronic circuit that interrogates each cell to determine which state or level the photocell is in. In other words, this scanning circuit will find out if the photocell is high (1) or low (0). We can then represent our object, in this case the letter A, as a

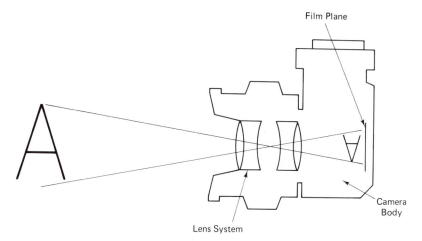

Figure 10-6   How a still camera forms an image.

series or mosaic of 1s and 0s arranged as in Figure 10-7. The resultant image is crude but nevertheless recognizable. Obviously, better resolution can be obtained by using more photocells. What is interesting about our vision system is that since we have digitized the image into 1s and 0s, much like the digitizing of computer data, we can easily store this data onto magnetic tape or disks just as we do with computer data.

   Now that we have a method of creating digitized images, we must look at a method for using this information in an automatic control system. Perhaps one of the most active areas of vision development right now in industry is the field of parts recognition. Figure 10-8 shows a typical example. Here we see a system

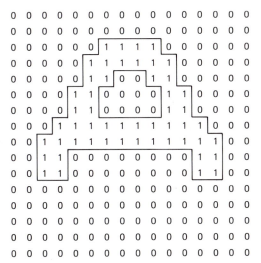

Figure 10-7   Digitizing an image. A bit map for the letter *A*.

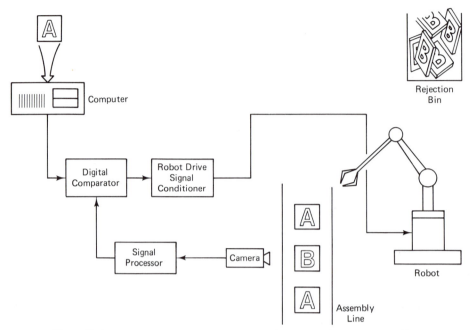

Figure 10-8 Automatic control system using vision system for parts recognition.

that has a capability to compare parts flowing past a checkpoint on an assembly. An overhead video camera takes a snapshot of each part, and the recorded image is sent to a processing center. At this location, the camera's image is compared to a prerecorded "standard" image of an identical part for a bit-by-bit comparison. If the two images compare, a signal is automatically sent to the robot telling the robot to let that particular part pass by. If, during the comparison process, the two images don't match, a signal is sent to the robot telling it to reach into the assembly line flow to pick out that part. The robot then places it in a holding area for further inspection.

In order to better understand the system, let's assume the following problem: Let's suppose that we are trying to separate As from Bs on an assembly line in a company that manufactures letter cutouts for an advertising firm. Again, we have a robot that will reach over to pluck out the As from the Bs when commanded to do so. Actually, our robot can be a simple pick-and-place type machine rather than the much more expensive articulated-arm type depicted in Figure 10-8. In any event, each letter is scanned by the camera and the image information is sent to a signal processing circuit. This circuit then converts the camera's electrical data into binary data and prepares the data for a direct match-up with the standard binary information stored in the computer. This comparison is done on a line-by-line, bit-by-bit basis, as mentioned earlier. If any of the bits don't match, as in the case of comparing an A to a B, or if the A is out of tolerance as a result of a portion or all of it being either too large or too small, an error signal is sent to the robot to reject the part. Looking at Figure 10-9, which shows the letter B and its bit map, it's fairly obvious that the third row of bits down from the top of the bit map doesn't match with the third row in Figure 10-7. The binary number is entirely different in value.

```
0  0  0  0  0  0  0  0  0  0  0  0  0  0  0  0  0
0  0  0  0  0  0  0  0  0  0  0  0  0  0  0  0  0
0  0  0 [1  1  1  1  1  1] 0  0  0  0  0  0  0  0
0  0  0  1  1  1  1  1  1  1  0  0  0  0  0  0  0
0  0  0  1  1 [0  0  0] 1  1  0  0  0  0  0  0  0
0  0  0  1  1  1  1  1  1  0  0  0  0  0  0  0  0
0  0  0  1  1  1  1  1  0  0  0  0  0  0  0  0  0
0  0  0  1  1  1  1  1  1  0  0  0  0  0  0  0  0
0  0  0  1  1 [0  0  0] 1  1  0  0  0  0  0  0  0
0  0  0  1  1 [0  0  0] 1  1  0  0  0  0  0  0  0
0  0  0  1  1  1  1  1  1  1  0  0  0  0  0  0  0
0  0  0 [1  1  1  1  1  1] 0  0  0  0  0  0  0  0
0  0  0  0  0  0  0  0  0  0  0  0  0  0  0  0  0
0  0  0  0  0  0  0  0  0  0  0  0  0  0  0  0  0
0  0  0  0  0  0  0  0  0  0  0  0  0  0  0  0  0
0  0  0  0  0  0  0  0  0  0  0  0  0  0  0  0  0
```

Figure 10-9 Bit map for the letter *B*.

In order for a proper image comparison to be made, it's necessary that the camera be placed directly overhead of the viewed object. This is so that there will be no visual distortion of the object's dimensions. Quite often, only silhouettes of the viewing object are used, since the outline represents the overall dimensional shape. The present technology is limited as far as being able to recognize details inside the object's profile lines; however, it will be just a matter of time before that will change.

The individual bit representing the light's intensity at a particular location in an image is called a *pixel*. Many times, it takes thousands of these pixels to compose just one image, depending on the image's physical size. It's possible, however, to use a completely analog system for parts recognition purposes. This means using ordinary video cameras. In some cases, they can be used just as effectively in an automatic control system as can the digital systems. It is obvious, though, that somewhere along the system's design layout it will be necessary to convert this information to digital information for computer analysis and storage.

10-4 CAMERA LEVELER SYSTEM

This next system we discuss is interesting because of its unusual design. In the entertainment field of movie-making and television, there is a need to maintain the levelness of the recording camera that has been mounted on a moving vehicle. Obviously, unless done for special effects, you don't want the camera to record any of the dips or rises that take place on not-so-smooth-or-level roads. Also, in the case of hand-held cameras, or cameras pushed on dollies, you want to be able to maintain levelness with the horizon at all times while shooting a scene.

One of the earlier solutions to this problem was strictly mechanical in its application. The solution involved using a gyroscopic device that spun rapidly to create the necessary opposing forces for maintaining a horizontal platform. However, considerable mass was necessary to produce the opposing forces, and consequently, the entire system was massive and cumbersome. In addition, some sort of on board power supply was needed to keep the gyroscope spinning. However, these systems were surprisingly effective, and some are probably still in use today.

A system that will enable us to maintain camera levelness is shown schematically in Figure 10-10. We see a camera-mounting plate that is pivoted in its

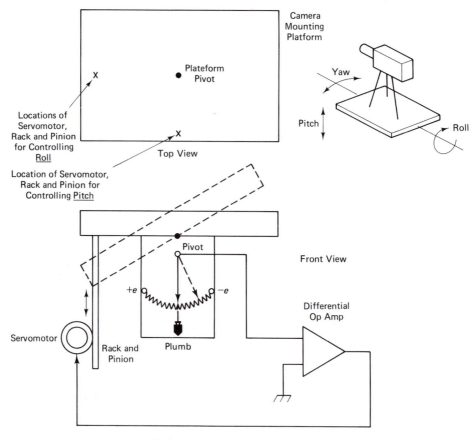

Figure 10-10 Camera leveling system.

center. The plate can rock in any direction about this pivot. Two servomotors are attached to the plate at the locations shown in Figure 10-10. Either or both motors, when energized, will cause the plate to tip. The plate's levelness is determined by two plumb sensing devices attached to the plate at right angles to each other. These devices are essentially nothing more than potentiometers whose wipers have become plumbs. When the plate is absolutely level, each pot outputs 0 volts to its respective servomotor. As the pots are tilted according to the tilt of the mounting plate, each pot puts out a proportionally varying DC voltage to its servo. This up-and-down tilting, by the way, is referred to as the object's *pitch*. An object's *roll* is the rolling over of the object; this is controlled by the pot which is at right angles to the first pot. Tilting in one direction causes either pot to output a positive DC voltage, while the opposite direction produces a negative DC voltage. The resulting actions of the two pots and their respective servomotors cause the mounting plate to remain level.

A system like this must be able to react quickly to sudden changes in its orientation. The system's components must have small time constants yet be able to move with virtually no hunting. The viewing audience watching the film or video tape will, otherwise, likely detect the slightest movement of any scene pitched relative to the horizon.

Up to this point, we have discussed a leveling system to accommodate a moving platform's orientation with respect to the horizon. But what about side-to-side motion relative to a fixed point out on that horizon? Is there any way to compensate for this side-to-side motion? One method involves using another potentiometer having a heavily weighted wiper that lies flat on the platform plate. Because of the massiveness of this wiper, its inertia will resist any sudden change in a sideways direction. Essentially, what happens is that the resistance windings of the pot will move beneath the wiper, which remains pointed straight ahead.

In order to accommodate both the up and down servomotor action for proper pitch and the sideways motion (this motion is often referred to as the *yaw* of a moving object), a modification to our camera platform has to be made. A second plate must be added that is mounted over the first plate and is free to rotate relative to the first plate. The sideways servomechanism is fixed to the first plate and pushes and pulls on the second plate to maintain its sideways orientation, as shown in Figure 10-11. The camera equipment is mounted on this second plate.

So far, we have discussed an all-electronic system using electrical servomotors. It is also possible to use a system consisting of either hydraulic or pneumatic actuators in place of the servomotors and their associated rack and pinion drives. One unique design using pneumatics involves the using of air bags mounted in place of the two servomotors under the bottom-most plate. However, for better reaction time and control, a third and fourth air bag are mounted opposite the first two. As one bag is inflating the opposite bag is deflating, and vice versa. However, in order to get this system to function properly, we have to have some sort of valve switching arrangement to shuttle air back and forth between the opposing bags. This is where the proportional-type servo-valve can be used. This is a

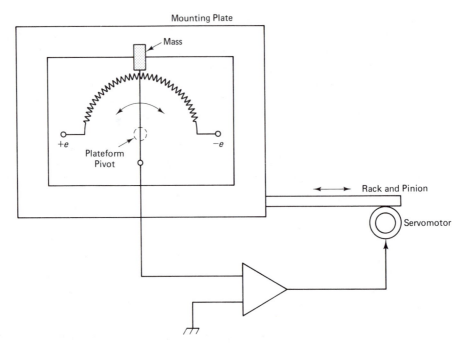

Figure 10-11 A system for controlling yaw. The mounting plate shown is mounted on top of the plate shown in Figure 10-10.

valve that is electronically controlled. It contains an internal shuttle or poppet whose position is precisely controlled by electronic pulses supplied from another circuit. The position of this shuttle determines the amount of air that is allowed to pass through the valve. The circuit producing the electronic pulses is told how many pulses to produce by the output of the attitude-sensing elements used earlier in our leveling system. The second plate that controls the camera's yaw can be controlled similarly. An added bonus of this type of system is that by using air bags, you obtain a high degree of shock absorbing capability that is not available with any of the other systems.

In place of the pneumatic air bags, hydraulic or pneumatic cylinders may be used. An advantage to using these types of actuators, especially the hydraulic actuators, is that tremendously heavy loads may be positioned using them. It is difficult, however, to obtain the high degree of positioning accuracy and control that you have with the servomotor. This is true in many automatic control systems. However, pneumatic and hydraulic systems tend to be somewhat less expensive, and with the advent of recent precision servo-positioning valves, hydraulic systems are becoming performance competitive. Keep in mind, though, that with any hydraulic or pneumatic system, you have to have a pump or compressor on board the system. This increases the noise, cost, and weight of the system.

10-5 PAPER PROCESSING SYSTEM

Like many manufacturing processes, the paper industry is heavily dependent on automatic feedback control systems. Automation has not only allowed the decrease in manufacturing tolerances on many products (resulting in higher quality), but automation has also removed operating personnel from undesirable or hazardous work situations which once required manual operation. Let's take a look at a typical paper manufacturing setup where we see the wet, newly formed paper entering its final stages of production. This particular manufacturing process is interesting because, unlike the others we have discussed, the response times (and therefore, time constants) are quite long. This is characteristic of any mechanical system using very large, massive system members for its operation. Because of this, considerable time may pass between the time the error signal is generated and the time a response is noted in the process. We may be looking at times measured in minutes or even portions of an hour. Nevertheless, we will still experience damping factors in the 0.4 to 0.7 range, hopefully, as we would like to have them. It's just that we will have to wait that much longer for our transient curve to develop. As a matter of fact, some system designers, when given the option of selecting a damping factor, will opt for values greater than 0.7 in the case of very slow systems such as this one, to reduce over-shoot from producing a high out-of-tolerance product in addition to a low out-of-tolerance product. (This practice is considered questionable by some system designers.) In reality, whenever a response is generated by a sensor, as we discuss shortly, the error signal is really not a step input. It is actually a gradually increasing or gradually decreasing error signal and is really considered to be a *ramp-type input*. This alters our overall transient curve form appearances from the ones we have been studying, but our overall system behavior is much the same. The study of ramp inputs is covered in advanced automatic controls texts. (Note: a ramp function is nothing more than the mathematical integral of a step function. And if you had a parabolic-type function for an input to your system, the parabolic function is merely the integral of a ramp-type input.)

Figure 10-12 shows the calender roller operation where the paper's thickness, moisture content, and weight quantities are all controlled automatically. Calender rollers are used for controlling paper thickness and humidity in the final manufacturing stages of sheet paper. In some instances, the final paper product is rolled onto a final take-up roller for storage before finally being shipped to the customer. What we are interested in are the feedback control systems used for controlling the paper's various aforementioned quantities.

Let's look at the first quantity, paper thickness. To control paper thickness, we must first have a transducer that can measure paper thickness while the paper sheet is moving rapidly from roller to roller. A roller caliper would do just that for us. This is a special caliper (Figure 10-13) having a set of rollers attached to each caliper end with each end positioned on either side of the moving sheet. The caliper rollers are adjusted so that the top roller rests on the bottom roller through

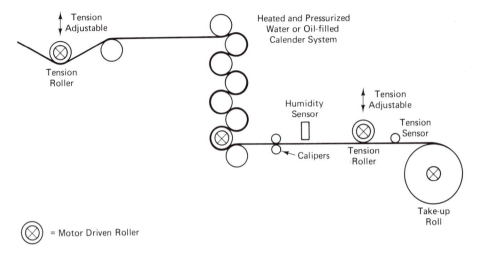

Figure 10-12 The calender control of a paper rolling operation.

the sheet. A spring tensioning mechanism is used to maintain proper press between the rollers at all times. The amount of caliper displacement due to the sheet's thickness is detected by a displacement sensor, consisting of a capacitive sensor, linear variable differential transformer, or optical encoder. The thickness is then automatically recorded and compared to the desired set point thickness. The calender roller pressure is then adjusted by varying the oil pressure inside the rollers. The rollers are actually flexible to allow their diameters to be changed.

Proper humidity must be maintained during the paper manufacturing process in order to maintain proper roller feed rates in the calender system. A too-wet condition will cause the paper sheet to become weak and to tear. A too-dry condition will cause static electricity to develop, setting up a potential for possible fire. Also, the curing of the paper sheet must be done at a specified rate to

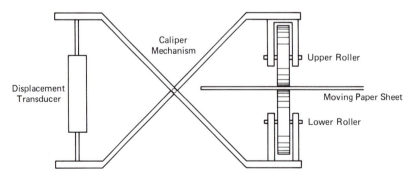

Figure 10-13 Roller caliper for measuring and monitoring sheet paper thickness during rolling operation.

maintain a certain fiber strength within the paper's material. Humidity affects this rate. Therefore, humidity sensors must be placed at strategic locations throughout the paper rolling process to monitor the relative humidity. Two types are used: One type consists of a horse-hair or human-hair hygrometer where the expansion and contraction of a hair strand is a measure of relative humidity (Figure 10-14). The other type is a chemical process where either the electrical resistance of a hygroscopic material (a material, such as calcium chloride, that absorbs water) is measured to determine the relative humidity, or there is a current flow generated due to a chemical ion exchange process (Figure 10-15), or a dielectric constant varies proportionally with humidity. The signal output information from the humidity sensors is sent by telemetry back to the paper driers, tension rollers, calender, and to the plant's environmental air conditioning systems for the proper humidity adjustments. In the case of the calender system, the calender rollers' temperature can be regulated by controlling the heating or cooling of oil inside the rollers. This, in turn, controls the relative humidity or moisture content of the paper sheet. It's obvious, at least in this particular control system, why response times could be considered slow compared to what we have been studying. Heating or cooling the volume of oil used in the calender could produce some rather large time constants.

The weighing of paper while still in motion coming off of the calender rollers can be an interesting measuring and control problem. One method is quite simple but disruptive to the operation of the manufacturing system. The system is simply shut down momentarily while sheet samples are cut from the paper stream and weighed. A less disruptive system has the take-up roller weighed as it revolves. It is necessary to know exactly how much length of paper has been taken up on the roll, but that figure is not difficult to obtain with the proper metering devices. What is difficult, though, is the adjusting of any needed paper weight. The response time for this operation is extremely long, and because of this, this particular operation is done manually rather than automatically. While the weighing operation is monitored regularly, the weight adjusting is usually done manually by

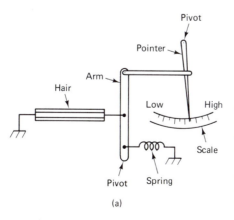

Figure 10-14(a)   Schematic of an hair hygrometer used for measuring relative humidity. (*Ronald P. Hunter, AUTOMATIC PROCESS CONTROL SYSTEMS: Concepts & Hardware, 2nd. Ed., copyright 1987, p. 103. Reprinted by permission of Prentice Hall, Inc., Englewood Cliffs, N.J.*)

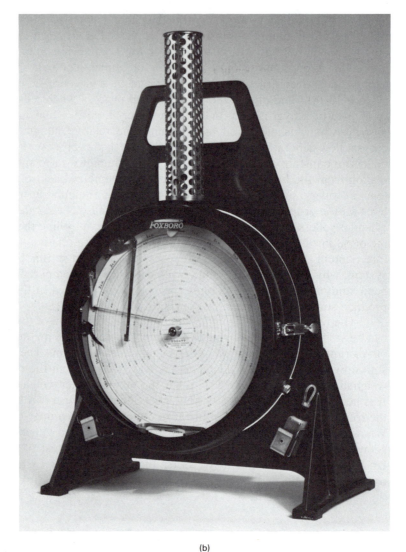

(b)

Figure 10-14(b)   Commercially constructed hygrometer. (*Photograph courtesy of The Foxboro Company, Foxboro, Mass.*)

adjusting the pulp mix back-up at the front of the manufacturing operation. This is assuming, of course, that the sheet is of the correct thickness. Otherwise, the weight variance could have been the result of a change in the paper's thickness.

There are areas in this manufacturing process that demand fast response times for the prevention of system failure. For instance, fast response times are needed to compensate for changes in sheet tension as the sheet flows through the

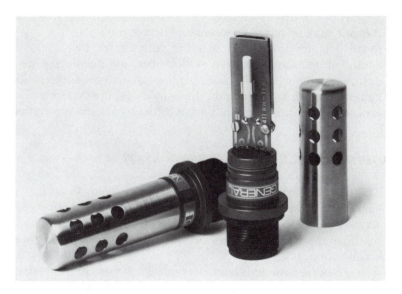

Figure 10-15   A sulfonated polystyrene ion-exchange sensor for measuring relative humidity. (*Courtesy of General Eastern Instruments Corp., Watertown, Mass.*)

calender rollers. If the tension is too great, there is a risk of sheet breakage. If the tension is not great enough, the take-up roller will become too loosely wound with the paper product. Sheet tension is adjusted and maintained by electrically or hydraulically driven tension rollers shown in Figure 10-12. A tension sensor in the form of a broad pressure-sensing roller that rides on the moving sheet detects any change in sheet tension. Any change is instantly translated to a compensating up or down movement of the tension rollers. The system behavior of this particular automatic control system closely resembles the classic systems that we discussed in earlier chapters.

Another important automatic control system has to do with variable roller speeds. As the paper sheet comes off of the calender onto the take-up roll, the take-up roller must be able to modulate its rotational velocity to match that of the sheet. Assuming that the linear velocity of the sheet remains constant coming from the calender, the take-up roller must be able to reduce its velocity proportionally as the roll becomes larger. This can be done with a sensor that senses the roll's increasing diameter. This can be a roller-caliper that rides on top of the take-up roll and whose displacement from the center of the roll is detected by a linear encoder. As this linear displacement increases as the roller becomes larger, the rpm of the take-up roller's drive motor is proportionally decreased. This is to compensate for the increase in linear velocity at the roller's outer edge where the sheet is coming onto the roll.

We have discussed only five automatic control systems of various types in this one area of paper processing. There are obviously many more that can be found operating elsewhere in other parts of the manufacturing system. Some control systems are so simple in their operation that during the design phase, the designer probably didn't take the time to analyze mathematically the system's behavior before having it built. In many cases, the system's behavior isn't all that critical, or it can be easily adjusted through trial-and-error after being built.

10-6 LIQUID FLOWRATE SUPERVISION SYSTEM

This particular topic covers a lot of territory in controls design. Flowrate supervision is used in controlling liquid flowrates or amounts in pipelines for chemical supplying or foods operation, or perhaps for controlling hydraulic fluids in a machine operation. Each installation of the automatic control system requires a unique control design.

To begin with, let's look at a typical flowrate control system shown in Figure 10-16. There are two important pieces of hardware in this system. One is the servo-controlled valve used to vary the liquid's flowrate in the supply pipe, and the other is the flowmeter used to monitor the liquid's flowrate. First, let's look at the servo-controlled valve. Figure 10-17 shows a diagram of a simple valve controlled by a servomotor. The valve's seat is designed so that for a given amount of rotation of the valve's stem produced by the servomotor and its associated gear transmission, a proportional flow of liquid is allowed to pass through the valve. In some valve designs, provision has been made for a clutch to allow the motor to

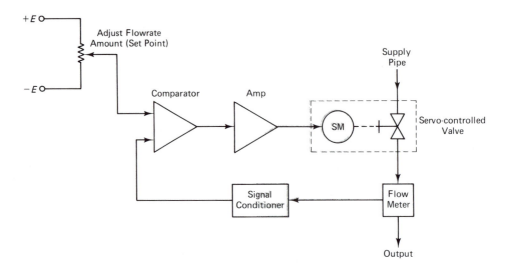

Figure 10-16   A typical flowrate controller system.

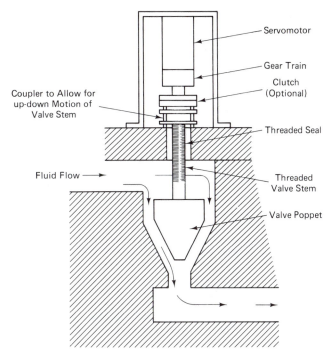

Servomotor

Gear Train

Clutch
(Optional)

Coupler to Allow for
up-down Motion of
Valve Stem

Threaded Seal

Fluid Flow →

Threaded
Valve Stem

Valve Poppet

Figure 10-17   A servo-controlled valve.

continue spinning if the valve seats completely or if the valve opens completely. The clutch mechanism protects the motor from becoming overheated should either travel extreme be reached. In the case of very small valve designs, the clutch is omitted and the motor is allowed to stall, having been designed to withstand the increase in operating temperature resulting from the stalled condition. In either case, there must also be a mechanical provision for accommodating the upward and downward movement of the valve stem as the stem rotates. Some valve designs use a linear motor rather than a rotational motor to move the valve stem. Each valve design has its own unique advantages and disadvantages. For the purpose of simplifying our discussion here, we use the valve design in Figure 10-17 to explain flow control. It's probably the easiest design to follow and to understand.

Our second important piece of hardware in the flow control system is the flowmeter transducer. There are many designs to choose from for this application, one of which is shown in Figure 10-18. This particular model contains a rotating turbine whose rotational velocity varies with the flowrate of the liquid inside the pipe. A variable reluctance transducer picks up the passage of the turbine's blades and transmits a variable frequency (FM) AC wave whose frequency varies directly with the turbine's speed. The FM signal is then demodulated to produce a variable DC voltage that is fed into a comparator for system controlling.

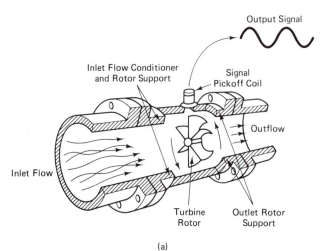

Output Signal

Inlet Flow Conditioner
and Rotor Support

Signal
Pickoff Coil

Outflow

Inlet Flow

Turbine
Rotor

Outlet Rotor
Support

(a)

Figure 10-18(a) Schematic of turbine flowmeter with installed sensor. (*Dale R. Patrick/Stephen W. Fardo, INDUSTRIAL PROCESS CONTROL SYSTEMS, copyright 1979, p. 122. Reprinted by permission of Prentice Hall, Inc., Englewood Cliffs, N.J.*)

(b)

Figure 10-18(b) Commercial turbine-type flowmeters. (*Photograph courtesy of The Foxboro Company, Foxboro, Mass.*)

Another frequently used flowmeter design is shown in Figure 10-19. This particular design utilizes a target that is suspended in the flow of the liquid or gas in the pipe. The force of the fluid on the target causes it to deflect backward. The amount of deflection is registered by a strain gage or other displacement-type

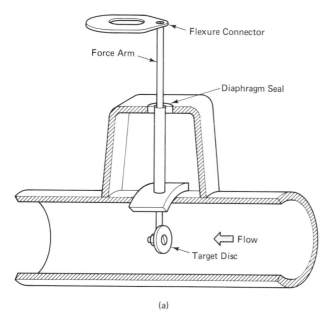

Flexure Connector

Force Arm

Diaphragm Seal

Flow

Target Disc

(a)

Figure 10-19(a)   Schematic of a target flowmeter. (*Dale R. Patrick/Stephen W. Fardo, INDUSTRIAL PROCESS CONTROL SYSTEMS, copyright 1979, p. 114. Reprinted by permission of Prentice Hall, Inc., Englewood Cliffs, N.J.*)

(b)

Figure 10-19(b)   Commercial target-type flowmeter. (*Photograph courtesy of The Foxboro Company, Foxboro, Mass.*)

sensor which, in turn, produces either an analog AC or DC voltage output, or a digital signal output if desired. In the case of our flow system in Figure 10-16, we would want an analog DC voltage output so that it would be compatible with the set point voltage at the comparator.

At this point, you may have recalled that we discussed a similar automatic control system back in Chapter 9. In Section 9-4 we discussed a remote valve positioner system that allowed a user to select a valve opening position that controlled a particular flowrate. The position had to be selected by some predetermined flowrate measurement. Once the valve opening position was selected, if that flowrate ever changed, the selected valve opening would still remain constant. In other words, the system was not designed to compensate for variable liquid flowrates inside the pipe. It was designed instead to position a valve stem and to readjust itself to hold that position if some outside force caused that position to change. Now, let's compare this system to the one shown in Figure 10-16. The system in Figure 10-16 does not have the capability to set a particular valve opening setting. Instead, for the system's set point, the user inputs a desired liquid flowrate. The control system then opens or closes the valve; that is, it self-adjusts until the desired flowrate is obtained. Notice that if the flowrate should change in the pipe due to a change in liquid supply conditions upstream in the pipe, the system can readjust or adapt its valve setting. The flowrate transducer would sense this change and command the valve to readjust itself in order to maintain the desired flowrate amount. This type of system is an *adaptive-type* automatic control system. We talk more about this type of system later in this chapter.

Now that we have become familiar with the hardware in our flow control system, let's go through its operation to make sure we understand how it operates. Also, we can then better appreciate the basic difference between this adaptive type of control system and the nonadaptive system in Chapter 9. Looking again at Figure 10-16, the system's operator sets the set point controller at the liquid flowrate level he or she wishes to maintain in the piping system. The controller's dial face on the control panel would most likely be calibrated in something like gallons per minute or cubic feet per second. However, in reality, the controller would be outputting a DC voltage which would be proportional to the dialed-in flowrate setting on the control panel. Let's assume that our system is just starting up with no flow in the piping system. With the set point data now dialed in, the output voltage goes to the amplifier where it is amplified and produces a proportional output power signal to the servomotor attached to the servo's valve. This amplified voltage causes the valve to open, allowing liquid to flow through. The flowrate transducer then senses the flow and generates a proportional output DC voltage, which is sent back to the control system's comparator. Depending on the amount of flow passing through the transducer, the magnitude of its output signal is determined. This, in turn, determines the magnitude of the error signal coming out of the comparator. The closer in magnitude the two voltage values are coming from the set point controller and the flow transducer, the smaller the error voltage supplied to the servovalve. The servovalve will continue to rotate until the error signal is zero; that is, until the transducer's output matches exactly the output of the set point controller. Notice what happens if, because of the servovalve's inertia, the valve opens too far beyond the time when the error signal becomes zero (as could actually happen in some system designs) and the error signal reverses polarity, causing the servomotor to reverse its rotation. In other words, the servomotor will always try to correct itself until eventually it settles at the proper flowrate setting. This is the same output characteristic we experienced in many of our other control systems, and so this system is fairly typical. Hunting should be kept to within the typical amounts associated with damping factors that are in the 0.4 to 0.7 range.

10-7 BOTTLE FILLING PROCESS

The food industry certainly has generated some very interesting challenges to the automatic controls engineer and technician. Automating mixes and batches of food ingredients can be extremely difficult where consistency is of utmost importance. Figure 10-20 is a schematic of a bottling operation for soft drinks. In this

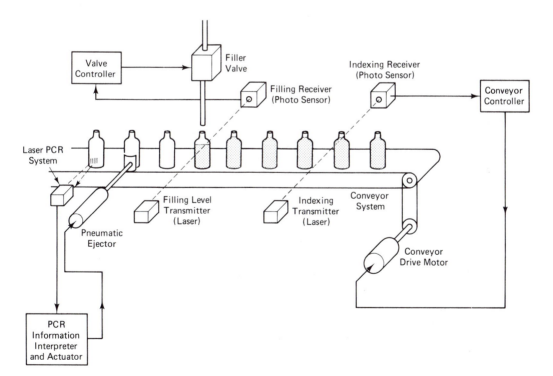

Figure 10-20 An automated bottle-filling operation.

schematic, we see bottles arranged in a straight line on a conveyor system. In reality, the bottles are usually on a circular rotating conveyor with an entrance provision for inserting empty bottles and an exit for filled bottles. The bottles can be filled by liquid level, by weight, or by a metered flow amount being supplied to the bottles. The system shown in Figure 10-20 uses a liquid level detector system. In this system, a light source projects a beam of light through the bottle being filled. This beam falls onto a photocell that produces a signal, which is sent to a valve controller. The valve controller responds to this signal by opening the supply valve over the bottle to be filled. When the filling liquid reaches a predetermined level inside the bottle, the light beam is cut off. The photocell, as a result, turns off, causing the valve controller to turn off, which, in turn, tells the supply valve to close. When the valve does close, the conveyor then automatically begins indexing. The indexing light source is then energized and the next bottle moves into position under the filler tube. As soon as the indexing light beam is cut by the next bottle, the conveyor is stopped. The filling light source is energized, causing the next bottle to be filled and thus repeating the operation.

The important design problem to overcome in this particular operation is the logic needed to tie together the two control operations so that they operate at the proper times. You certainly don't want the filling operation to be taking place while the conveyer is moving, or vice versa. Also, provision must be made for the likelihood that a bottle may be missing from the conveyor line. You will want to be able to insert a control circuit that detects that possibility. You may also want to be able to tell if the correct bottle is being filled. If you are running a 12-ounce bottle operation and a 16-ounce bottle inadvertently got placed in the bottle line on the conveyor, you will want to be able to recognize and cope with that situation. This is where a vision system may be used, although simpler systems could be devised to do the same bottle inspection. This depends on how radically different the various bottle types may be that may appear on the conveyor lines. The system shown in Figure 10-20 uses a Product Code Recognition or PCR laser system. This system is virtually identical to the system often seen at the checkout lanes of supermarkets for ringing up grocery prices on the cash register. The bar code printed on each bottle is scanned by this system, and if the improper bottle size data is detected, a pneumatic cylinder or ram ejects the bottle from the conveyor line.

In place of the level detection system used for the bottle filling operation, a weighing system may be used. As each empty bottle is positioned beneath the filling tube, the bottle is made to be positioned on a small platform containing either a strain gage or some other deflection-type transducer. This can be a linear differential amplifier or an optically encoded displacement transducer. When the proper signal voltage is generated by this device representing the proper filling weight, a voltage comparator circuit then closes the filling valve and indexes the conveyor to the next bottle.

Using a metered volume filling operation would be very similar to the system described in Section 10-6. In this system, a flow transducer would be used to measure out a specified volume of liquid. Preferably, this transducer would be a *positive displacement* type. That is, the transducer is designed to measure a specific volume of fluid before discharging the fluid from its outlet. An example is the rotating-vane meter (Figure 10-21). This is the type of meter used in gasoline

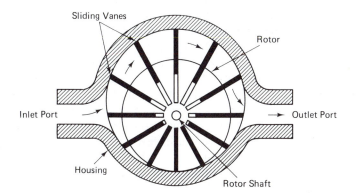

Figure 10-21   A positive displacement metering device used for filling operations. (*Dale R. Patrick/Stephen W. Fardo, INDUSTRIAL PROCESS CONTROL SYSTEMS, copyright 1979, p. 127. Reprinted by permission of Prentice Hall, Inc., Englewood Cliffs, N.J.*)

pumps for metering dispensed gasoline. These transducers are extremely accurate. Since the exact volume of liquid that can be trapped between the rotating vanes of this device is known ahead of time, it's merely a matter of counting the vane's number of revolutions to determine the exact volume of metered liquid. Accuracies greater than 0.1% are not uncommon for this device. Knowing the number of revolutions ahead of time needed to produce a certain volume, the total revolution count can be converted to a DC voltage and sent to a comparator circuit for controlling the filling valve. This voltage conversion may be easily handled by a digital-to-analog converter.

The filling operation, the conveyor indexing system, and the bad bottle ejection system may all be thought of as bang-bang servo systems, since they behave much like the one we analyzed in Section 9-8. That is, these control systems are either all the way on or all the way off. This is a fairly typical characteristic of the many control systems used in process control applications. They are generally cheaper than the more complex linear servo systems and are easier to maintain.

10-8 TELESCOPE TRACKING SYSTEM

This next example of a process control automatic control system is a fairly recent innovation in astronomical research. It is a unique control system and has saved labor and increased the accuracy of telescopic tracking.

We are all aware that the earth rotates daily on its axis creating the illusion that the sun, the stars, and the planets rotate around the earth. Whenever the astronomer wishes to photograph a portion of the sky using an optical telescope, lengthy time exposures are required. These time exposures may run several hours, and in some instances, even days. The reason for these long exposures is to gather enough light in order to record an image. Many of the objects that are photographed are hundreds of thousands or millions of light years away, and by the time that light reaches earth, it is understandably quite weak. (One light year is approximately 5.87 trillion miles.) What we may not be aware of, however, are the intricate mechanisms needed to keep the telescope with its installed camera film pointed at the object or objects being photographed. These mechanisms essentially allow the earth to rotate beneath the telescope while the telescope remains fixed, continuously pointing at the desired object. While this description is somewhat flawed since the earth really doesn't rotate beneath the telescope, because the scope is rigidly attached to it, it is partially true. The way most telescopes are presently mounted on their earth-fastened mounting bases allows this to happen. These mounts are called *equatorial mountings*. In order to appreciate the control system we study next, it's necessary to understand these mounting systems.

Figure 10-22 shows a telescope mounted on its equatorial mount. Because all objects in the sky appear to rotate around a point located close to the north star (since this is the point in the sky where the earth's rotational axis is pointed), it becomes a matter of designing a two-axes mount, one of which points at the north

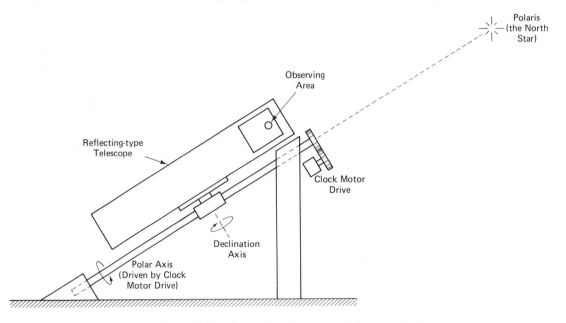

Figure 10-22   The motions of an equatorial-mounted telescope.

star. This axis is called the scope's *polar axis*. A second axis, called the *declination axis,* allows the scope to swing away from the equatorial axis and point to any other spot in the sky. In order to track an object automatically at any of these points, it becomes a matter of attaching a drive motor to the polar axis to allow the scope to slowly rotate about this axis. Of course, the motor's speed must be precisely synchronized with the rising and setting time of the objects in order to keep up with their movement. This motor is called the scope's *clock-drive*. While this type of system works acceptably well, it is still an open-loop control system that is vulnerable to outside influences—influences such as accidental bumping of the telescope frame by operating personnel, variable refraction of the object's light reaching earth due to sudden variations in the earth's atmospheric density, and rapid motion of the object in the sky due to its being near the earth or because of very high orbital velocities. All of these variations can be compensated for using present control designs, but not without requiring complex electromechanical system designs and requiring constant, almost minute-by-minute, supervision. Typically, a technician is required to make small minor touch-up corrections to the clock-drive using what is called a *slew motion pad.* He or she uses a smaller guide scope, similar to a target scope of a rifle, to follow a bright star, or guide star, which is usually located near the photographed object. This guide scope is attached to and is parallel with the main frame of the larger scope. In other words, the scope is slewed in whatever direction is necessary to bring it back on target should either it or the object wander off.

The following system is a simplified version of a closed-loop automatic tracking system now being used in some observatories around the world (Figure 10-23). It is comprised of two sets of photocells mounted in the guide scope. Each pair of photocells has its outputs sent to a differential amplifier, power amplifier, and motor. A second closed-loop system, which is a positional system, is also shown which is used for initially pointing the telescope at the correct object. The entire system works like this: The object to be photographed is often invisible to the scope's operator. However, its coordinates are usually precisely known. (These coordinates are like the coordinates of an object located here on the earth's surface where two x–y coordinates are needed to form an intersection denoting where the object is located.) A guide star is then selected that's known to be near the object of interest. That object's positional coordinates are then entered into the computer's control system. The computer corrects the coordinates for the time of day (converting the right-ascension coordinate to what is referred to as the object's *hour angle*) and sends that information to the scope's two setting circles. This information is compared to the existing setting circle values and, if an error voltage is present due to different setting circle values, the scope is automatically moved by means of that error voltage until the error voltage drops to zero. This is done on both axes. Because of the massiveness of the scope, little if any hunting has been designed into the system. Damping factors are usually fairly large for this size of system. Response time is also quite slow. Once the scope is positioned on its target, a guide star is found and centered on the guide

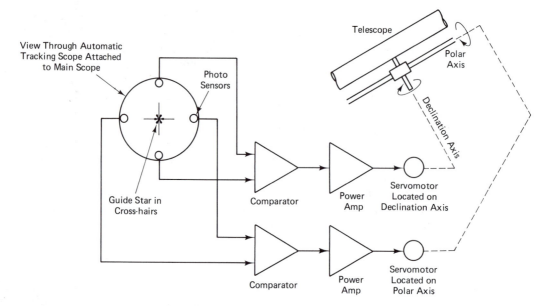

Figure 10-23   An automatic guiding system for telescopes.

scope's cross-hairs. At this point, since the opposite photocells are receiving equal light from the guide star, the comparators for each photocell pair will be generating no error signals. However, should the guide star be off center from the cross-hairs only slightly, one of the photocells will receive more light than the other. Consequently, an error signal is generated in the otherwise balanced comparator, and the appropriate drive motor is energized to bring the guide star back on center or until the error once again drops back to zero voltage. Most likely, during an automatic correction both comparators will be affected.

It is interesting to note that some telescope drive manufacturers have done away completely with the polar axis. Instead, they have mounted the scope on what is called an *altazimuth* mount. This type of mount is very simple to construct and is similar to the mount used on surveying equipment. Here, a computer continuously drives *both* what are now the altitude axis and the azimuth axis of the scope in an effort to keep the scope on its target. This is a somewhat revolutionary design and has worked out very successfully. It eliminates the difficult task of constructing a polar axis that is exactly parallel to the earth's rotational axis. Automatic slewing works very well with this system and is almost mandatory, since the polar axis is no longer used.

10-9 DIE CASTING PROCESS CONTROL SYSTEM

Die casting is a very old and interesting process. While the process has been around for a long time, it has only been since the 1960s that the die casting industry has developed an automated process for producing castings. And it was as recent as the early 1980s that it was proven that molten aluminum, a material often used in die casting, required a certain insertion velocity into the die in order to produce a good die-cast part. Let's look at a typical die casting machine and see how it is presently being automated.

Figure 10-24 shows a typical die casting machine and all its associated parts. What we are particularly interested in is referred to as the machine's *shot end*. This is the end where the molten metal is ladled in by a robot and is then "shot" or rammed into the die. The robot ladles a measured amount of molten aluminum into the pour-hole (not shown in Figure 10-24) of the machine's cold chamber. This is similar to the loading of a cartridge into the breach of a cannon, except in place of a solid cartridge material being inserted, we're inserting a liquid material. After the ladling, a hydraulic ram immediately shoots the metal through a hole in the die casting machine's front plate and into the die cavity, located inside the two die-halves situated between the front plate and the machine's traveling plate. The traveling plate is clamped against the die halves and front plate by means of a mechanical toggle system that can generate tremendous clamping forces anywhere in the range of 200 to as much as 3000 tons, depending on the size of the machine. The toggle system, during this clamping procedure, pushes against the machine's back plate. The back plate in turn stretches the four tie bars (only two are seen in Figure 10-24) that are fixed to the front plate. This stretching causes the two die-halves to be further "squeezed" together between the traveling plate and the front plate while the "shot" is being made. After a short curing or freezing time, the toggles are released and the traveling plate moved back, causing the die-halves to part. A second robot then reaches in between the two die-halves and extracts the newly made part.

Now that we have discussed the overall action of the die casting process, let's look at some of the automation techniques being used. First, let's go back to the machine's shot cylinder. Earlier, we suggested that the molten aluminum being shot into the die must be delivered to the die at a particular velocity by the shot cylinder, or ram, in order to make a good part. This velocity is called the *critical velocity*. To obtain and hold this critical velocity, we have to employ a method that enables us to monitor and to adjust automatically the shot cylinder's ramming velocity to maintain this velocity. To monitor the velocity, an optical encoder can be attached to the shot cylinder's piston rod to measure its forward speed. Since the optical encoder generates digital information in the form of pulses, this can be converted to a DC voltage and compared to the desired dialed-in velocity, which is the desired critical velocity. If an error voltage is created at the comparator, this voltage can readjust a hydraulic proportional valve to either decrease or increase the shot cylinder pressure for correcting the next shot. Analog-type transducers can also be used in place of the digital optical encoders; however, using digital signals wherever possible within the system will make the control system's signals as immune to electrical noise as possible. Unfortunately, electrical noise is extremely prevalent around many

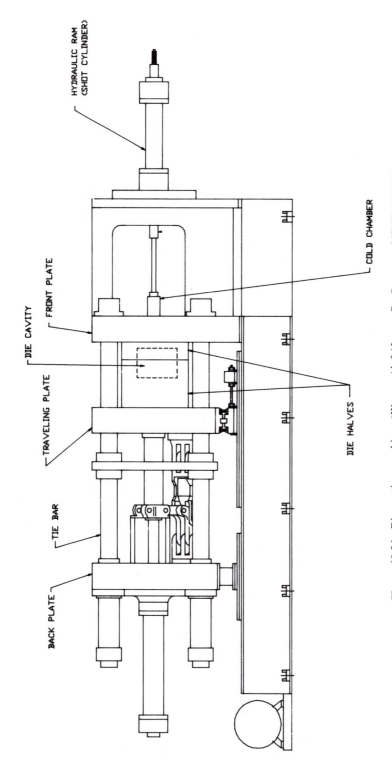

Figure 10-24 Die casting machine. (*Wayne Alofs/James R. Carstens, MECHANICAL MAINTENANCE AND EVALUATION OF DIE CASTING MACHINES, copyright 1987. Reprinted by permission of the Society of Die Casting Engineers, River Grove, Ill.*)

die casting shops because of electrical welding and heavy reliance on electro-mechanical relays.

Now, let's look at the so-called closing end of the die casting machine. This is the opposite end of the machine where the rear end of the toggle system is anchored to the machine's back plate. In order to monitor the amount of force being generated within the tie bar system and to insure that all four tie bars are evenly loaded, we need to install transducers that can sense deflection in the tie bars. The logical choice would be to use strain gages; however, it is difficult to maintain calibration with strain gages in an industrial atmosphere, not to mention their temperature susceptibility (of which there is plenty in a die casting shop). The linear variable differential amplifier or the optical encoder would be good choices instead. The optical encoder produces a digital signal and would have that particular advantage over a linear variable differential amplifier system. Both systems are shown in Figures 10-25 and 10-26, respectively. With these systems, a particular locking tonnage can be dialed into a comparator and the machine then locked up on a die. The stress in each of the four tie bars can then be compared to one-fourth of the inputted desired tonnage. The outputs of the four tie bar indicators can then be compared to the dialed-in data and the machine adjusted accordingly. This adjustment is performed by either tightening or loosening the appropriate tie bars to obtain the desired balance loads. This tightening or loosening

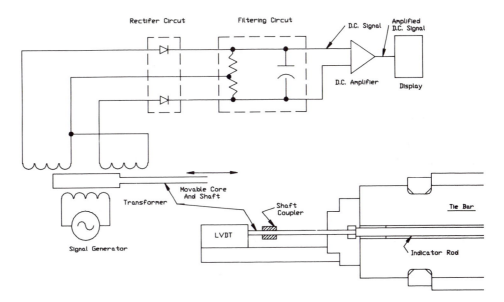

Figure 10-25 Using the LVDT as a stress indicator. (*Wayne Alofs/James R. Carstens, MECHANICAL MAINTENANCE AND EVALUATION OF DIE CASTING MACHINES, copyright 1987. Reprinted by permission of the Society of Die Casting Engineers, River Grove, Ill.*)

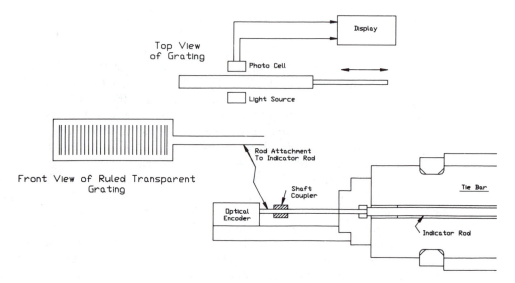

Figure 10-26 Using the optical encoder as a stress indicator. (*Wayne Alofs/ James R. Carstens, MECHANICAL MAINTENANCE AND EVALUATION OF DIE CASTING MACHINES, copyright 1987. Reprinted by permission of the Society of Die Casting Engineers, River Grove, Ill.*)

process can be fully automatic and can be done by energizing a gear drive system that screws the tie bars into or out of threaded mounts. One or all the tie bars can be moved, depending on which tie bar requires this adjustment.

Much development is needed yet in automating die casting operations. This is a relatively new field and is of particular importance to the automotive industry, since they depend heavily on the die casting process for the manufacturing of their components.

10-10 A PROPORTIONAL PNEUMATIC CONTROLLER

Proportional pneumatic controllers are used to create proportional pneumatic signals to actuators requiring intermediate positioning capabilities. An example would be the controlling of a damper setting on an air conditioning system, where the damper would have to report to an infinite number of intermediate positions for mixing air in the duct work depending on the heating or cooling demands of the system (Figure 10-27). The damper's setting would be controlled by an attached pneumatic cylinder whose plunger extension would be controlled by a proportional pneumatic controller.

Proportional pneumatic controllers have, to a great extent, been replaced with electronic proportional controllers. However, there are numerous places in industry where there is still demand for them. Pneumatic controllers are used in

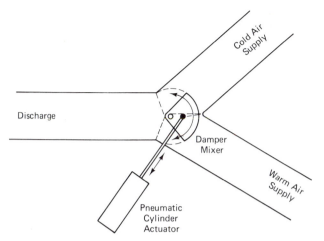

Discharge

Damper
Mixer

Pneumatic
Cylinder
Actuator

Figure 10-27 A proportional damper
mixing system in an air conditioning
system.

volatile atmospheres where operating electrical devices would be too dangerous.
Also, the cost of one of these controllers is considerably less than a comparable
electronic controller.

Figure 10-28 is a schematic of a typical proportional pneumatic controller
being used in conjunction with a pneumatic temperature sensor. The system
works like this: A temperature sensor is installed in a room having ductwork that
supplies both heated and chilled air to that room. The amount of heating or
cooling air is determined by a damper blade that is actuated by a pneumatic
cylinder. If the blade is caused to rotate counterclockwise, more cooling is allowed into the supply ducts. If the blade rotates clockwise, more heating is supplied to the supply ducts. The temperature sensor is a bimetallic temperature
element attached to an air flow control valve. If the temperature happens to be
rising, the bimetallic strip expands, therefore restricting the airflow valve and
allowing less air to flow into the bellows. A falling temperature causes the bimetallic strip to contract, therefore opening the air valve and allowing more air to
flow into the bellows. The desired temperature, or set point controller, is a spring-
loaded screw adjuster. The system is calibrated so that when the bellows is exactly offset by the spring-loaded screw adjustment, that particular temperature
exists in the supply duct. Let's assume now that the temperature rises above the
set point. This will cause less air to enter the bellows, causing the blocking lever
to move left in the diagram and blocking more of the air-bleed port. This, in turn,
causes more air to be bypassed into line *A*. At the same time, though, more air is
also bypassed into line *B* and into the fluidic amplifier. The amplifier's diaphragm
expands, pushing the actuator control valve to a more closed position. The actuator cylinder causes the piston to move downward in the diagram which, in turn,
causes the damper to rotate counterclockwise, thereby increasing the cooling.

Let's assume now that the temperature sensor senses a falling temperature
in the supply duct. The falling temperature causes the bimetallic strip to contract,

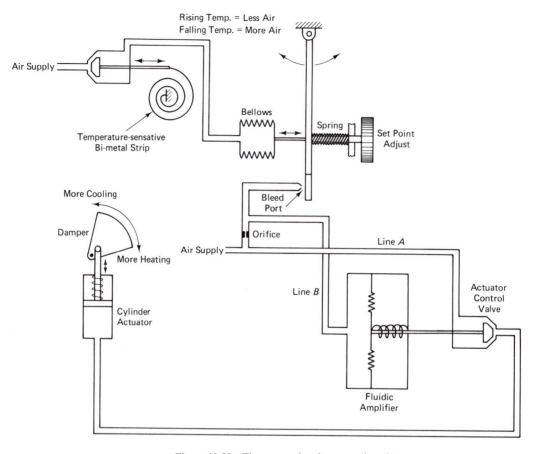

Figure 10-28   The proportional pneumatic valve.

therefore opening the air supply valve and causing the bellows at the blocking lever to expand. This causes the blocking lever to swing right in the diagram. This creates a greater bleed-port bypass. As a result, less air is delivered to the fluidic amplifier through line *B* and less air also to be delivered by line *A* to the actuator control valve. The fluidic amplifier's diaphragm now relaxes, therefore opening the actuator control valve further. More air will now flow to the actuator, causing the piston to move upward in the diagram and moving the actuator clockwise. This action increases the heating supply to the supply duct.

   This type of closed-loop temperature control is very effective. It's inexpensive and quite rugged. Its only major disadvantage is having to have a continuous air supply along with extensive piping. Periodic maintenance is needed to assure a leak-free system.

10-11 A REMOTE CONTROL ANTENNA ROTATOR

The final process control system that we discuss here is a remote control antenna rotator system. This is a system used to enable an operator to position a communications antenna from a remote location so that the antenna can be directed at a distant receiving antenna. With the closed-loop servo system, the antenna will self-correct itself if gusts of wind or any other disturbances should happen to attempt to move it off course. Figure 10-29 shows a diagram of this system. The system uses a CX–CT 400 Hz synchromechanism for the input command signal and checking the output position of the antenna. This is a very precise system and has many applications in other fields of control. It is quite similar to the remote valve positioner example discussed in Section 9-4. The reason for the 400 Hz CX–CT system is only because this system was used frequently in aircraft control. The 400 Hz frequency is used extensively aboard aircraft and by the military in general.

Taking a look at the system, we see that a DC motor is used for turning the antenna. The reason for this is that DC motors are capable of much higher torque

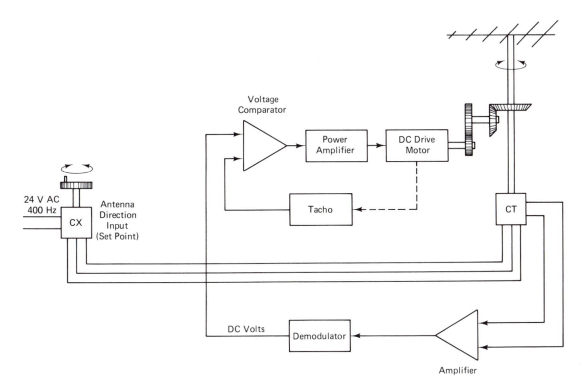

Figure 10-29   An antenna rotational controller.

outputs as compared to AC induction motors. The disadvantage in using this type of motor, however, is that some means must be used to convert AC to DC to make the motor control compatible with the AC CX–CT system. This is the reason for the demodulator. The output of the CT is demodulated so that its DC equivalent voltage can be eventually compared with the DC output of the tachometer mechanically coupled to the antenna's drive motor.

This is how the system works: It must be assumed that the system has been initially calibrated. That is, whatever bearing heading is dialed into the CX back at the control station, the antenna system located at its remote site is guaranteed to be pointing at that same bearing heading. In other words, there is zero error signal being generated within the control system. Initial calibration would probably involve a rather simple procedure similar to placing the CX at 0° heading and then adjusting antenna so that it points due north, 0°. To analyze the system's operation, let's assume that the operator adjusts the CX for a compass bearing of 30° east of north. The antenna, however, is pointing at, say, 150° from a previously commanded heading. (All compass headings in this example are measured clockwise from north.) Because of the difference in headings of the CX and CT units, an error signal is generated at the CT in the form of a certain amplitude AC waveform that is either in phase or out of phase with the supply voltage AC waveform. Remember that AC control systems always use the supply voltage waveform for referencing phase relationships. In this case, the system is designed so that out-of-phase signals at the CT produce a counterclockwise rotation of the antenna. The amplitude of the error signal causes the drive motor to rotate at a proportionally related speed. The higher the amplitude, the faster the rotation. In this example, an out-of-phase signal will be produced that will cause the servosystem to rotate the antenna counterclockwise. There will be a relatively large amplitude error signal that will be produced due to the differences between the set point

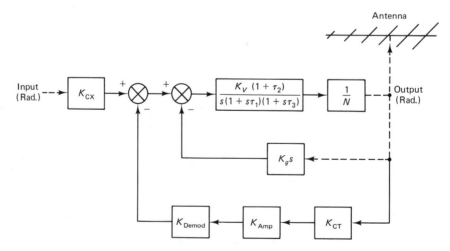

Figure 10-30 Block diagram of antenna system shown in Figure 10-29.

signal and the original resting point of the antenna (150° southeast). To decrease the likelihood of over-shoot and to dampen an otherwise lively system because of the components being used, a tachometer is used to convert velocity into position. This system's block diagram is shown in Figure 10-30.

SUMMARY

The systems presented in this chapter are but a very few of the many workable systems that are being used today. The ones described here have been simplified so that their basic operating concepts can be better seen. Despite these simplifications, if you understand what has been presented in this chapter, you will have little difficulty understanding the more detailed systems encountered in industry. If there are concepts that you are still struggling with in understanding automatic control systems, reread the earlier chapters. Chapter 10 represents a parting from the theoretical and an entry into the practical real world systems. Unless the theory is adequately "nailed down" at this point, you may have difficulty continuing with the practical information from this point onward.

REVIEW QUESTIONS

10-1. What advantage does an analog automatic control system have over a digital automatic control system?

10-2. What two advantages does a digital automatic control system have over an analog automatic control system?

10-3. Why was it necessary to develop the gray code to replace the binary code in certain encoding situations?

10-4. What are the three major types of power sources used in robotics today? List advantages and disadvantages of each.

10-5. Explain what is meant by the term *bit map* as applied to optical control systems.

10-6. Explain the function of a digital comparator circuit. What are its similarities and differences when compared to an analog voltage comparator or differential amplifier?

10-7. Explain what is meant by the term *adaptive control* in a process control system.

10-8. Explain the purpose of a tachometer as used in an automatic control system.

10-9. What would be the advantage of using an optical encoder for the measuring of very small increments of displacement versus using a linear variable displacement transformer for the same application?

10-10. Why is adaptive control so important in process control systems? In other words, why would you want to use it in a process control design?

REFERENCES

HUNTER, RONALD P., *Automated Process Control Systems,* Englewood Cliffs, N.J.: Prentice-Hall, Inc. 1987.

JOHNSON, CURTIS D., *Process Control Instrumentation Technology,* New York, N.Y.: John Wiley & Sons, 1977.

PATRICK, DALE R., *Industrial Process Control Systems,* Englewood Cliffs, N.J.: Prentice-Hall, Inc., 1985.

POTVIN, JEAN, *Applied Process Control Instrumentation,* Reston, Va.: Reston Publishing Company, Inc., 1985.

11

System Communications

11-1 THE PURPOSE OF SYSTEM COMMUNICATIONS

Up to this point, we have more or less taken for granted the fact that control signals, by some means or another, traveled trouble-free from one portion of a control circuit to another. We gave little thought to the means or methods used in transmitting these signals between circuits. We assumed that they arrived at their destination intact and unaltered. In this chapter, we study the various methods used in conditioning the control signals for travel and why this preparation is done. There are definite advantages to using certain means of preparing signals for transmission and for using certain decoding methods on the data that are imbedded in these transmissions.

Some of the discussion here involve a light dosage of radio theory and computer data transmission. To make the discussion a little easier to digest, it has been broken down into three major categories:

1. analog data transmission,
2. digital data transmission, and
3. carrier data transmission.

Each category is discussed, and advantages and disadvantages are presented in each. In reality, carrier data transmission techniques may be considered an analog form of communication. As a matter of fact, as we soon find out, carrier transmissions are used primarily to transmit analog-type data. However, in an effort to make the discussions a little easier to understand and to be more concise, the carrier data transmission discussion was separated from the analog data transmission material.

Perhaps you are asking yourself, why not just send the output of one circuit directly into the input of the next? Why bother to transform the output into some form of data signal before transmitting it to the next input? Unfortunately, things aren't quite that simple in hooking circuits together. A problem that could develop with the direct hook-up of circuits is this: The transmission of original variable voltages or currents is fine for small distances measured in, say, inches or a few feet. But when you are talking about hundreds of feet or miles, there are problems with outside electrical noise or interference, not to mention the problems with DC or AC voltage losses due to resistance in the wires. In addition, there are problems with lack of compatibility between circuit component inputs and outputs. One may require an AC voltage while another may require a DC current. There are also problems with impedance matching between these inputs and outputs in order to obtain maximum power transfer between circuits. All of these factors can really complicate an otherwise simple direct hookup. Therefore, in this chapter we attempt to unravel some of the mystery and jargon used in describing the various data communications systems used in automatic control systems.

11-2 ANALOG DATA COMMUNICATIONS

In the beginning days of automatic control systems, analog data communications was the only practical means of sending signals from one portion of the system to another. However, as time passed and control systems became much larger and more sophisticated, other means of data transmission had to be contrived. A good example of this problem is the remote antenna rotator system described in Section 10-11. It's conceivable that in this installation the antenna could be located several thousand feet from the control station as depicted in Figure 11-1. This is a common situation where the separation is necessary to reduce the problem of radio frequency interference (RFI) in the control station, especially if the antenna should be pointing directly at the station. In addition, to further reduce this problem and to increase the range of the antenna's radio transmissions, the antenna may be mounted on top of a tower hundreds of feet high. This adds to the path length of the controlling signals and to the control cable itself. Using an analog control system for this application would be somewhat difficult. However, we discuss a possible solution in a later section.

Analog systems (Figure 11-2, for example) are comprised mainly of those system components whose outputs are a variable amplitude voltage or current. The amplitude of this current or voltage is a linear representation of that component's input. Many components, such as transducers, synchromechanisms, amplifiers, signal conditioning circuits, etc. are designed so that their inputs and outputs are relatively compatible; that is, the output of one can be fed to the input

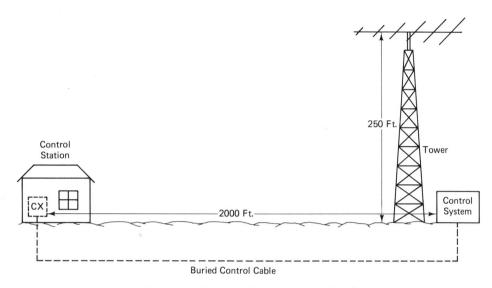

Figure 11-1 Typical communications antenna installation.

of the other with only minor circuit modifications. Many process control devices are designed to put out a current in the range of 4 to 20 mA. In other words, many circuits are designed so that 4 mA represents the lowest expected output signal, while 20 mA represents the highest expected output. Some form of circuit scaling is needed to insure the meaning of this range. Obviously, this very low current

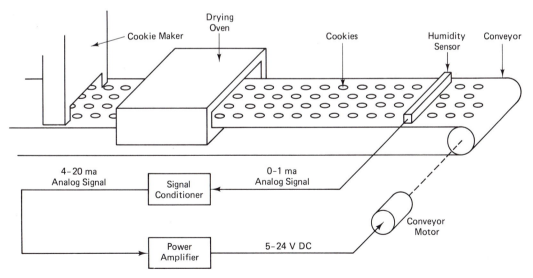

Figure 11-2 A process control system using a 4-20 mA analog circuit automatic control components.

range cannot be expected to drive very many output devices by itself. It would first be necessary to boost the power by using a power amplifier.

Figure 11-2 is an example of an analog-type automatic control system. The humidity sensing element used to detect the proper amount of moisture content in a packaged cookie product has a 0–1 mA output. This output signal is sent to a signal conditioner where it is amplified and the sensing signal "scrubbed" of any extraneous electrical noise that might create false data. This signal, which is now in the 4–20 mA range as a result of amplification, is sent to a power amplifier where the current signals are converted to an analog drive voltage to drive the cookie's conveyor motor. Because of the relatively short control lines needed in this system, the analog signals may be transported by ordinary wire or cable, although shielded cable is highly recommended. (See also the discussion in Section 12-7.1.)

11-3 CARRIER MODULATED CONTROL SYSTEMS

Carrier modulation (sometimes referred to as CW, meaning continuous wave) is seldom used for transmitting data in control circuits. It was used extensively many years ago, but not for this purpose. It was used instead for the transmission of Morse code for communications purposes. The transmitter's carrier was interrupted to form combinations of dashes and dots, each group forming an alphanumeric character or punctuation. It is still used in limited applications for the transmitting of data strictly for informational purposes, such as in emergency communications, but not for circuit control. There are applications, though, where the systematic turning on and off of a transmitted carrier is used in the controlling of bang-bang servo systems, but these cases are somewhat limited. An example would be the controlling of a garage door opener system. A hand-held transmitter is turned on momentarily to energize a receiver at the garage door. The receiver, when receiving this transmitted signal, turns on a motor which lifts the door. The door stops when limit switches installed in the door's track path detect the door's maximum travel. Another short burst of transmitted carrier from the transmitter then reverses the procedure and closes it.

11-4 AMPLITUDE MODULATED CONTROL SYSTEMS

Amplitude modulation data communications refers to the transmitting and receiving of radio frequency signals that contain data represented by the varying amplitude of the carrier signal. The transmission and reception of AM radio and television signals are two good examples. In the case of television, the data being transmitted by amplitude modulation is the color information for the TV picture. The changing amplitude of the carrier represents the varying intensities of each frame or picture being transmitted. In the case of AM radio, the sound is transmitted by the same means. The sound's amplitude and pitch is represented by the amplitude and frequency of the modulation "envelope" surrounding the carrier. This is demonstrated in Figure 11-3.

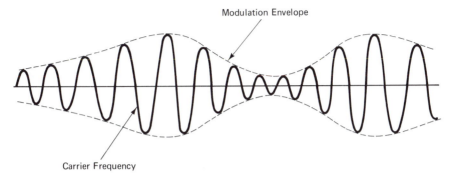

Figure 11-3   An amplitude-modulated (AM) carrier wave.

To understand how an amplitude modulation data system works, let's refer to the block diagram for the remote antenna controller described earlier in Figure 10-32. Let's assume that we want to adapt this system to an AM control system having the system measurements shown in Figure 11-1. The distance between the CX and the remaining control system components, which are all located in a control shed at the antenna tower site in this figure, is 2,000 + 250 feet. It should become obvious that transmitting low-amplitude voltages in an analog system may result in too-weak control signals by the time the transmitted signal voltages appear at the tower's base. Long lengths of transmission lines tend to be susceptible to stray voltages from outside sources that are inductively and capacitively coupled into the line. Using shielded cable reduces the problem but doesn't eliminate it entirely. The result is gross distortion of the control signals because of these stray induced voltages.

We use an amplitude modulated carrier system. This is a system borrowed from radio frequency transmission technology and involves transmitting a radio signal at a specific radio frequency. The transmitted signal is comprised of an *rf carrier* that is *amplitude modulated* by the CX's three output phase voltages. We have each of the three control phase voltages vary the amplitude of three transmitted tones, called *subcarriers,* having three different frequencies of, say, 400 Hz, 730 Hz, and 960 Hz. We could have picked any three frequencies as long as they are sufficiently separated from each other. These three values were picked, though, because their harmonics and subharmonics will cause a minimum of interference with each other. Each of these tones will then modulate the transmitted carrier frequency signal, as seen in Figure 11-4. In our system, the 400 Hz control voltage line frequency (or whatever other power supply line frequency happens to be used, such as 60 Hz) is removed through *rectification*. In other words, the AC control signal is converted to a DC voltage whose varying ampli-

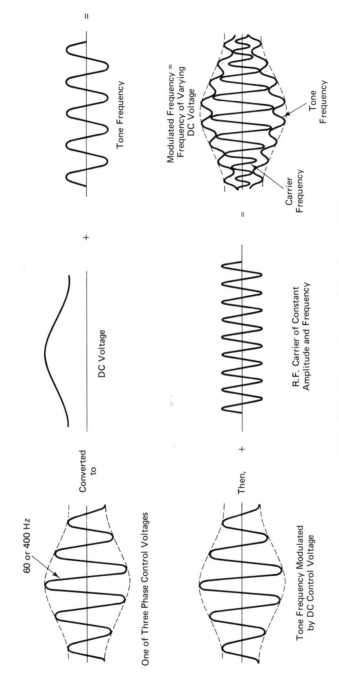

Figure 11-4 Transmitting control signals using AM.

352

tude coincides with the varying amplitude of the AC signal. The DC signal is then allowed to modulate, that is, vary the amplitude of one of the audio tones. Then, the audio tone with its varied amplitude is allowed to modulate an r.f. carrier. The AM (amplitude modulated) carrier, as it is now called, is of very low power, typically less than 500 milliwatts. Not much power is required to transmit a signal down the 2,250-foot path of cable. Besides, this reduces any likelihood of interference with other services in the area of the system. The frequency of this carrier is also usually quite low compared to normal communications frequencies, typically in the area of 10 to 500 KHz. The reason for this is to discourage radiation of the signal through the air over long distances and to discourage interference with the much higher communications frequencies normally being used by the other services. Since shielded cable is used for control signal transmission and because the cable is buried in the ground for the most part, little radiation will be lost from the cable.

Each of the three phase control signals from the CX are now converted to audio tones and transmitted simultaneously down the control cable to the waiting control system at the antenna tower base. We need a means of deciphering the voltage and phase information being transported by the modulated carrier. Therefore, a radio receiver must be used to *demodulate* the carrier and "strip off" this information. If this receiver had a speaker attached, you would hear a chorus of three amplitude-varying tones, but they would still be of no use to the control circuit at this point. Specially tuned filter circuits are needed to separate the tones back into three separate discrete channels, just as they were back at the CX. Then, following this filtering process, they must be mixed with the original control voltage frequency (400 Hz in our case) to get them back into their original forms for sending on to the CT. Figure 11-5 shows the complete remote control system now modified for the automatic controlling of our antenna.

11-5 FREQUENCY MODULATED CONTROL SYSTEMS

Figure 11-6 shows an example of a *frequency modulated* control signal. Instead of modulating the amplitude of a transmitted carrier as we did in Section 11-4, we modulate the carrier's *frequency*. In other words, again using the example of the antenna rotator system, the only changes we make to our system's design are to change the type of modulator used on the transmitter and the type of receiver used at the tower site. Both the transmitter and receiver must now be designed for frequency modulation and demodulation, respectively.

The advantage of frequency modulation over amplitude modulation of control signals is that FM is much more noise immune as compared to AM. Very little man-made or nature-made electrical noise is FM. It's for this very reason that FM is quite popular for commercial broadcast applications.

There is some room for debate here as to whether or not a transmitted carrier is actually needed to send and receive tone encoded information. Why not simply

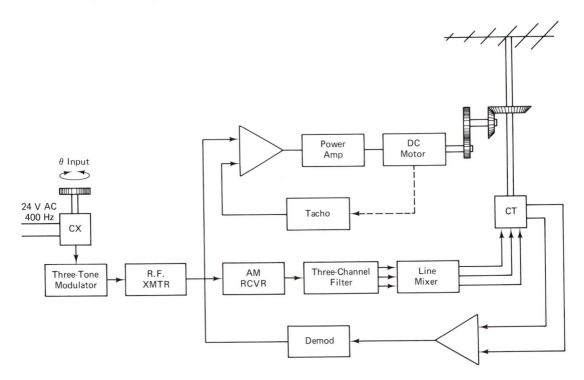

Figure 11-5 Complete remote control antenna controller system.

transmit the tones themselves without an rf carrier, must like a transmitted telephone signal? It can be argued that audio tone frequencies are nothing more than very low frequency radio waves having different propagation characteristics as compared to the higher frequency carriers. Experimentation is still being conducted on which frequencies propagate better over lines having lengths similar to the example we discussed here. One advantage to simply transmitting the tones is that you won't be troubled quite as much with the signal leaving the wire. The wire has less of a tendency to act as an antenna as compared to the case of the rf carrier transmission setup. Another advantage is that the circuitry is somewhat less complicated. Figure 11-7 shows a schematic of a simple but effective means of converting a variable amplitude AC voltage to a variable frequency tone. The

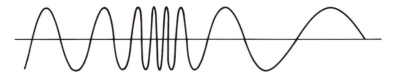

Figure 11-6 A frequency-modulated (FM) carrier wave.

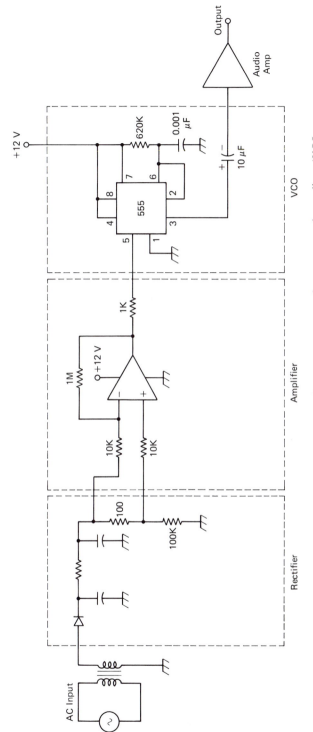

Figure 11-7 A variable voltage-to-frequency converter for tone control encoding. (*VCO circuit reprinted with permission from THE 555 TIMER APPLICATIONS SOURCEBOOK, by H. M. Berlin, published by Howard W. Sams & Co., copyright 1976.*)

355

circuit is comprised of four main subcircuits. The first circuit converts the incoming variable AC voltage signal from the CX into a variable amplitude DC signal. This, in turn, is sent to an amplifier to increase its amplitude to a point where it will drive a voltage-to-frequency oscillator (or VCO). This circuit converts the DC signal into a variable frequency. From there, the signal is sent to an amplifier whose job is to boost the line's transmitting power to prepare the signal for transmission down the control cable to the receiving site. Virtually any type of low-power amplifier in the range of a watt or so would work for this application. About the only thing that is somewhat critical about this circuit is the matching of the control line's impedance to that of the amplifier's output. Even that is not a difficult job, since what minor mismatch may occur at this point can be compensated for by merely increasing the gain of the amplifier somewhat.

The decoding of the received tone at the opposite end of the control line is equally fairly simple. Figure 11-8 illustrates a schematic of such a system. A Schmitt trigger is used to "strip-off" any line noise in the control cable. The 555 IC circuit does the actual variable frequency decoding, whereas the field effect transistor (FET) produces a very stable and reliable output voltage that is independent of any fluctuations in the circuit's supply voltages. The DC-to-AC inverter converts the output variable DC voltage to an AC voltage whose frequency is 400 HZ once again.

11-6 PHASE MODULATED CONTROL SYSTEMS

The discussion of phase modulation is presented here only to complete the discussion on modulation techniques. There is little advantage in using this form of modulation over that of FM. One possible advantage is the somewhat better noise immunity as compared to FM. The method of detecting a PM (phase modulated) signal is very similar to that of FM methods. Essentially, phase modulation is a method of varying the *phase* between the transmitted carrier and the modulating data. The modulating data is in the form of a frequency-varying AC signal. The phase variations are then detected by a receiver which, for all practical purposes, can be an FM-type receiver.

11-7 DIGITAL DATA COMMUNICATION

Digital signals are in the form of pulses or square waves. They are either transmitted in groups to form coded alphanumeric data or they are transmitted in continuous but variable numbers allowing their frequency to be counted for the purpose of deciphering the data (Figure 11-9).

With the advent of the microcomputer in the early 1970s, it became evident that control circuits would also be affected. Circuits were redesigned so that direct interfacing with the computer would become possible. This then would

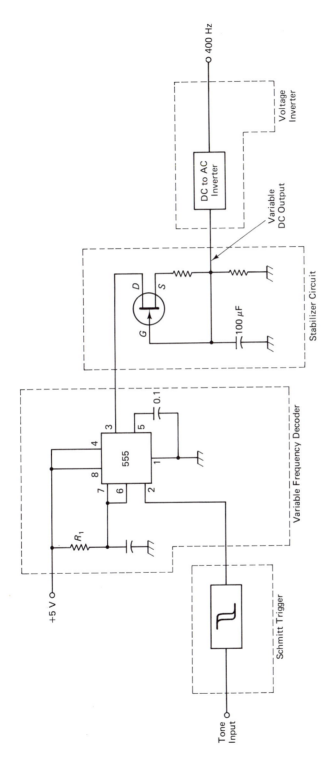

Figure 11-8 A tone decoding (frequency-to-voltage) converter. *(Reprinted with permission from THE 555 TIMER APPLICATIONS SOURCEBOOK, by H. M. Berlin, published by Howard W. Sams & Co., copyright 1976.)*

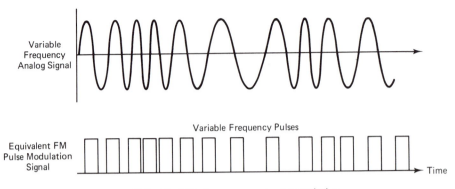

Figure 11-9   Variable frequency pulse transmission.

allow the computer to "talk" directly with the circuit components with no intermediate data conversions necessary. But there were other reasons for going to
digital signal control. Digital signals were far more immune to electrical noises
than were some analog signals. By far the worst affected control signal was the
analog signal. As mentioned earlier, most man-made and naturally-made electrical noise is amplitude modulated. Consequently, the receiving circuits that were
designed to respond to this type of transmission could became easily confused
with this noise. Little such noise problems existed with digital-type circuits, however. But it was soon discovered that because of increased complexities of circuits, increased gain, and because much lower voltage values were being used for
output signals, the digital circuits become more susceptible to ground loops and
stray r.f.-type noise. (Ground loops are undesirable electrical currents flowing in
the ground and through wire commons or cable shielding, behaving like uncontrollable feedback signals, producing undesirable electrical circuit noise in systems.)
Solid state components, unfortunately, have a habit of acting as miniature radio
detectors in that they often and inadvertently pick up radio transmissions. But
despite this major flaw, there was an added bonus. Because of using digital signals, much of the control circuitry operated only during the duration of the pulse.
In other words, the duty cycle for the circuit was considerably less as compared to
the continuously operating analog circuits. As a result, less heat was dissipated
and less energy was consumed by the circuits. Less power was needed, too, to
run these circuits, only because of the elimination of power-hungry filaments
associated with vacuum tubes.

Another major potential flaw associated with digital communications is that
it generates its own share of electrical noise. Since, mathematically, a square
wave or pulse is composed of an infinite number of sine waves, a square wave or
pulse can generate tremendous amounts of interference because of the infinite
number of frequencies being generated. Despite this problem, and because its
advantages far outweigh the disadvantages, digital control circuitry is becoming
extremely popular.

11-8 PULSE MODULATED CONTROL SYSTEMS

There are several methods used for pulse modulation, all of which fall under the heading of digital communications. The ones that we discuss here are the following:

1. pulse amplitude modulation,
2. pulse width modulation,
3. pulse position modulation, and
4. pulse code modulation.

These four methods are the most basic pulse modulation methods used in industry at the present.

11-8.1 Pulse Amplitude Modulation (PAM)

Figure 11-10 shows a picture of a pulse at different periods of time demonstrating how its amplitude varies with whatever data is being used to vary its height. The pulse's height is modulated by the input signal or data. This is called *pulse amplitude modulation, or PAM*. As an example, an input voltage of 3 volts would produce a pulse height that would be twice as high as an input voltage of 1.5 volts. This is assuming that the pulse's height varies linearly with the input. This type of digital system requires a very quiet no-signal or background noise condition in order for it to function properly. It is susceptible to the same noise problems as an amplitude modulated system.

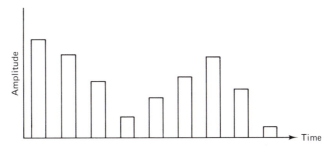

Figure 11-10   Pulse amplitude modulation.

11-8.2 Pulse Width Modulation (PWM)

Figure 11-11 represents a pulse whose duration or width is modulated by an input signal. The amplitude of the pulse remains the same. This type of system is called *pulse width modulation, or PWM*. The receiver for this type of transmission must

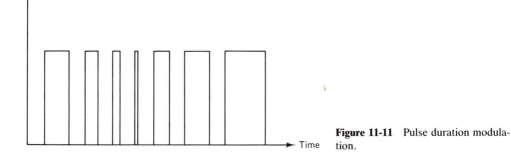

Figure 11-11 Pulse duration modulation.

have the capability of measuring the duration of the received pulse and translating the measured amount into an equivalent voltage, current, or other data form. This system has a better noise immunity than the PAM system. As a matter of fact, there are similarities between the PAM system and FM as far as how each performs under noisy conditions.

11-8.3 Pulse Position Modulation (PPM)

This type of modulation is shown in Figure 11-12. Pulse position modulation is somewhat more complicated than the other methods already mentioned. It consists of a pulse of fixed width and amplitude. However, its position constantly shifts relative to a known time or position reference. The pulse's position relative

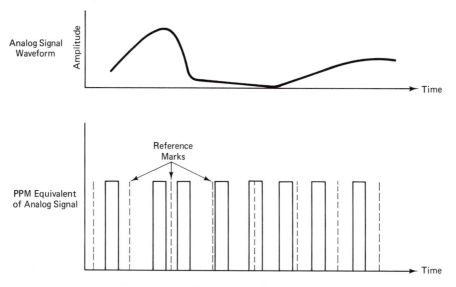

Figure 11-12 Pulse position modulation.

to this reference is varied by the input data and determines the magnitude of that input. Its noise immunity is somewhat like that of a PM system, meaning that it is somewhat better than PWM and certainly better than PAM.

11-8.4 Pulse Code Modulation (PCM)

Of the four listed methods of pulse communications, pulse code modulation is the one modulation type most extensively used today. It also has the most variations. One form of PCM is seen in Figure 11-13. In this type of pulse modulation,

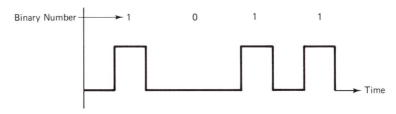

Figure 11-13   The decimal number 10 represented by the 8–4–2–1 binary code.

a group of pulses are combined to form a coded group to represent an alphanumeric byte. Out of the group of four pulses in Figure 11-13 (the third pulse from the right is missing), the binary number 1011, or decimal 11, is formed. This grouping is based on the so-called 8–4–2–1 binary system. Which pulses show up in which four possible locations in the grouping determines the magnitude of the decimal number. Additional groups of four binary *bits* can be formed to create larger magnitude decimal equivalent numbers. Table 11-1 is the truth table for the 8–4–2–1 system. (A truth table is a listing of all possible logical states that a

TABLE 11-1   TRUTH TABLE FOR THE 8–4–2–1 CODED BINARY SYSTEM

Decimal no.	8	4	2	1
0	0	0	0	0
1	0	0	0	1
2	0	0	1	0
3	0	0	1	1
4	0	1	0	0
5	0	1	0	1
6	0	1	1	0
7	0	1	1	1
8	1	0	0	0
9	1	0	0	1

coded group of symbols can have for a possible set of conditions.) As an example, if we wanted to represent the decimal number 359 using the 8–4–2–1 code, we would write the binary equivalent as 0011 0101 1001. The numbers 8–4–2–1 are the *weights* for this code. That is, noting which columns have 1s occurring in them, and then merely adding the weighted values of those columns, will produce the equivalent decimal number value. Notice also in Table 11-1 that we used only 10 of the 16 possible states that exist using an 8–4–2–1 coding scheme. The remaining 6 states are referred to as forbidden states, and should these binary numbers crop up in a digital automatic control system, erroneous data has obviously been generated. A digital system can be made to look for these errors to prevent system errors.

There are numerous other binary coded systems that are used in industry for digital control. Each has its own particular advantage for usage. Table 11-2 lists some of the more common *five-bit codes,* and Table 11-3 lists some of the more-than-five-bit codes being used.

TABLE 11-2   FIVE-BIT BINARY CODED DECIMAL (BCD) CODES

Decimal	2-out-of-5	63210	Johnson code	51111
0	00011	00110	00000	00000
1	00101	00011	00001	00001
2	00110	00101	00011	00011
3	01001	01001	00111	00111
4	01010	01010	01111	01111
5	01100	01100	11111	10000
6	10001	10001	11110	11000
7	10010	10010	11100	11100
8	10100	10100	11000	11110
9	11000	11000	10000	11111

TABLE 11-3   COMMON MORE-THAN-FIVE-BIT CODES

Decimal	50	43210	9876543210
0	01	00001	0000000001
1	01	00010	0000000010
2	01	00100	0000000100
3	01	01000	0000001000
4	01	10000	0000010000
5	10	00001	0000100000
6	10	00010	0001000000
7	10	00100	0010000000
8	10	01000	0100000000
9	10	10000	1000000000

The 2-out-of-5 code has a built-in *parity check* provision. Parity checking is a method of counting all the 1s and 0s to make certain there is the required total number or required odd or even number of 1s or 0s present for that particular character. Parity checking is often used in automatic control systems to check data validity. In the 2-out-of-5, there are always two 1s. The same is true for the 8–6–4–2–1 code. In the 5–1–1–1–1 code, the code is *self-complementing.* That is, the complement of a decimal number has a binary complement for its binary equivalent. For instance, the 9's complement of decimal 2, from Table 11-2, is decimal 7 (9 − 2 = 7). (The term 9's means to subtract the decimal number from 9 to find that number's 9's complement.) The 1's complement of binary 00011 (decimal 2) is binary 11100 (decimal 7). (The 1's complement of any binary number is generated by merely changing all 0s to 1s and all 1s to 0s in that number.) Therefore, the complement of the decimal using this coding scheme has a binary 1's complement equivalent. Another example: The 9's complement of decimal 4 is 5 (9 − 4 = 5). The complement of binary 01111 (decimal 4) is binary 10000 (decimal 5). Because of this self-complementing feature, error checking is made easier and the electronics needed to decode the binary numbers is also made easier.

Looking at another coding scheme, the 50 4321 code, we have another built-in parity check system. The 01 and 10 bits are even parity checkers. That is, when all the 1s are added in the entire binary number including the parity bits themselves, we always come out with an even number of 1s. As a matter of fact, we always come out with two 1s in every number. This makes another fast and accurate system error check.

Another popular form of PCM is the ASCII code. The letters ASCII (pronounced askey) stand for *American Standard Code for Information Interchange*. This is the code that virtually all personal computers use today for communication purposes between computer systems and their peripheral equipment. Figure 11-14 shows the truth table for the ASCII code. Notice that this coding scheme is referred to as an eight-bit code, whereas only seven bits are explained in the truth table. The reason for this is that the eighth bit is used as a parity check for the other seven.

11-9 FREQUENCY SHIFT KEYING

Frequency shift keying (FSK) has been around for many years in the field of communications. This system uses two audio tones that are keyed alternately off and on. One tone represents a 1 while the other represents a 0. Teletype systems have used this system extensively for transmitting messages over telephone lines and over radio. FSK is probably used more today than ever before because of the advent of the computer. Computer information is transmitted by this method over both telephone and radio using a device that is called a *modem*. The word is a composite of the two words *modulator* and *demodulator*. The tone frequency pairs often used for this application are 1,270 Hz and 1,070 Hz, and 2,225 Hz and 2,025 Hz. Other tone combinations are also used as long as the receiver has been designed to receive and separate those tones.

For radio transmission, the tones are fed directly into the transmitter's modulator audio input. The major disadvantage to this method is that all data must be transmitted serially. In other words, in order to transmit a "word" of information, the binary equivalent must be transmitted bit-by-bit, one at a time in succession. This is called *serial transmission*. The problem with this method is that it is very slow. However, data can be transmitted over relatively long distances using this process. The other mode of transmission, called *parallel transmission,* allows

BIT 4	BIT 3	BIT 2	BIT 1	BIT 7: 0 / BIT 6: 0 / BIT 5: 0	0 / 0 / (0 1)	0 / 1 / 0	0 / 1 / 1	1 / 1 / 0	1 / 0 / 1	1 / 0 / 0	1 / 1 / 1
0	0	0	0	NULL	DC$_0$	b	0	@	P	↑	↑
0	0	0	1	SOM	DC$_1$	!	1	A	Q		
0	0	1	0	EOA	DC$_2$	"	2	B	R		
0	0	1	1	EOM	DC$_3$	#	3	C	S		
0	1	0	0	EOT	DC$_4$ (STOP)	$	4	D	T		
0	1	0	1	WRU	ERR	%	5	E	U	UNASSIGNED	UNASSIGNED
0	1	1	0	RU	SYNC	&	6	F	V		
0	1	1	1	BELL	LEM	' (APOS)	7	G	W		
1	0	0	0	FE$_0$	S$_0$	(	8	H	X		
1	0	0	1	HT / SK	S$_1$	)	9	I	Y		
1	0	1	0	LF	S$_2$	*	:	J	Z		
1	0	1	1	V$_{TAB}$	S$_3$	+	;	K	[		
1	1	0	0	FF	S$_4$	, (COMMA)	<	L	\		ACK
1	1	0	1	CR	S$_5$	−	=	M	]		(1)
1	1	1	0	SO	S$_6$	.	>	N	↑		ESC
1	1	1	1	SI	S$_7$	/	?	O	←	↓	DEL

Legend:

NULL	Null/Idle
SOM	Start of Message
EOA	End of Address
EOM	End of Message
EOT	End of Transmission
WRU	"Who Are You ?"
RU	"Are You . . . ?"
BELL	Audible Signal
FE$_0$	Format Effector
HT	Horizontal Tabulation
SK	Skip (Punched Card)
LF	Line Feed
V$_{TAB}$	Vertical Tabulation
FF	Form Feed
CR	Carriage Return
SO	Shift Out
SI	Shift In
DC$_0$	Device Control Reserved for Data Link Escape
DC$_1$–DC$_3$	Device Control
DC$_4$ (Stop)	Device Control (Stop)
ERR	Error
SYNC	Synchronous Idle
LEM	Logical End of Media
S$_0$–S$_7$	Separator (Information)
b	Word Separator (Space, Normally Nonprinting)
<	Less Than
>	Greater Than
↑	Up Arrow (Exponentiation)
←	Left Arrow (Implies/ Replaced By)
\	Reverse Slant
ACK	Acknowledge
(1)	Unassigned Control
ESC	Escape
DEL	Delete/Idle

Example: Character "R" is represented by 0100101

Figure 11-14 ASCII eight-bit code truth table. (*James Martin, TELECOMMUNICATIONS AND THE COMPUTER, 2nd ed., copyright 1976, p. 308. Reprinted by permission of Prentice Hall, Inc., Englewood Cliffs, N.J.*)

the user to transmit all bits of an entire "word" simultaneously. However, this requires using many channels of transmission media. That is, one channel for each bit is required. If wire is going to be used for the conducting media, then multiconductor wire has to be used. We discuss this and other transmission media in the next section.

11-10 TYPES OF TRANSMISSION MEDIA

There are five major types of transmission methods and media that are presently being used to transmit and receive control signals. These are:

1. wire conductors,
2. radio waves,
3. fiber optics,
4. light waves, and
5. pneumatic transmission lines.

As pointed out in the previous section, the faster method of digital data transmission is parallel transmission. If that is going to be the selected mode of data control transmission in an automatic control system design, then either multiconductor wire cable must be used or multistrand fiber optical cable. Otherwise, serial transmission can be employed, using any of the four aforementioned methods. Keep in mind, though, the much slower speed of transmission. Despite this one limitation, serial transmission of PCM information has had widespread use in automatic control applications.

Since multichannel transmissions for PCM systems are cumbersome and not all that convenient to design (in many instances, you may have to convert analog data to digital data and then convert digital data back to analog data, requiring A/D and D/A circuits), much simpler systems avoiding digital circuitry altogether may be used. The transmission of data by means of carrier transmission may be used instead. An example of this was illustrated in Figure 11-1 where the three phase control voltages were converted to AM signals (an FM or PM signal conversion would probably have been better to take advantage of the superior noise immunity characteristics). You can see the problem that would be created if it had been decided instead to convert each of the three phase voltages to say, eight-bit binary data and then to transmit that data to the receiving site at the tower. Each of the three phases would require a *minimum* of eight channels of transmission, not to mention the A/D and D/A conversion circuitry needed. Furthermore, to transmit parallel information over such long distances and to expect it to arrive at the receiver all at the same time to maintain proper phase alignment, is extremely difficult with the present state of the technology. With all of this in mind, let's take a look at our four basic transmission media.

11-10.1 Wire Conductors

Wire conductors used for instrumentation and control signal applications in industry are usually shielded and heavily insulated to withstand abrasion. A wide variety of these conductors are on the market today. In most cases, an outside shield surrounding the inner signal conductor is usually kept at ground potential to prevent interference from occurring from outside stray unwanted signals. This type of cable is often referred to as *coaxial cable,* or simply *coax.* Coax is manufactured with different impedances, the two most common impedances being approximately 50 ohms and 75 ohms. Of these two, the 50-ohm coax is the more popular.

Recently, flat cable, or ribbon cable as it is more popularly called, has become popular for control signal applications. The reason for this is due to the popularity of the microcomputer. The computer has made it necessary to transmit and receive digital data using parallel transmission techniques over short lengths. Ribbon cable can also be purchased with shielding if desired; this is recommended especially for industrial environments.

11-10.2 Radio Waves

A common means of transmitting data is by radio transmission. In recent years, some spectacular results have been obtained using this method in the science of satellite technology. The automatic control of satellites and extraterrestrial robotic vehicles over immense distances has proven beyond all doubt that this is a viable means of data transmission. In some cases, more than one frequency is used in order to handle more than one channel of data as seen in Figure 11-15. One frequency is the command frequency for transmitting the setpoint information, while the other is the response frequency. In some cases, elaborate schemes are used where only one frequency would be used but several subcarriers would be transmitted and received on that same frequency. The subcarriers would

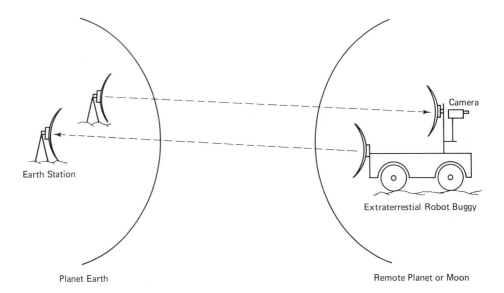

Figure 11-15   Long-distance robotic remote control.

either be audible frequencies or at lower radio frequencies below the carrier's frequency. The earth-based transmitting power needed for satellite or robotic control is in the hundreds or thousands of watts. The onboard transmitting power beaming back to earth would only be several or tens of watts due to power supply energy and weight limitations.

11-10.3 Fiber Optics

Fiber optics represents the newest developed media for the transmission of data. Fiber optics involves the using of glass-like fibers that act as optical light pipes. A modulated light source, usually a laser, is used for transmitting light beams that act as carriers through the fiber. At the other end of the fiber cable, a photocell circuit is used to demodulate the light beam carrier to remove the information. The advantage of fiber optics over radio or telephone transmissions is that little cross-talk or interference between channels is experienced. Different colored lasers can be used to create different channels in the same fiber. At the receiving end, the receiver can filter out its particular color of data and exclude all others, much like the audio tone filtering system described earlier. In addition, many fibers can be used to handle many different channels. It's not unusual to see one cable carrying many hundreds of fiber channels.

11-10.4 Light Waves

The transmission of data by light waves is another relatively new technology. The theory of operation has been around for a good many years, but it has been only recently that practical systems have been worked out. Basically, the only difference between this system and the one just described involving fiber optics is that no fiber optics are used. Instead, the modulated light beam information is transmitted directly either through the atmosphere or through space to the photosensitive receiver. Again, lasers are used to generate the powerful narrow beams of light necessary for this kind of system to work. The disadvantage with this system is that it is susceptible to atmospheric and man-made interference. The fiber optic system, on the other hand, is virtually interference-free. Fiber optics require a considerable installation investment (cost of cable, installation labor, etc.) whereas the light communications system has zero installation costs between the transmitters and receivers. At the present time, there are very few installations using this method of remote automatic control; however, as time goes on, this system will be used more and more.

11-10.5 Pneumatic Transmission Lines

The method of transmitting control signals by using pneumatic lines is the only practical nonelectrical method presently being used in industry. This is perhaps one of the simplest methods also.

As the name implies, an air line, usually made of copper tubing or other durable tubing material, is filled with pressurized air. The tubing acts as a conduit to transmit control signals from a transmitting device, capable of modulating the air's pressure inside the tube, to a receiver at the opposite end of the tube, used for the purpose of interpreting the modulated signals. The transmitter is generally comprised of a "flapper" valve whose open and closed position is controlled by a positioning lever or member. The position of this lever could be in direct response to a change in a position, a temperature, pressure, or flowrate of a fluid. Figure 10-28 in Chapter 10 shows just such a system. The amount of bypassed air created by the flapper causes a proportional decrease or increase of air pressure inside the air tube. The receiver at the tube's opposite end then interprets the rise and fall of air pressure as a proportional change in the measurand at the other end. In the simpler control systems, the pneumatic controls are set up strictly for an on-off or bang-bang servo application rather than a proportional application.

There are two inherent drawbacks to a pneumatic control transmission system. For one, the maximum practical length of pneumatic transmission line can be restrictive. Transmission line lengths in excess of 600 feet are rarely used. Pressure signals become smoothed or very poorly defined beyond this length. Also, pneumatic systems are limited by relatively low transmission rates because of the air's compressibility. Transmitted pressure fronts comprising the signal information can travel no faster than the speed of sound, thereby limiting the response time of the overall system. The major advantages of this kind of system, in light of the overwhelming disadvantages just mentioned, are low cost, low maintenance, and that the systems can operate in volatile atmospheres. Because of these three reasons, pneumatic systems are still in popular use today.

SUMMARY

The study of control system communications is extremely important, with the increasing sizes of these systems. One of the biggest problems confronting the design of such systems is the speed of communications. Obviously, digital serial communications is the only practical method that can be used for long-distance data communications. On the other hand, digital parallel communications is fairly fast but very limited in the distances it can travel. (This may change in the near future, however, with the advent of multichanneled laser space transmissions.) This is way carrier-type transmissions have found many applications in this area. Basically, carrier transmissions were developed to transmit the analog information of the earlier automatic control systems and to do it reliably. Considerable development is needed to make fiber optics cost-competitive with the wire conductor systems for long-distance digital control system applications. However, it will be only a matter of time before this happens. Already, many wire conductor communications systems are rapidly being replaced with fiber optics. Because of the much wider bandwidth and multichannel capabilities of fiber optics, so much more digital information will be able to be transmitted and received.

REVIEW QUESTIONS

11-1. List the three major categories of data transmissions and give a characteristic of each.

11-2. Describe two examples of bang-bang servo systems. What main feature differentiates a bang-bang system from a proportional control system?

11-3. Why is FM used more frequently than AM for the transmission of analog data signals?

11-4. Explain how a Schmitt trigger works in a noise filtering system.

11-5. Explain the difference between an FM data signal and a PM data signal. Are there any advantages in using PM signals versus using FM signals in a communications system?

11-6. What is a major concern in transmitting pulsed or square-wave-type signals? Explain in detail making references to sine-wave and square-wave theory.

11-7. List the four basic forms of pulse modulation and describe characteristics of each.

11-8. Describe what is meant by the term *self-complementing* in reference to certain binary-type codes.

11-9. Cite an advantage and disadvantage of serial data transmission; do the same for parallel data transmission.

11-10. Cite an advantage for using pneumatic data transmission versus using electrical transmission methods. Cite at least two disadvantages for using pneumatic transmissions.

REFERENCES

American Radio Relay League, *The Radio Amateurs Handbook,* Newington, Conn.: ARRL Inc., any recent annual edition.

BUCHSBAUM, WALTER H., *Practical Electronic Reference Data,* Englewood Cliffs, N.J.: Prentice-Hall, Inc., 1978.

VERGERS, CHARLES A., *Handbook of Electrical Noise: Measurement and Technology,* Bike Ridge Summit, Pa.: TAB Books, 1979.

12

Computers and Automatic Control Systems

12-1 INTRODUCTION

In Chapter 11 we discussed methods of transmitting and receiving digital information being used for automatic control systems. These systems were obviously using some form of computer or hardware digital devices to generate and handle the many bits of data. Because of the impact that computers are having in industry and on automatic control systems in particular, it's necessary to understand how the digital computer works. In Chapter 12 we pick the computer apart, using a typical microcomputer as our subject, to see how control data is processed. With a basic understanding of how computers work, you won't feel the intimidation that many feel when they first approach a computer or any other kind of digital system. In order to work with digital automatic control systems, it's necessary to understand how computers work and how they work with control systems.

12-2 WHAT IS A MICROPROCESSOR?

In order to understand what a microprocessor is, we must first understand what a CPU is. The letters CPU stand for *Central Processing Unit*. Before the advent of the microcomputer, the microprocessor didn't exist. All computers, instead, had CPUs. The CPU was the brain of the computer. Its purpose was to perform all of the computer control, all the handling of input and output data transfers between circuits, all the arithmetic and logical operations, and it performed all the instructions obtained from the various memory banks within the computer.

As time went on, the first computers began evolving into more powerful and compact machines. This was due to the integrated circuit chips becoming more sophisticated and more dense as more circuits were crowded into their limited containment areas. Medium-scale integrated circuit chips, or MSI chips, gave way to large-scale integrated (LSI) chips, until finally, what was originally called a CPU and was spread out over several square inches of circuit board became a single LSI chip, dubbed the *microprocessor* chip.

A typical 40-pin microprocessor containing all the circuitry just mentioned, along with its general size, is illustrated in Figure 12-1. The processor chip itself

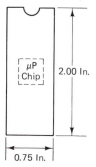

Figure 12-1   A close-up look at a microprocessor.

most likely covers an area no greater than the dashed-line area shown inside the figure's sketch of the IC's housing itself. Because of the rapid development of computer chips and the computers themselves, the standardization of the nomenclature has been left behind in this flurry of activity to survive on its own. As a result, what comprises the CPU and what comprises the microprocessor for a microcomputer nowadays is a little confusing. As it turns out, they are one and the same. The microprocessor chip *is* the microcomputer's CPU. The circuit board on which the microprocessor is installed is called the *mother board*. The box or cabinet that contains the mother board and all the other peripheral devices is called the computer's *CPU box* (Figure 12-2).

Let's look more closely now at the organizational structure of a microprocessor chip. Figure 12-3 shows a box diagram of a rather simplified version of a typical microprocessor. It really isn't necessary to understand how this chip works to use a computer, but it does become necessary to have some idea as to what goes on inside these chips if you become involved with circuit design or with some aspect of trouble-shooting. Besides, it gives you a much better feeling for what is going on inside computer-controlled circuitry such as digitally controlled automatic control circuits.

The ALU (or arithmetic-logic unit) of a microprocessor performs all the mathematical and logical chores on the data that is supplied to it. It has two inputs: One input is from the accumulator, the other from the sequence controller. Generally, the ALU does only two forms of math; it adds binary numbers and

Figure 12-2 Complete personal computer showing CPU. (*Courtesy of HiTech International, Inc., Milpitas, Calif.*)

subtracts them. Technically, it only adds binary numbers. Subtractions are done through what is called 2's complement addition. Here is an example, except we use 10's complement addition (for reasons we explain in a moment): To subtract, say, 4 from 7, we instead *add the 10's complement of 4 to 7*. The 10's complement of 4 is found by taking 9 − 4 (not 10 − 4) which is 5, and then adding 1. In our case, the 10's complement of 4 is 6. Then, adding 6 to 7 we get 13. We then drop the 1 in front of the 3 to get our answer. We used the 10's complement only

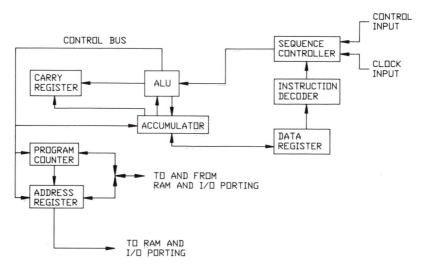

Figure 12-3   Flowchart for a typical microprocessor.

because we were working with decimal numbers. In binary numbers, we would use the 2's complement. It would work like this:

$$
\begin{array}{r}
7 = \quad 0111 \\
\text{Normal subtraction:} \quad -\ \underline{4} = -\ \underline{0100} \\
3 = \quad 0011
\end{array}
$$

Using 2's complement: The 2's complement of 0100 is 1011 (the 1's complement or reverse of 0100) + 1, which is 1011 + 1 = 1100. Then,

$$
\begin{array}{r}
0111 \\
+\ \underline{1100} \\
10011
\end{array}
$$

↳ drop the "carry"

The answer is 0011, or decimal 3.

This routine seems a bit complicated and perhaps somewhat nonsensical at first, but it is so easy to design digital logic circuits to perform the 2's complement for subtraction that it is used extensively. The binary numbers that are operated on (called *operands*) by the ALU are supplied by the accumulator. The results of an ALU arithmetic procedure are dumped back to the accumulator. Notice that attached to the ALU is a carry register. Its purpose is to handle the carry routines needed to properly manipulate the binary numbers, such as the foregoing routine needed to discard the carry to obtain the final proper answer.

The purpose of the accumulator is to accumulate data for the ALU and to temporarily store completed data from the ALU. The accumulator's instructions come from the control bus that originates in the ALU. Essentially, the ALU tells the accumulator when to operate and when not to operate.

The data register is another temporary holding area for data while an instruction is being decoded. It also acts as an overflow for the accumulator. Usually, the register can hold only one eight-bit data word. The data register then goes into an instruction decoder to a sequence controller where the decoded instructions coming from the ALU (these are usually math instructions such as $\times$, $+$, $\div$, etc.) are cleared for operation and synchronized with other control input data coming from outside the microprocessor. Notice that there is a clock input also going into the sequence controller to do the actual synchronizing of all commands. The clock insures that all data is in pace with their control signals and that everything arrives at a particular location in the data processing at the correct time.

Let's look at the program counter now. This is another memory area that contains the address of the next command or instruction that is to be run in a program. The program counter's main job is to sequence the order of each programming instruction. The numbering sequence may be rearranged however, if the programmer wishes to "jump around" within the program. This is done by issuing branching instructions within the program's routine. The program counter is controlled by the aforementioned control bus. Its output goes into the address register.

The address register is used to temporarily store the address of a data word's memory location. What is interesting about this feature is that this gives the programmer the option to alter the data's address, thereby changing the routine or sequence of which data is to be processed. Again, this area of the CPU is controlled by the control bus. The output of the address register goes to the CPU's temporary memory, called random access memory (RAM), and to the I/O porting. The I/O porting refers to the circuitry that controls the inputting and outputting of the floppy drives and hard disks. This porting also controls the sending of data to the monitor and printing devices, not to mention receiving data from the keyboard.

The microprocessor, or processor as it is often shortened to, "talks" to the outside world by way of three different routes or paths. One path is called the data bus, another is the address bus, and the third path is called the control bus. It's the data bus that distinguishes one microprocessor chip from another. The data bus is a group of connections or pins on the IC chip that allow for the two-way exchange of data or for the inputting of instructions to the processor. There may be 4 lines, 8 lines, 16 lines, or 32 lines. This group of lines represent the number of digits that can be made up to form a binary "word" that can be manipulated by the microprocessor. That is, these lines represent the parallel processing capacity of the processor itself. Therefore, these microprocessors are referred to as 4-bit processors, 8-bit processors, 16-bit processors, etc. Table 12-1 lists some of the more popular microprocessors that have been used in the past

**TABLE 12-1   POPULAR MICROPROCESSOR
CHIPS PRESENTLY BEING USED
IN MICROCOMPUTERS**

Chip number	True data bus width	Manufacturer
Z80	8	Zilog
6502	8	MOS Tech.
8086	8	Intel
8088	8	Intel
80286	16	Intel
80386	32	Intel
6800	8	Motorola
68000	16	Motorola

and are presently being used in microcomputers today, along with the names of their manufacturers.

In Table 12-1, the reason for the term *true* data bus length is because some computer designers can take, say, an 8088 chip and simulate a 16-bit processor through manipulation of the data bus configuration, but in reality it is still only an 8-bit processor chip. The chips that have been listed here are those that comprise the greater part of the microcomputer market. The term *microcomputer* is really what is now more commonly referred to as a personal computer or PC. Many PCs, as a result of containing large amounts of RAM (Random Access Memory) and having several 10's of megabytes of hard-disk memory available, are approaching the sizes of *minicomputers*.

12-3 THE COMPUTER AND ITS SOFTWARE

Let's look at a personal computer a little more closely, especially at the mother board contained by its CPU box, as illustrated in Figure 12-4. On this mother board, the CPU (which, by the way, is an 80286 16-bit processor) is installed, with a heat sink, in the upper center on the board, alongside the hard drive enclosure. Slightly above and to its left we see the 80287, a 40-pin chip. This is the microprocessor's matching math coprocessor. A math coprocessor is an auxiliary chip that speeds up the math computation speed of the computer. It's been especially designed to perform high-speed calculations and other related routines. If you're planning to use a computer that's going to be used for math-intensive applications for automatic control applications, or for any other applications, you should consider installing one of these. Just be sure that it is the matching coprocessor for your particular microprocessor and that it's designed to work with your microprocessor's operating speed.

Near the lower right-hand corner of the mother board we see several rows of ICs which are the computer's RAM chips. RAM is the computer's temporary

LITHIUM BATTERY 80287 COPROCESSOR CHIP 80286 UP CHIP HOUSING FOR HARD DISK

EXPANSION SLOTS (8) BIOS ROM CHIPS RAM CHIPS

Figure 12-4 A mother board.

memory for holding data that is being worked on. The memory is temporary, in that when the computer is turned off, the memory information is lost forever. However, when energized, information can be read from this memory and also "written" into this memory. This is an important point to realize, because not all memory has this read/write capability. For instance, the two chips seen with the foil labels at the lower edge and left-of-center of the mother board are ROM (Read Only Memory) chips. These chips have data information that has been permanently implanted into the chips and can be read at any time. However, you can't write to these chips. These particular chips have a unique function in the computer. They comprise what is referred to as the BIOS for the computer. This is an acronym which means *Basic Input/Output System.* In other words, these chips contain operating commands that manage the operation of the computer's several input/output ports, which are attached to peripheral equipment such as floppy disk drives, hard disk drives, back-up tape drives, etc.

In the left-hand portion of our mother board, we see several empty *card slots.* These are the slots that accept additional printed circuit boards that are dedicated to the operation of our computer's peripheral devices. One slot may contain a video control board that allows the operation of a monitor. Another board may be the controller board for the floppy disk drives or the hard disk. Another board may be a board for allowing the computer to "talk" to another computer over the telephone. (This board is called a modem.) It's important to realize the differences between these two kinds of memory devices. A floppy disk drive is an electromechanical device that allows the computer operator to read and write data information to a disk for permanent storage applications. They are quite slow in operation as compared to a RAM chip. However, relatively large amounts of data can be stored and accessed in this way. Typical capacities range from 360 kilobytes to as much as 1.4 megabytes. A hard disk is also an electromechanical device, but is much faster in operation as compared to a floppy disk drive, and can hold tremendous amounts of data. Typical storage capacities range from 10 megabytes to several hundreds of megabytes. Hard disks are still considered to be slow compared to RAM storage, but at least the stored information is *nonvolatile,* like the floppy. That is, the memory is not affected by loss of power to the memory device as is the case with RAM chips. (The hard disk, on the other hand, *is* susceptible to stray magnetic fields much like any other recording magnetic media, and also to dust or dirt particles that may be present between the media and the reading and writing heads of the hard disk mechanism.) The RAM chip is considered to be a *volatile* memory device. A nonvolatile RAM chip memory system is made possible by using batteries to hold up the memory. This is a fairly common practice to prevent memory loss during power outage conditions.

Now that we have identified the major parts of a microcomputer, let's look at the different forms in which data may be written and programmed so that it can be stored and processed by a computer. These different forms are the computer's *software* languages. Essentially, there are only three major categories of programming languages:

1. binary or machine code language,
2. assembly language, and
3. the higher level, or structured languages.

The most basic of software languages is *machine language* or code. This is a language form comprised of nothing but binary bits or numbers. This is an extremely tedious and lengthy method of writing and is not often used anymore. Machine language is referred to as a low-level language. As a matter of fact, it's the lowest possible level one can use in writing software, since it is *the* language of the computer. All data manipulations and all control signals are handled in binary form on the computer's mother board and peripheral boards. Sometimes when using machine language, rather than enter binary numbers, often what was done was to enter this binary information in hexadecimal form. This was done to reduce the number of keystrokes and to reduce keystroke errors. The keyboard was in hexadecimal form, and some form of light-emitting diode (LED) display was used on the earlier computers to keep track of programming progress.

The next lowest level software language is *assembly language* or code. This is a language form comprised of mnemonic statements; that is, statements that could be pronounced by the user for easy memorization. Each mnemonic statement represents a binary coded statement or several binary words that the user would never see. However, should that mnemonic statement be used, a series of binary codes would be generated. An *assembler* is used to actually perform the conversion of the assembly language into actual binary words. In order to program in assembly language, the user must have a thorough understanding of the computer's hardware and architecture.

The category of programming languages covers a wide territory. The programmer has a wide selection of programs to choose from. Higher level programming languages encompass those languages that have the user enter certain English words of instruction that perform a particular action or manipulation of the data. Perhaps one of the better examples of such a language is BASIC. By entering the proper sequence of these commands or instructional words, the program will do what the programmer has instructed it to do. The programmer is unaware that for each of his or her instructions in the program, that instruction has been automatically converted to machine language for the computer to run on. That portion of the high-level program that does the high-level language-to-binary conversion is called an *interpreter,* or *compiler*. The major difference between an interpreter and compiler is found in the methods used in generating the machine code. A compiler reads the entire high-level program first and then generates the machine code; the interpreter, on the other hand, reads one high-level instruction or statement at a time, producing the machine code as it goes along.

Some high-level programs are higher level than others. For instance, there are several statements in BASIC that occur frequently together to perform a particular function. Another programming language will take those statements and combine them into one statement so that all one has to do is type that one word instead of having to type the several words as had to be done in BASIC. Much time in programming can be saved using a higher level language. However, programming versatility is lost when using such a language. This used to be one of the major drawbacks; however, this is changing with more sophisticated programs now being marketed. This allows the user to have greater flexibility in programming and still maintain the greater programming speed characteristics of higher level languages. It's interesting to note that some programming languages are approaching using common everyday English expressions for the programming commands. As a matter of fact, more than one common expression will perform the same command. This means that the programmer, without having to resort to a memorized command set of words as is now done, can instead issue a command word using his or her intuition. The programming language will then refer to a dictionary of terms that will interpret equivalent words and issue what it interprets to be the desired command.

12-4 *AN EXAMPLE OF AUTOMATIC CONTROLLING USING A COMPUTER*

Now that we have some understanding of how computers work and what software is all about, our next task is to understand how all of this information is applied to a computerized digital automatic control system. To begin with, we have already gained some insight on how the controlling and analyzing of data is done. But we back away now and take a look at the computer itself and at the surrounding hardware that it can send information to and receive information from.

Let's begin with analyzing a computer-controlled automatic control system used for process control. As an example, the process we discuss here will be the cold-chamber aluminum die casting process we discussed earlier in Section 10-9. We wish not only to *monitor* the various important parameters that go into the making of a good aluminum die casting part, but we also wish to *control* these parameters and to make any necessary corrections, automatically. In other words, we want to make an adaptive control system for our die casting machine. Each process control system has its own particular set of parameters it must monitor and control in order to insure producing a good part. For the die casting process, these parameters are (refer to Figure 12-5):

1. metal temperature;
2. shot velocity (this is the velocity of the hydraulic ram or plunger used for injecting the molten aluminum into the die);

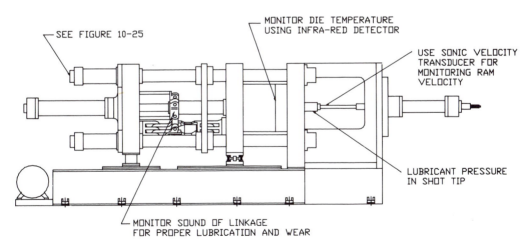

Figure 12-5 Parts of a die cast machine. (*Wayne Alofs/James R. Carstens, MECHANICAL MAINTENANCE AND EVALUATION OF DIE CASTING MACHINES, copyright 1987. Reprinted with permission by the Society of Die Casting Engineers, River Grove, Ill.*)

3. lubrication of shot tip (this has to do with the molten metal injection portion of the cold chamber and is critical for keeping wear to a minimum between the shot plunger and the shot sleeve itself);

4. location of ladle used for pouring metal into cold chamber shot hole;

5. dimensions of finished part;

6. proper part ejection;

7. tie bar tension on the die casting machine to maintain proper closure tonnage on die during metal injection and cooling of part;

8. die temperature;

9. lubrication of the die casting machine's toggle system (This is the mechanical portion of machine that mechanically amplifies the hydraulic pressure used in creating the proper closing pressures on the die); and

10. electrical energy consumption of the die casting machine.

It takes at least one or perhaps two minutes to produce one part out of a die casting machine. In other words, this is the cycle time for the process. Within this cycle time, all 10 of the foregoing parameters must be monitored and any corrections to the system made. In order to monitor and control a complicated operation such as this using a computer, transducers must be installed in appropriate areas surrounding the process. Ideally, these transducers should have the capability of producing digital signals. This would eliminate any analog-to-digital conversions that would otherwise be needed for interfacing to the computer. In addition, you would have the maximum circuit noise immunity needed for a safe and reliable operating system.

In developing the software needed to supervise this system, a program will have to be written that will scan each of our 10 listed areas. The outputs of our transducers will have to be wired so that each output can be accessed at the computer's input/output ports. This is done so that each signal can be compared to a prerecorded signal value that has already been considered to be an acceptable value. If any of the transducers' outputs are considered to be unacceptable, appropriate corrections will have to be made automatically.

The best way to develop and understand an automatic control and monitoring system is to make a flowchart of block diagram of the system. Figure 12-6 illustrates such a diagram for our die casting system. Block diagrams tend to be uninteresting, but they do contain a wealth of information. Be sure to study Figure 12-6 closely and make sure you understand each line and notation. If you do this, you'll have little problem understanding any other systems you may have to work with in the future.

Figure 12-6 shows what appears to be a mechanical scanning system where a switch travels from one block system to another, making an electrical contact. This is shown for illustrative purposes only. In reality, all electrical switching is done electronically with no moving parts. As each portion of the die casting machine is scanned, the computer sends back corrective information in the form

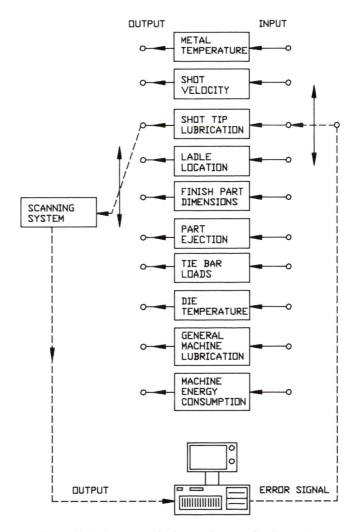

Figure 12-6   A computerized control system for die casting.

of an error signal to that same area. As a matter of fact, both the scanning and error corrections may be done several times a second, with the exception of the following: shot velocity, finished part dimensions, ladle location, tie bar load, and part ejection; these particular items can be checked only during or immediately following the making of the part.

To illustrate the preceding remarks, let's assume that our software has told the input/output (I/O) port on the computer to read the metal temperature in the ladle as it is being scooped from the nearby furnace. Thermocouples in the ladle read the metal's temperature and the resultant generated voltage. This voltage signal is sent to an A/D converter circuit. The circuit, in turn, sends a digital signal to the computer, where its newly converted binary equivalent value is compared to a desired temperature (probably around 1,250°F to 1,300°F). Any difference in temperature is immediately sent back in the form of an error signal to the furnace's temperature controller, where any necessary adjustments are made. Because of the massiveness of the furnace and the contained aluminum charge, the response time, and consequently its settling time to the set point, is going to be very slow, measurable in minutes.

The die casting machine's hydraulically controlled shot producing system is probably one of the more complicated and difficult systems to control automatically on the machine. Its description was given in Section 10-9. Because of the speed with which a shot is made, it is difficult to make speed corrections with generated error signals *during* the shot. It is much easier to make corrections on the *next* shot. This is a typical problem for many mechanical devices having large masses. Their response times are very slow compared to their total travel time, and consequently, they must settle for next-cycle correction only.

The next item, shot tip lubrication, involves having to apply lubrication to the tip of the ramming device that pushes the molten metal into the die. If this part is not lubricated, excessive drag and wear is created in the shot sleeve. This is the chamber in which the metal is poured. If the computer software detects any slowing down of the ram during successive shots, this could be an indication of lack of shot tip lubrication. A normal closed-loop system would create an increasingly larger error signal as the lubrication decreases, causing the velocity to become compensated with increased speed. Unfortunately, this would cause the entire shot system to eventually destroy itself if no corrective steps were taken to shut down the control system. However, with the adaptive control system that we have described up to this point, additional lubrication would be automatically added to the shot tip area to prevent its self-destruction from taking place.

Ladle location must be determined in order for the ladler or robot to accurately hit the pour hole in the shot sleeve as it pours the molten metal. This means that the ladle's position must be continuously monitored and corrected using the methods outlined in Section 10-2.

The checking of finished part dimensions is performed by a vision system similar to the one described in Section 10-3. This particular system is a go-no-go system. If the part is within specifications, it is saved; if it isn't, the part is recycled.

The part ejection system checks to make sure that the mechanical ejection system built into the die has in fact ejected the part out of the die. This check may be made by a displacement transducer built into the robot's gripper that is used to remove the part from the die. If the part is properly grasped by the gripper, the transducer will transmit the proper displacement of the gripper. Otherwise, any other signal value coming from the gripper will indicate a problem with the ejection procedure. This system is basically another go-no-go type system rather than a proportional feedback system.

The tie bar loading on the die casting machine is monitored and controlled according to the description given in Section 10-9.

For the controlling of die temperature, the die contains temperature conditioning coils that have water circulating through them. Temperature mixing valves are used to mix heated and cooled water to obtain the desired die temperatures. Because of the massiveness of the steel dies used, the temperature response time is quite slow, on the order of several minutes typically. This implies that the settling times for any temperature corrections will be quite long. Installed on the mixing valves are servovalves that adjust the valve stem positioning. The die's temperature is monitored by a scanning infrared detector that measures temperatures over its entire face. These temperatures are then compared to the desired temperature values stored in software. The software then calculates any arithmetic differences in these values and sends that difference back to the die's water mixing facilities in the form of an error correction to operate the appropriate valves for temperature correction.

General machine lubrication may be checked by sonic transducers (microphones) placed at the important bearing points on the machine. The sound output of these detectors can then be compared to acceptable output levels. If any levels are exceeded due to plugged or broken lubrication lines, alarms may be sounded or lights flashed to get the attention of the machine supervisor. This system is another go-no-no control system.

Machine energy consumption can be checked by comparing the electrical current draw of the machine and comparing that information to what is considered a normal current draw. The current may be monitored by a digital ammeter circuit. Any increase in current draw may be an indication of binding members on the machine due to member failure or a failed lubrication system.

As was pointed out during the describing of each of the machine operating parameters, some of the parameters listed in Figure 12-6 obviously don't require an analog automatic feedback control. The finished part dimensioning, part ejection, shot tip lubrication, general machine lubrication, and machine energy consumption areas are all go-no-go decision-making operations. Instead, these may be thought of as bang-bang automated systems. As in any control system, however, we want our computer control system to coordinate all of the various operations of our die casting process; this means writing a rather extensive software program to do all of this.

Again, we use a flowchart to see how we might write the software program to automatically control our die casting operation. This is a very generic program. That is, it is merely to be an overview for a far more detailed program. The choice of programming language used for this project is a choice left up to the programmer and the computer that will be used to run the program. Regardless of the language used, the programming logic remains the same. Figure 12-7 shows a flowchart that could be used in aiding the programmer in writing the software for controlling our die casting process. It must be pointed out that this is a very crude chart. No safety interlocks are shown, and many other auxiliary control circuits

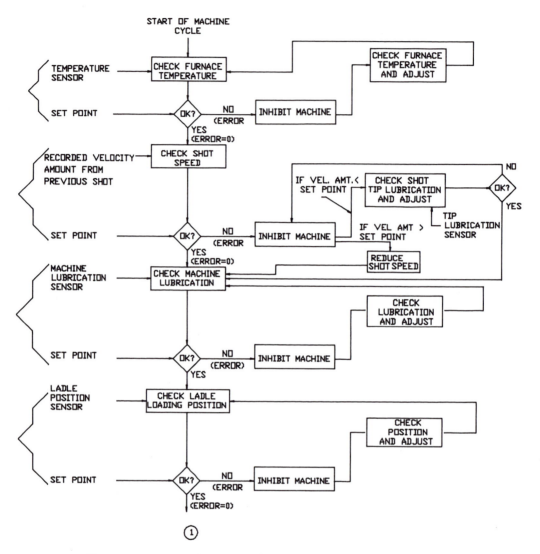

Figure 12-7 Flowchart for programming the automatic control of a work cell of a die cast machine operation.

are also missing. Also, there are many other ways to construct the programming flow to create the same end results.

Looking at Figure 12-7, the flowchart begins with checking the furnace temperature to make certain that the aluminum which is to be used in the die casting machine is at the proper pouring temperature. A sensor on the furnace supplies the temperature data and is compared to the set point temperature. If an error signal results, the machine's operation is stopped so that the furnace's tempera-

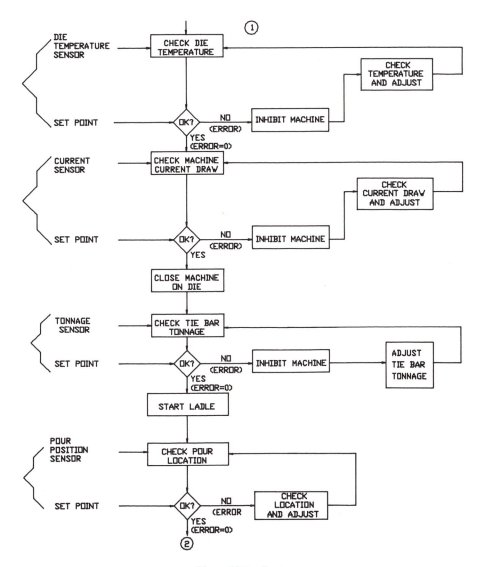

Figure 12-7   *Cont.*

ture can be readjusted. During this readjustment period, the temperature sensor continues taking readings until the set point is reached, or until the error signal is zero. When this occurs, the next parameter is analyzed, which is check shot speed. Again, a sensor's reading (from the previous shot) is compared to the set point. If the shot speed sensor detected a greater velocity from the shot just completed as compared to the set point velocity, the shot controller is told to reduce the shot velocity for the next shot by an amount equivalent to the error

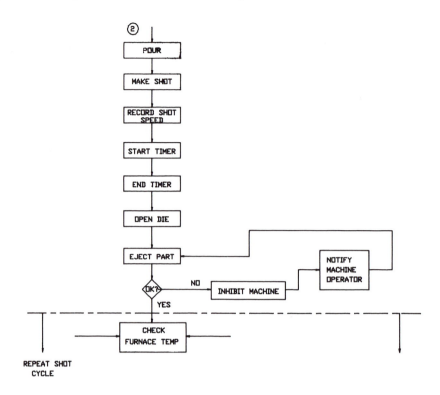

Figure 12-7 *Cont.*

value. If the shot speed sensor detected a smaller velocity as compared to the set point velocity, the shot controller is told to increase the next shot velocity an amount equivalent to the error value. However, in addition to this, the tip lubrication circuit is also checked to make sure proper lubrication is present. This additional check is made because a decreased shot velocity could be a result of lack of tip lubrication. Following these checks, the software proceeds to check machine lube.

As we continue down the flowchart, we encounter a series of functions beginning with the pour command and continuing down through the start timer and end timer commands. These two commands control a timer that determines how long the die must be kept closed in order to allow the part to completely freeze before reopening the die and ejecting the part.

In each of the parameter check categories listed in the flowchart, notice that each parameter is confronted with a YES/NO routine. This is a common programming routine where checks have to be made on circuit conditions. These checks are logical functions carried out by software routine statements that ask questions similar to, "If *x*-axis ladle position is ≤0.010 inch, then go to next statement (i.e., YES), else, (i.e., NO) correct *x*-axis position to reduce error and recheck." All the statements in the flowchart to the right of each parameter check are *loop statements*. These statements perform a correction and then cause a loop to occur so that the parameter can be rechecked for error correction before proceeding to the next parameter.

The foregoing process control system is admittedly rather detailed and specific. It wasn't intended to be a dissertation on the die casting process. However, the analysis of the work-cell software routines is quite representative of many automatic process control systems encountered in industry and is definitely worth studying.

12-5 *USING THE STEPPER MOTOR IN AN AUTOMATIC CONTROL SYSTEM*

The stepper motor is an ideal actuator for interfacing with the computer for control. Since the motor itself responds to digital-like signals, it's a natural device for accepting commands and converting them into rotary motion. The theory of operation of this motor was discussed in Section 4-17. What we discuss here is how to control this motor with a computer and how the motor is used in an automatic control application.

A typical stepper motor control system appears in Figure 12-8. This is an entirely digital control system. A positional transducer is attached to the stepper motor's load to create the necessary error signal back at the comparator. The directional control logic circuitry in Figure 12-8 is circuitry that converts the mathematical sign of the error signal into the logic needed that will rotate the stepper either clockwise or counterclockwise. (Be sure to review Section 4-17 if you have forgotten how a stepper's rotational direction is changed.) The software

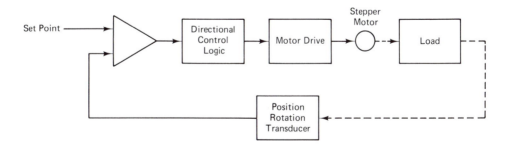

Note: Dashed Lines Indicate a
Mechanical Coupling

Figure 12-8   An automatic control system using a stepper motor.

needed to interpret the stepper's position is laid out in the form of a flowchart in Figure 12-9. This program is a continuous looping operation, since it must catch any changing of the set point at any time made by the operator of the system. The number of updates of the stepper's position is controlled by the frequency of the loops made in the software. This, in turn, is determined by the clock rate of the computer's CPU.

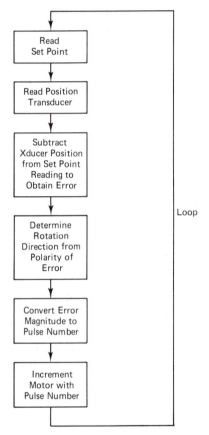

Note: The Number of Loops Per Second Made by the Software Will be Determined by the Computer's Clock Speed.

Figure 12-9   Flow-charting the software for a stepper motor positional system.

A typical application of a system using a stepper motor for positional control would be in the designing of a graphics pen plotter (Figure 12-10). In this system, the plotter's pen location will be continuously changing, while at the same time in the background the pen's position would be continuously monitored and updated by the algorithm seen in Figure 12-9. More than likely, this program would be found in a ROM chip where it would then be protected from accidental alterations.

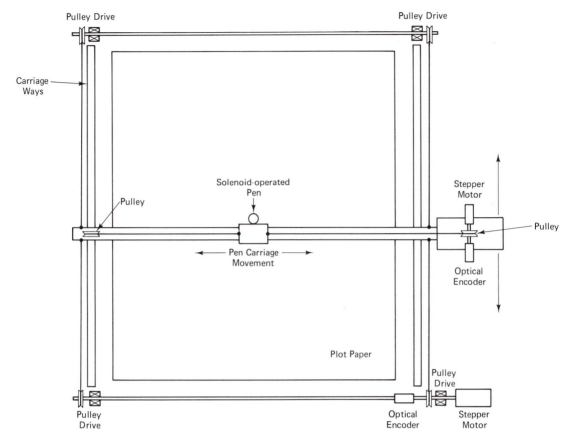

Figure 12-10   A stepper motor pen plotter.

12-6 USING A PROGRAMMABLE CONTROLLER
FOR AUTOMATIC CONTROL

A programmable controller (or PC as they are frequently called) is a device that in many respects resembles a personal computer (also called a PC). The first programmable controller was produced in the middle 1960s for the purpose of replacing electrical relays. Up to that time, the electromechanical relay was used extensively to control industrial process control systems. All logical functions (the AND, OR, NOT, NOR, etc. statements) were done in this manner. However, the PC was a solid state device that used transistors, and later integrated circuits, to perform all switching logic. A modern industrial PC is illustrated in Figure 12-11. In addition to using solid state switching for relays, the PC is programmable. The programming language most commonly used closely resembles the relay language used by electricians and control engineers to draw out the control cir-

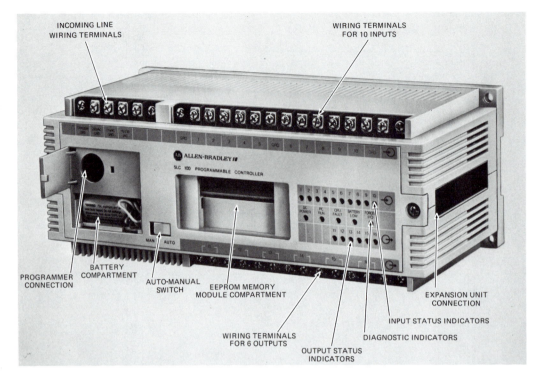

Figure 12-11   A programmable controller. (*Courtesy of Allen–Bradley Co., Milwaukee, Wis.*)

cuitry schematics, called *ladder diagrams* (Figure 12-12). These diagrams are comprised of all electrical relay symbols, and because of the familiarity that electricians have with this relay logic, the relay ladder diagram programming method with the PC has been very popular. As a matter of fact, to this day most of the PCs being manufactured still use this form of programming. There are at least two other forms of languages used, however, with the PC. These are Boolean algebra-based languages and mnemonic-based languages (similar to the computer assembly languages discussed earlier). Boolean algebra is a form of algebraic logic that uses special math symbols in which the logic AND statement is represented by a dot (·), the logic OR statement is represented by a plus (+), and equivalent statements are represented by the equal sign (=). This type of programming is more electronic in its approach as compared to the more conventional electrical approach.

The major advantages in using a PC is that, like a microcomputer, it is programmable and it is also equipped to respond to the input signals coming from switches and transducers (on the more elaborate models). It is also built to withstand the hostile environment of an industrial installation. Most computers can't survive in such areas. Unfortunately, most PCs lack the versatility of a micro-

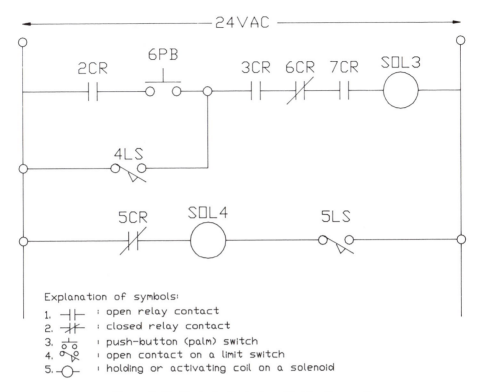

Figure 12-12   Typical relay logic or "ladder diagram."

computer as far as being able to process data, handle large memory requirements, and run complex programs using the usual programming languages. Also, a microcomputer allows the user to monitor the information being processed by using the appropriate software. This is especially important when interpreting data coming from a transducer. The major components of a PC are:

1. microprocessor,
2. nonvolatile memory,
3. input/output interfacing circuitry, and
4. programmer.

DEFINITIONS

MICROPROCESSOR   The microprocessor has the same function as the one found in microcomputers, except it usually doesn't have the processing capabilities as the ones found in computers. The reason for this is to keep the PC manufacturing costs down. (This is changing, however. Some PC manufacturers are beginning to install actual microcomputer chips to take advantage of the chip's additional computing power.)

NONVOLATILE MEMORY The nonvolatile memory has stored in it all the commands and logic information needed to control the process. In case of a power failure, the program wouldn't be lost, since the memory would be "held up" with batteries. This can become a serious problem with volatile memory when power is restored and the controlled process has no homing instructions to begin new cycling. Damage to equipment could result, and personnel safety could be jeopardized.

INPUT/OUTPUT INTERFACING CIRCUITRY This is the circuitry that allows the PC to control and to gather information from the process. The input circuits and the output circuits are separated from each other through a modular construction so that they can be segregated and moved around. Each I/O module may handle over a dozen different circuits and is designed to handle AC, DC, pulse, and BCD data in the more elaborate models.

PROGRAMMER The programmer is a system that allows the user to program the PC. It may consist of a CRT (cathode ray tube) for following the programming procedure, and a keyboard. The keyboard may be either attached to the PC through a detachable cable, or the programming may be done through a hand-held keyboard. Some hand-held models may even have a small liquid crystal display screen built into them for programming interpretation.

There are also auxiliary devices that can be used in conjunction with the PC to allow the user to monitor process information. Figure 12-13 shows a device designed to do this very thing. It's called an Industrial Graphics Terminal. It has the capacity to be programmable much like a computer so that it can display high-resolution graphics. The graphics give a visual representation of the process being monitored. It is interesting to note that some of the later models of PCs are beginning to look and behave more and more like computers. They can handle large numbers of input/output signals, perform complex math, create graphics, and automatically control large process systems. In addition to all of this, they are programmable in BASIC-like languages. One such system is called the PCAM, or Programmable Controller and Monitor. It is specifically designed to be used in automatic control systems where it can both monitor and control a process. A complete system hook-up is shown in Figure 12-14. Notice that one PCAM unit is assigned to a particular work cell area or machine. A personal computer is then used to gather data from each PCAM through a networking system. This means that a supervisor can be in a remote location such as an office with his or her computer and have the capability to control and monitor each of these areas. Several dozen work cells can presently be controlled in this manner. As an added side note, the user can also gather management and diagnostic data from those areas with this system.

Figure 12-13   Graphics terminal for monitoring process data. (*Courtesy of Emory–Anderson, Inc., Comstock Park, Mich.*)

Because this type of controller is programmable in a language similar to BASIC, it is possible to write fairly complex programs. In other words, it is possible to write a program similar to, or as extensive as, the one described by the flowchart in Figure 12-7. This type of controller is especially designed to work with transducers and control devices that are commonly used in automatic control system applications. As an example, referring back to our flowchart of the die casting operation, an ordinary PC could not handle and interpret the information necessary for the adaptive controlling of the shot-making process. A PCAM device, on the other hand, could not only interpret this information easily and make the necessary adjustments automatically, it could also generate any necessary monitoring information on a high-resolution graphics screen built into the unit.

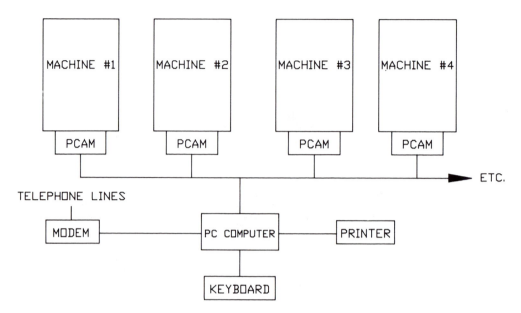

Figure 12-14 A PCAM network. (*Courtesy of Emory–Anderson, Inc. Comstock, Park, Mich.*)

Factory automation is becoming more and more prevalent these days. As a result, the automation of processes has become increasingly important not only from an economic standpoint but also from a safety standpoint. One of the great concerns for a manufacturing process firm is how to remove its workers from dangerous work situations. Automatic control of these processes is the answer.

12-7 METHODS OF COMPUTER INTERFACING

In working with electronic circuits, we learned that in order to hook up the output of one circuit to the input of another, it was necessary to match their impedances for maximum power transfer. Also, we had to be careful that proper voltage protection existed so that we weren't feeding, say, a 440 VAC three-phase circuit into a 12-volt CMOS circuit. Obviously, we would have an instant problem developing here. In hooking up circuits to a computer, we have much the same problem. We are dealing with sensitive high-density solid state components inside the computer which lack the ruggedness of the old vacuum tubes. We have to be especially careful when designing control circuitry and wiring these circuits to computers. Fortunately, the computer industry has standardized some of the hook-up circuitry to reduce some of the difficulties just mentioned. Some of the standardization procedures are still being formulated, but for the most part, enough has been done to eliminate incompatibility problems between circuits and computers.

Every industry has created its share of buzzwords for the English language, and the computer industry is certainly no exception. When hooking up a circuit to a computer, the term *interfacing* is used instead of the term *hooking up*. As a matter of fact, the circuit being interfaced just may be a *peripheral device* for the computer. In other words, it may be a circuit which assists the computer in performing a particular function. What we are interested in here are the methods used for computer interfacing. We have to be able to select the proper connectors, the cabling, and be able to understand the circuit enough to be assured that the signals we want to swap back and forth between the computer and circuit are digestible by each other.

Assuming that we are going to interface a digital automatic control circuit to a computer so that the computer can monitor and control the circuit, we have to ask ourselves, is the data being sent to the computer serial-type or parallel-type data? This will determine the standard interfacing method to be used. If the data coming from the circuit is serial, we would most likely use the RS-232C Interface. If the data is parallel, we would probably want to use the IEEE-488 Interface. An explanation of each of these along with others follows.

12-7.1 The 20 mA and 60 mA Current Loops

This is the oldest of existing methods used for interfacing electronic and electrical equipment together. It was originally used for open-loop control of electromechanical equipment in communications (such as teletypewriters) and industrial data processing. Distances between the control transmitter and the controlled circuit at the receiving site can be as much as 2,500 feet. However, at these distances considerable voltage-drop losses could result. These systems were used for transmitting and receiving serial data. A notable disadvantage with this system is the very low rate of data transmission. This was due primarily to the electromechanical switching devices that were used at the time. The highest rate attainable was approximately 150 baud. (The term *baud* refers to the rate of data transmission in units of bits/sec.) Two systems were popular; one involved using current flows of 20 mA, while the other used 60 mA (Figure 12-15). The two systems were identical otherwise. At the receiving end of the data line was a solenoid that actuated a mechanical system for printing characters. In the later systems, the solenoid was replaced with amplifiers and sensors. These were used for process monitoring at remote distances. However, even though the electromechanical switching devices were eventually replaced with much faster vacuum tube circuits, and later still, with solid state devices, the current loop systems were beginning to be replaced with the RS-232C systems.

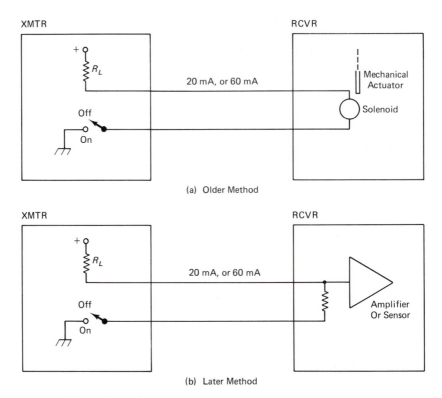

Figure 12-15   Methods of using 20 mA and 60 mA current loops.

12-7.2 The RS-232C Standard

This standard was developed by the Electronic Industries Association (or EIA) and was originally designed for use in communications involving printers, terminals (i.e., CRTs), and modems. The standard was originally referred to as the EIA RS-232 standard but was almost immediately revised, consequently becoming known by the present designation of the RS-232C standard. This standard is another serial data system. It's interesting to note that even though serial transmission of data requires only two lines, one for the data and another for ground return, the RS-232C standard specifies 25 pin connections in all for the connectors at each end of the cable that must be used. The pins are identified in Table 12-2.

Figure 12-16 shows the pin configuration on the D-type connector commonly used for RS-232 interfacing. In reality, most of the pins are not used for most applications. Pins 1, 2, 3, and 7 are the only pins really needed in many installations. As few as two pins, pins 3 and 7, can be used in some installations for receiving data only. Figure 12-17 depicts a typical RS-232C installation between a computer and a process station. The maximum *baud rate* for an RS-232C system

**TABLE 12-2   EIA RS-232C INTERFACE
STANDARD SHOWING PIN CONNECTOR
ID**

Pin	Description
1	Protective ground
2	Transmit data
3	Receive data
4	Request to send
5	Clear to send
6	Data set ready
7	Signal ground (common return)
8	Received line signal detector
9	(Reserved)
10	(Reserved)
11	Not used
12	Secondary rec'd line signal detector
13	Secondary clear to send
14	Secondary transmitted data
15	Transmit signal timing
16	Secondary received data
17	Receiver signal timing
18	Not used
19	Secondary request to send
20	Data terminal ready
21	Signal quality detector
22	Ring indicator
23	Data signal rate detector
24	Transmit signal element timing
25	Not used

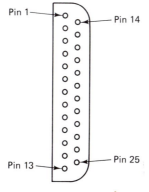

Figure 12-16   The 25-pin D-type connector.

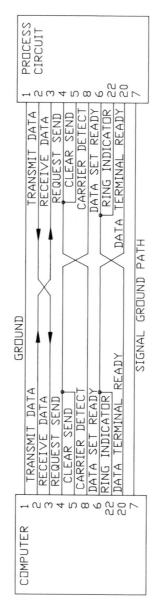

Figure 12-17 RS-232C data communications between computer and remote process circuitry.

is 20,000 baud. That is, only 20,000 bits per second of data can be handled with the serial interface.

12-7.3 The Centronics System

The Centronics system is really not an interface scheme used for control systems as such. It is presented here only because of its extensive use in computer systems and to make this discussion on interface systems complete. The Centronics system is used primarily for interfacing printers to computers (see Figure 12-18). It is a parallel interface that is probably the most popular system used today for printer communications. It can transfer data at tremendously high rates compared to the serial interfaces just discussed, which accounts for the high-speed printing rates that are common today. In Figure 12-18 we see two connector pinouts illustrated. One is for the connector that goes into the printer device, while the other is the connector that goes into the computer or CPU. The CPU's connector is constructed in such a manner that it is not possible to mistakenly plug the wrong end of the interconnecting printer cable into the computer's printer connector. The connector on the printer is a 36-pin connector containing eight lines that carry the eight-bit wide word in parallel. The transmission of the binary data is controlled by a computer-generated *strobe* signal. The flow control of the binary data is done through handshaking; this is the turning on (causing a line to go high with 1 bit) or turning off (causing a line to go low with a 0 bit) of the ACKNOWLEDGE or BUSY leads, or a combination of both these lines. The connector on the cable's other end is a 25-pin connector similar to the D-type connector used for the RS-232C system. The pins' wiring connections, however, are radically different. Also, the 25-pin connector for the RS-232C cable is a female connector, whereas the 25-pin Centronics cable connector is a male connector.

12-7.4 The IEEE-488 General Purpose Instrumentation Bus

The International Electronics and Electrical Engineering Society standardized a parallel interface scheme called the IEEE-488 General Purpose Instrumentation Bus, otherwise known as the IEEE-488 GPIB. The system was actually developed by Hewlett Packard Corporation for their instrumentation needs but was later adopted and standardized by the IEEE organization. The 488 bus allows two-way communications between as many as one dozen or so different instruments equipped to be attached to the system. However, the spacing of these instruments from one another is rather limited, for reasons that were mentioned earlier in our discussion on communication techniques. The circuitry needed for this system is quite complicated requiring specially designed ICs for its implementation. Basically, the 488-bus system allows instruments to transmit, receive, and control data flow between themselves in a very ordered and error-free fashion. In general, the 488 interface is comprised of three kinds of devices: (a) The *talker:* This is a device designed to place data on the bus, allowing it to be transmitted to

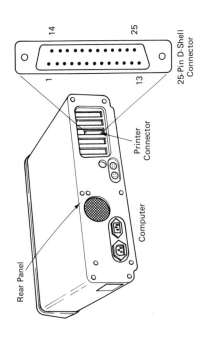

Signal Pin No.	Return Pin No.	Signal	Direction (with ref. to printer)	Description
1	10	STROBE	In	STROBE pulse (negative going) enables reading data.
2	20	DATA 1	In	1st to 8th bits of parallel data. Each signal is at "HIGH" level when data is logic "1" and "LOW" when logical "0".
3	21	DATA 2	In	
4	22	DATA 3	In	
5	23	DATA 4	In	
6	24	DATA 5	In	
7	25	DATA 6	In	
8	26	DATA 7	In	
9	27	DATA 8	In	
10	28	ACKNLG	Out	"LOW" indicates that data has been received and that the printer is ready to accept other data.
11	29	BUSY	Out	"HIGH" indicates that the printer cannot receive data.

Note: Pins 12, 13, 14, 15, 18, 31, 32, 34, 35 and 36 vary in function depending upon application; they are commonly used for printer auxiliary controls, and error handling and indication. Pins 16 and 17 are commonly used for logic ground and chassis ground, respectively.

Signal Designation / Pin Number

+5 V — 18
Chassis GND — 17
Logic GND — 16
OBCXT — 15
Supply GND — 14
Select — 13
Paper End — 12
Busy — 11
Acknowledge — 10
Data Bit 8 — 9
Data Bit 7 — 8
Data Bit 6 — 7
Data Bit 5 — 6
Data Bit 4 — 5
Data Bit 3 — 4
Data Bit 2 — 3
Data Bit 1 — 2
Data Strobe — 1

Pin Number / Signal Designation

36 — Undefined
35 — Undefined
34 — Undefined
33 — Undefined
32 — Fault
31 — Input Frame
30 — (R) Input Frame
29 — (R) Busy
28 — (R) Acknowledge
27 — (R) Data Bit 8
26 — (R) Data Bit 7
25 — (R) Data Bit 6
24 — (R) Data Bit 5
23 — (R) Data Bit 4
22 — (R) Data Bit 3
21 — (R) Data Bit 2
20 — (R) Data Bit 1
19 — (R) Data Strobe

(R) Indicates Signal Ground Return

Figure 12-18 The Centronics connector and the pin-out used both at the computer and at the printer. *Connector and pin-out information reproduced by permission of Black Box Corporation, 1987 ed. of Black Box® catalog, Pittsburgh, PA.*

one or more connected devices. As a rule, only one talker is allowed to talk at a time. (b) The *listener:* This device is designed to accept data from the bus and is allowed to do so through a system called *handshaking.* There can be more than one operating listener on the bus at any given time. (c) The *controller:* This is a device whose job is to issue commands to both the listeners and the talkers. There may be more than one controller on the bus, but only one controller can operate at a time.

The connector style, pin-out, and additional pin descriptions for the connectors used in this system are all shown in Figure 12-19. Because some of these

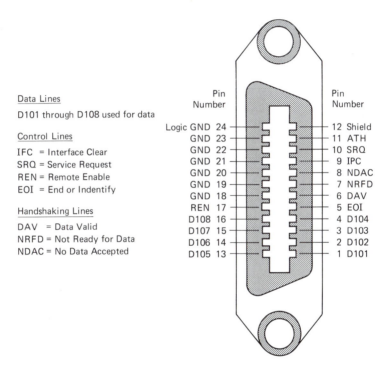

Data Lines

D101 through D108 used for data

Control Lines

IFC = Interface Clear
SRQ = Service Request
REN = Remote Enable
EOI = End or Indentify

Handshaking Lines

DAV = Data Valid
NRFD = Not Ready for Data
NDAC = No Data Accepted

Pin Number		Pin Number	
Logic GND	24	12	Shield
GND	23	11	ATH
GND	22	10	SRQ
GND	21	9	IPC
GND	20	8	NDAC
GND	19	7	NRFD
GND	18	6	DAV
REN	17	5	EOI
D108	16	4	D104
D107	15	3	D103
D106	14	2	D102
D105	13	1	D101

Figure 12-19 The IEEE-488 General Purpose Interface Bus. *Connector and illustration and pin-out information reproduced by permission of Black Box Corporation, 1987 ed. of Black Box® catalog, Pittsburgh, PA.*

devices must both receive and transmit data and yet not interrupt the flow of data on the bus, a provision must be made in the connector design to allow for the "stacking" of connectors, much like the stacking of appliance plugs at an AC socket outlet. Further information on this system can be obtained from the references listed at the end of this chapter.

12-8 PLUG-IN CIRCUIT BOARD INTERFACING METHODS

Up to this point, we have discussed methods for interfacing circuits with other circuits that are external to the computer. When designing an automatic control system to operate around a custom-built microprocessor-based system, the problem develops of interfacing internal circuit boards with other circuit boards designed to operate directly with a mother board containing the CPU. This is a custom systems design problem that is solved by using one of the several existing circuit board interfacing standards. This is a discussion for advanced digital circuit design, and we don't get into it here at this level. However, it's important to be able to recognize the names of these interface standards if only for reference purposes. The IEEE-696/S100 standard (sometimes referred to simply as the S100 bus) is used extensively for designing plug-in circuit boards, as is the STD-BUS system. The term *bus* merely refers to the circuit paths that exist between two or more devices in a computer system that are used for the transferring of data, control signals, and operating power. The S100 bus card is shown in Figure 12-20(a); the STD bus card is shown in Figure 12-20(b). Other systems exist also, and their combined totals probably outnumber the two just mentioned. But you can be assured that these other more numerous systems will simply be minor variations of the S100 and STD bus systems. As to which of the two systems is the more popular, the S100 bus would probably be the choice. While the STD bus is a better defined system (it has a 56-pin bus with 8 data lines, 16 address lines, 22 control lines, and 10 power line connections), the S100 bus has 16 data lines, 24 address lines, 11 interrupts, and is capable of multiprocessing (i.e., it can support the operation of two or more CPUs within the same system). Address lines carry

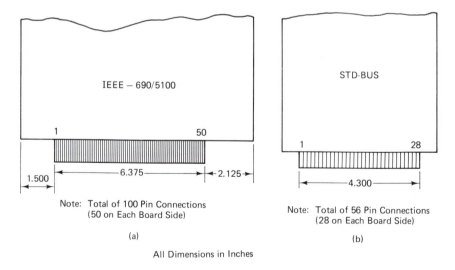

Figure 12-20   Two popular edge connector codes.

the data placement information so that the data get stored in the proper location within a register or other computer-related storage area. Interrupts are signals used to temporarily halt the operating system. These signals come from either within the computer itself for system supervisory reasons, or they originate from outside the computer, possibly coming from an I/O device indicating a completion of data transfer.

Because of the increasing usage of computer systems and components in automatic control systems, it makes sense for the design technician or engineer to become familiar with these various bus systems. It's becoming increasingly difficult for mechanical engineers and technicians to remain purely mechanical, and for electrical engineers and technicians to remain purely electrical. The area of automatic controls almost demands that there be a cross-pollination of ideas and information between these two fields in order to keep up with the increasing complexity of control system development and design.

SUMMARY

Unquestionably, the single biggest change in automatic control design has been the adaptation of computer control. This change has forced many of the control designers to rethink their design philosophy and to learn new hardware and software technology. In some cases, especially in Chapter 12, you may not have seen the relevancy of all the emphasis on computer technology with automatic controls. That's easy to do, since there is so much to learn in this subject. People in industry have had an especially difficult time with this. More often than not, they have had to adapt and learn computer science on their own time because of the magnitude of information they needed to consume. Then, having done that, they have had to figure out how to integrate all that information into their particular application. The use of programmable controllers in process control is just one of these adaptations industrial users are having to cope with. Certainly, the communications industry has had to do the same thing. A good example of this is the use of computer remote control in communications satellites orbiting the earth and the remote control of extraterrestrial robots. Every one of these systems requires some form of automatic control for its proper operation. But, in order to understand the control's system of operation, it's necessary to understand how the computer operates.

REVIEW QUESTIONS

12-1. Explain the various functions of the microprocessor chip. What does the ALU do in this chip?

12-2. Explain the purpose of a data bus in a microcomputer.

12-3. Explain the difference between RAM and ROM memory chips. Why is it necessary to have both?

12-4. Explain why we can't depend solely on floppy disks and hard disks for memory in a computer. In other words, why can't we do away with RAM chips entirely?

12-5. What is the purpose of the BIOS chips in a microcomputer?

12-6. Explain the difference between volatile and nonvolatile memory.

12-7. What is meant by the language level of a computer language? Why is assembly language considered a lower level language as compared to BASIC?

12-8. What is the function of a compiler? What is the basic difference between a compiler and an interpreter?

12-9. Explain the function of a PLC. What is a PLC's I/O?

12-10. Explain what is meant by Centronics connector, parallel connector, and RS-232 connector.

REFERENCES

ARTWICK, BRUCE A., *Microcomputer Interfacing,* Englewood Cliffs, N.J.: Prentice-Hall, Inc., 1980.

BOYCE, JEFFERSON C., *Digital Computer Fundamentals,* Englewood Cliffs, N.J.: Prentice-Hall, Inc., 1977.

BREY, BARRY B., *Microprocessor/Hardware Interfacing and Applications,* Columbus, Ohio: Charles E. Merrill Co., 1984.

GAONKAR, RAMESH S., *Microprocessor Architecture, Programming, and Applications,* Columbus, Ohio: Charles R. Merrill Co., 1984.

GILMORE, CHARLES M., *Introduction to Microprocessors,* New York, N.Y.: McGraw–Hill Book Company, 1981.

GUNN, THOMAS G., *Computer Applications in Manufacturing,* New York, N.Y.: Industrial Press, Inc., 1981.

STREITMATTER, GENE A. AND VITO FIORE, *Microprocessors Theory & Applications,* Reston Va.: Reston Publishing Company, Inc., 1979.

TOCCI, RONALD J. AND LESTER P. LASKOWSKI, *Microprocessors and Microcomputers: Hardware and Software,* Englewood Cliffs, N.J.: Prentice-Hall, Inc., 1979.

WAGNER, T. J. AND G. J. LIPOVSKI, *Fundamentals of Microcomputer Programming,* New York, N.Y.: Macmillan Publishing Company, 1984.

Answers
to Odd-Numbered
Problems

The following are the answers to the odd-numbered problems in the listed chapters. Chapter 1 is a review chapter; therefore there are no answers listed. Chapter 9 contains design problems; therefore, because of the variety and complexity of solutions that are possible, no answers are offered there either.

CHAPTER 2:

2-1: 25.7   **2-3:**

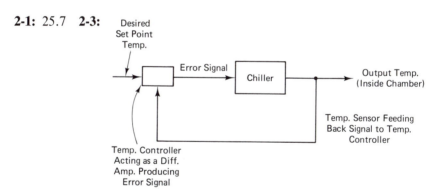

2-5: 20.68 in./gal.   **2-7:** *a.* 60 dB   *b.* Not definable since one can't calculate the log of a negative number. Also, negative voltage gains are not definable.   *c.* 23.2   *d.* −3.74   **2-9:** −3.02 dB

CHAPTER 3:

3-1: $107/s$ **3-3:** $6/s^3 + 9/s^2 - 1/s$ **3-5:** Assume that $10 = 1/\tau$, then $\dfrac{2.5}{s(0.1 + 1)}$

3-7: Assume $1/4 = a$, then $\dfrac{5}{s + 0.25}$ **3-9:** Assume $\tau = 3$, then $\dfrac{1}{s(3s + 1)}$

3-11: $\dfrac{3.2\omega}{s^2 + \omega^2}$ **3-13:** $\dfrac{2}{(s + 0.5)^2}$ **3-15:** $f(t) = 22te^{-t}$ **3-17:** $f(t) = 6.2e^{-3t}$

3-19: $10\sin10t$

CHAPTER 4:

4-1: T.F. $= 0.305$ VDC/in.   **4-3:** $\tau = 0.018$ sec   **4-5:** $\dfrac{E_{\text{out}}}{E_{\text{in}}} = \dfrac{1 + s43.33 \times 10^{-6}}{1 + s173.3 \times 10^{-6}}$

4-7: T.F. $= \dfrac{48}{s(1 + 0.032s)}$ **4-9:** T.F. $= \dfrac{50(1 + 0.06j\omega)}{j\omega(1 + 0.1j\omega)(1 + 0.008j\omega)}$

CHAPTER 5:

5-1: $\dfrac{E_{\text{out}}}{E_{\text{in}}} = 35{,}000$ **5-3:** $\dfrac{E_{\text{out}}}{E_{\text{in}}} = \dfrac{10}{s + 15}$ **5-5:** $\dfrac{E_{\text{out}}}{E_{\text{in}}} = \dfrac{G_1G_2 + G_3}{1 + H(G_1G_2 + G_3)}$

5-7: $\dfrac{S_{\text{out}}}{S_{\text{in}}} = \dfrac{G_1G_2}{1 + G_1G_2\left[\dfrac{H_1 + G_3H_2}{G_3}\right]}$ **5-9:** $\dfrac{S_{\text{out}}}{S_{\text{in}}} = 3363$

CHAPTER 6:

6-1:

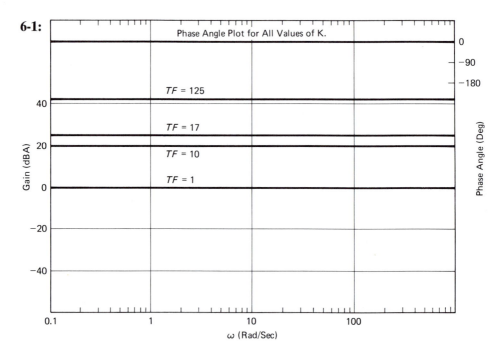

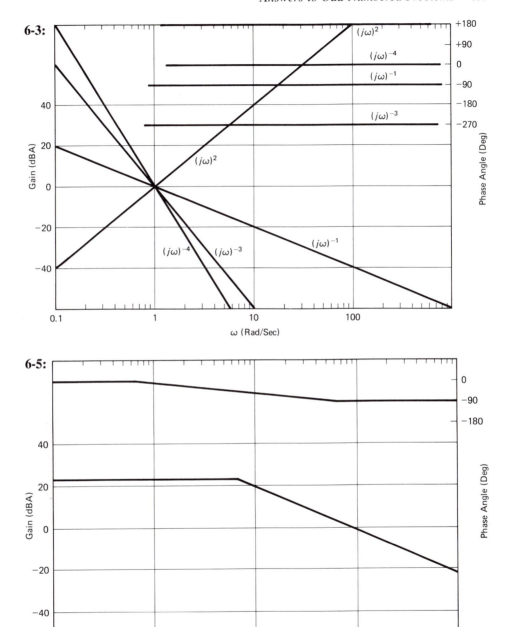

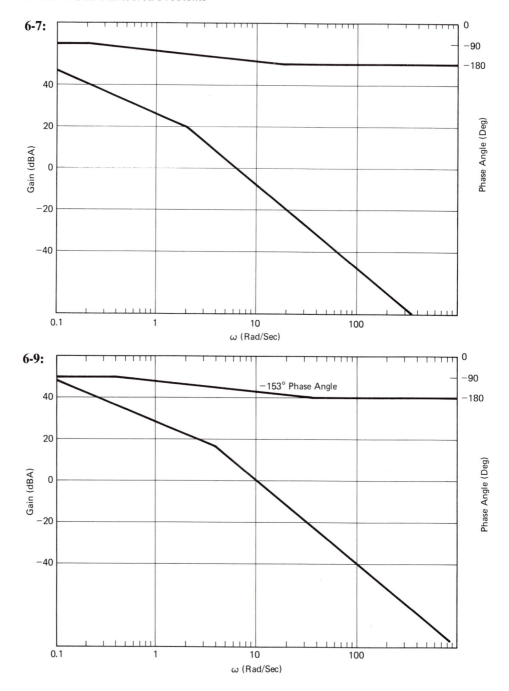

CHAPTER 7:

7-1:

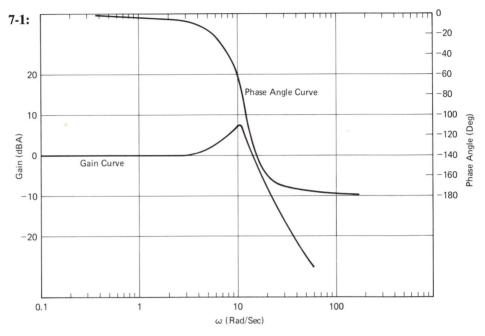

7-3: The open-loop bandwidth is 13.5 *r/s.* The closed-loop bandwidth is 16 *r/s.* In general, going to a closed-loop system usually increases a system's bandwidth.

7-5:

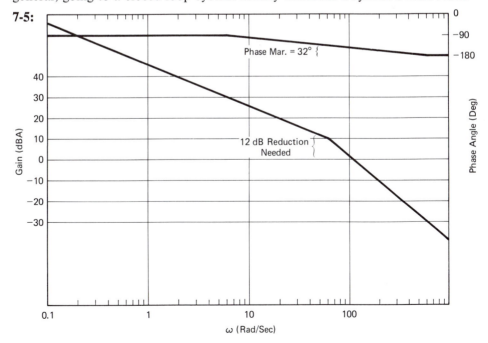

7-7: The Nichols plot should resemble very closely the closed-loop curve of Problem 7-1.   **7-9:** Approximately -20 dBA in gain reduction is needed for stabilizing this system.

CHAPTER 8:

8-1: *a.* zeros: $= -5$ *b.* zeros: $= -0.167$
      poles: $= 0, -3.33, +0.33$ poles: $= -0.33, +1, -1$
  c. zeros: $=$ none   *d.* zeros: $= -0.33$
      poles: $= -10, -2$ poles: $= 0, -2, -0.1$
  e. zeros: $= -4$ *f.* zeros: $= 5$
      poles: $= -1.01, -0.04$ poles: $= 2, -0.167$

8-3:

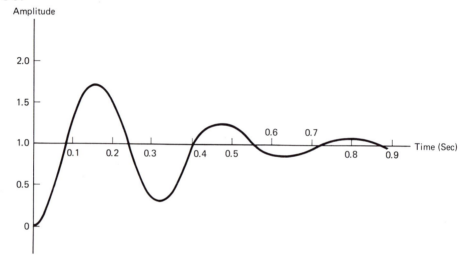

8-5: $\omega_n = 7.746$ *r/s*   **8-7:** $\omega_n = 25$ *r/s*
      $\omega_d = 7.484$ *r/s* $z = 0.125$
      % overshoot $= 43.2\%$ $\pm 5\%$ settling time $= 0.96$ *s*
                                         number of osc. $= 3.79$
                                         $\omega_m = 24.6$ *r/s*

8-9: Eq. (8-41) states the following: asymptote location pt. $= \dfrac{50 - 0}{2} = 25$

CHAPTER 10:

10-1: An analog automatic control system has the advantage of not having as much of a resolution problem in set-point control as compared to a digital automatic control system since there are no incremental (digital) settings to contend with. Theoretically, an analog system is capable of an infinite degree of resolution because of the infinite number of set-point opportunities. **10-3:** Because of the uncertainty existing with the output of a binary-type encoded wheel or slide during the transition from one binary position to the next, the Gray code was necessary in order to eliminate this uncertainty or ambiguity of switching. The switching problem arises from the encoder's sensors detecting this change-over at slightly different times as the boundaries separating the binary-generating areas sweep beneath the sensors, thereby creating a series of random, unwanted binary number readings. **10-5:** A bit map is a picture of an image that is to be digitized showing the locations of the binary bits representing the various light intensities on that image. **10-7:** Adaptive control is an automatic control system that, through its feedback, can make corrections to an operating system. The one distinguishing feature of this type of control system as compared to any other automatic control system, is that an adaptive system responds to the environmental changes surrounding the main automatic control system. The purpose of an adaptive control system is to prevent an automatic control system from destroying itself in its attempt to make an error correction due to some sudden or unanticipated change within its operating environment. **10-9:** An optical encoder can be made to have a much larger travel displacement to accommodate larger measurand values as compared to an LVDT. Because of this, much greater resolution is possible. Any attempts to increase or to amplify the displacement of an existing LVDT using mechanical means usually results in back-lash and/or hysteresis.

CHAPTER 11:

11-1: The three major categories of data transmission are: (1) analog, (2) digital, and (3) carrier transmission.

Analog transmissions are characterized by a continuously changing or modulated voltage or current whose amplitude is a linear representation of the measurand being sensed at the transmission site.

Digital transmissions are characterized by a transmission of rectangular or pulse-like voltages or currents whose amplitudes, positions, widths, or frequencies of occurrence represent the measurand at the transmission site.

Carrier transmissions are characterized by the systematic interruptions of a transmitter carrier wave, the duration and frequency of the interruptions containing the transmitted data or information.

11-3: FM transmissions are more often used for data transmissions as compared to AM transmissions because of less electrical interference present with FM transmissions. This is because most electrical interference, either man-made or natural, is of an AM nature rather than of an FM nature. **11-5:** In an FM system the carrier's frequency is modulated by the transmitted data whereas in a PM system, the electrical phase existing between the carrier and the transmitted data is varied. There is little advantage in using the one method over the other, although, it is believed that PM may have a slight advantage over FM in electrical noise immunity.

11-7: The four basic forms of pulse modification are: (1) pulse amplitude modulation, (2) pulse width modulation, (3) pulse position modulation, and (4) pulse code modulation.

Pulse amplitude modulation is characterized by its varying pulse heights containing the transmitted data. Pulse width modulation is characterized by its varying pulse widths containing the transmitted data. Pulse position modulation is characterized by the positions or locations of each pulse relative to a reference location, the position containing the transmitted data. Pulse code modulation is characterized by the presence or absence of each pulse sent in groups representing coded binary numbers.

11-9: An advantage of serial data transmission is that it requires fewer transmission channels (a minimum of two, one for the signal and one for ground return) as compared to a parallel transmission system which may require a minimum of 4 (8 being more typical), depending on the "word" length being transmitted. Another advantage is that serial transmissions can be transmitted over great distances as compared to parallel transmissions. A disadvantage of a serial transmission is that it is very slow compared to a parallel transmission.

Parallel transmissions have the great advantage of being very fast as compared to serial transmissions. However, their disadvantages are, very limited transmission distances, greater complexity in circuitry and many more transmission channels required.

CHAPTER 12:

12-1: The function of the microprocessor chip is to control the flow of traffic, so to speak, of all the data that needs to be processed. The chip contains data registers, instruction sets, binary arithmetic functions for data conversions, data address registers, and program counters to properly manipulate the data's flow and processing.

The function of the ALU, or arithmetic-logic unit, is to perform the necessary math functions and logic needed for processing the input data.

12-3: RAM (random-access memory) refers to a type of memory chip that acts as a temporary storage location of data. However, RAM is volatile; that is, the memory capability is lost when power is removed from these chips. The ROM (read-only memory) chip, on the other hand, is a non-volatile chip that contains data or information that can be read from it. That is, the information can be extracted at any time; however, it is not possible to store user data within this chip. ROM chips are used for storing programs that can be used at any time when desired. When power is removed, the ROM's contents still remain intact. **12-5:** BIOS (basic input-output system) chips contain a computer's operating commands that control the computer's many input/output ports which, in turn, allow the computer to send and receive data from peripheral devices attached to these ports.   **12-7:** The ''level'' of a computer language refers to that language's ability to perform a command or series of commands with a minimum number of command statements and symbols. The higher level languages require fewer statements and symbols as compared to the lower level languages that require many statements. However, the higher level languages tend to lose versatility in their applications as compared to the lower level languages.   **12-9:** The function of a PLC (programmable logic controller) is to act as a sequencing or logic device, and sometimes as a data processor, for an industrial machine used in a manufacturing facility. The I/O on a PLC refers to the input/output ports which connect to, and receive data from, the machine that the PLC is controlling. Typical devices attached to a PLC's I/O are, the machine's control switches, sensors attached to the machine, and annunciators.

APPENDIX A

Rectangular/Polar
Polar/Rectangular
Conversions and How
to Handle Math Routines
Using These Forms

When working with electronics and in automatic control system design, it's necessary to be able to convert quickly from rectangular expressions to polar expressions and back again. The reason why we have to contend with both systems of expression is because adding and subtracting requires one expression form while multiplication and division requires another form. Also, depending on the application, one form is probably easier to understand than the other. For the purposes of automatic control applications, you should know both systems and be able to switch from one to the other.

Figure A-1 illustrates the trigonometric relationships between the two systems for locating a point, m. Point m can be given the position (A, jB) using the

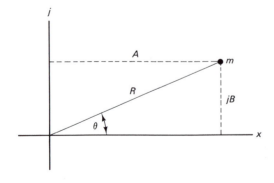

Figure A-1 Graphical relationship between rectangular and polar coordinates.

rectangular coordinate system, or it can be given the position ($R \angle \theta$) using the polar coordinate form of system. To convert from the rectangular to the polar form, use the following conversion equations:

Rectangular → Polar: $R = \sqrt{A^2 + B^2}$ (A-1)

$$\theta = \arctan \frac{\pm B}{A}$$ (A-2)

EXAMPLE

Convert ($2 + j5$) to polar form.

Solution:

Use Eq. (A-1) to find R:

$$R = \sqrt{2^2 + 5^2}$$
$$= \sqrt{29}$$
$$= 5.385$$

Use Eq. (A-2) to find θ:

$$\theta = \arctan \frac{5}{2}$$
$$\theta = \arctan 2.5$$
$$= 68.2°$$

The final answer is expressed as $5.385 \angle 68.2°$.

To convert from the polar form to the rectangular form, use the following equations:

Polar → Rectangular: $A = R\cos\theta$ (A-3)

$B = R\sin\theta$ (A-4)

EXAMPLE

Convert $2.87 \angle -56.8$ to the rectangular equivalent.

Solution:

Use Eq.(A-3) to find A:

$$A = (2.87)(0.548)$$
$$= 1.573$$

Use Eq. (A-4) to find B:

$$B = (2.87)(-0.837)$$
$$= -2.402$$

The final answer is expressed as ($1.573 - j2.402$).

Handling mathematical manipulations using the two coordinate forms can be easy if you choose the right form for the right math process. For instance, you will want to *use the rectangular expressions only for addition and subtraction.* You will want to *use the polar forms only for multiplication and division.* You can violate these rules, but the math can really become a problem. The following are examples of how to perform these four math functions:

EXAMPLE

Add (4.7 ∠ 30°) to (−6.4 − j3.7).

Solution:

Since we are adding, we will want to use all rectangular forms. Therefore, convert the one polar form that was given to rectangular form, then add the two expressions as you would normally do to any algebraic expression. Converting (4.7 ∠ 30°) to rectangular form:

$$A = 4.7\cos30° \quad \text{(Eq. A-3)}$$
$$= 7.070$$
$$B = 4.7\sin30° \quad \text{(Eq. A-4)}$$
$$= 2.350$$

Therefore, (4.7 ∠ 30°) = (7.07 + j2.35)
Now adding the two rectangular values:

$$\begin{array}{r} 7.07 + j2.35 \\ (+) \quad -6.34 - j3.70 \\ \hline \text{Our answer} \longrightarrow \quad 0.73 - j1.35 \end{array}$$

EXAMPLE

Subtract (3.87 + j8.33) from (1.19 − j6.98).

Solution:

No conversions are necessary, since both given expression are already in the desired rectangular form.

Solution:

$$\begin{array}{r} 1.19 - \ j6.98 \\ (-) \quad 3.87 + \ j8.33 \\ \hline \text{Our answer} \longrightarrow - 2.68 - j15.31 \end{array}$$

EXAMPLE

Multiply the preceding two expressions together.

Solution:

First, we must convert both expressions to polar coordinates. (We won't show the conversion math this time, just the results.):

$$1.19 - j6.98 \rightarrow 7.081 \angle -80.3°$$
$$3.87 + j8.33 \rightarrow 9.185 \angle 65.1°$$

The rule for multiplying polar expressions is this: The *R* values of the expressions are multiplied as usual, but the angle values (i.e., the *θ*s) are *added*. Therefore,

$$(7.081 \angle -80.3°) \times (9.185 \angle 65.1°) = 7.081 \times 9.185$$
$$= 65.039 - 80.3° + 65.1° = -15.2°$$

Our answer $\rightarrow 65.039 \angle -15.2°$

EXAMPLE

Divide
$$\frac{64.8 \angle -124.0°}{4.99 \angle 23.6°}$$

Solution:

The rule for division of polar expressions is this: The *R* values of the expressions are divided as usual; however, the angle values (i.e., the *θ*s) are *subtracted*. Therefore,

$$64.8 \div 4.99 = 12.986$$
$$(-124.0°) - (23.6°) = -100.4°$$

Our answer $\rightarrow 12.986 \angle -100.4°$

APPENDIX B

A Computer Program for Plotting Transient Waveforms

Being able to obtain a visual image of the transients generated by a closed-loop feedback system is a great step toward understanding the system itself. Provided that you have the equation describing the transient response curve, the BASIC program that follows will give you such a plot. This program was designed to run on an AT IBM computer. However, it will most likely run on any IBM personal computer with little or no modification. If it is run on a *compatible* IBM, the BIOS chip(s) have no BASIC program and therefore no IBM graphics capability which this program requires. Therefore, you must use either GW BASIC, BASIC, or BASICA along with a graphics routine enabling you to print graphics to your hi-res monitor. The program will not function without the graphics routine.

Follow these instructions:

1. Type the code in Figure B-1, being very careful to duplicate every character and symbol used.

```
5 SCREEN 0
6 REM This program is called PLOTFUN.BAS.  It was originally published in
7 REM MACHINE DESIGN magazine, Feb. 26, 1987, J. Perkins, author.  The
8 REM program has been modified to suit this particular application.
10 PRINT "PROGRAM PLOTFUN . . . PLOTS FUNCTION ON SCREEN"
20 PRINT " "
30 NL = 1
40 OPEN "FUNCTION.BAS" FOR OUTPUT AS #1
50 PRINT "ENTER THE EQUATION TO BE PLOTTED, BEGINNING WITH 'Y = ' AND HAVING"
51 PRINT "EQUATION EXPRESSED IN BASIC NOTATION AND A FUNCTION OF 'T'."
52 PRINT "WHEN COMPLETED, HIT <ENTER>."
55 INPUT A$
60 L = 990 + 10*NL
```

Figure B-1 BASIC code for PLOTFUN.BAS. (*Reprinted with permission from MACHINE DESIGN magazine, Vol. 59, number 4.*)

418

```
65 NL = NL + 1
70 LNUM$=STR$(L)
90 A$ = LNUM$ + " " + A$
100 PRINT #1,A$
120 PRINT #1, "2000 RETURN"
130 CLOSE #1
140 CHAIN MERGE "FUNCTION.BAS",200,ALL,DELETE 1000-2000
200 REM
210 INPUT "ENTER TIMESCALE;TMAX",TMAX
220 INPUT "ENTER VERTSCALE;YMIN,YMAX",Y1,Y2
230 X1=0:X2=TMAX:DELT=TMAX/640
240 REM
250 REM PLOT THE SECTION ON THE SCREEN
260 CLS:SCREEN 2: KEY OFF
270 XS=639/(X2-X1):YS=197/(Y2-Y1)
280 REM
290 PSET(0,0)
300 LINE-(0,199)
310 LINE-(639,199)
320 LINE-(639,0)
330 LINE-(0,0)
340 REM
345 REF=200*Y2/(Y2-Y1)
350 PSET(0,REF)
360 LINE-(640,REF)
370 PSET(0,REF-(Y*YS))
380 REM
390 T=0
400 XX=Y
410 GOSUB 1000
420 UU=DELT*XS + UU
430 VV=REF-(Y*YS)
440 LINE-(UU,VV)
450 T=T+DELT
460 IF T<(TMAX+DELT) GOTO 400
470 END
480 REM
1000 REM
2000 REM
```

Figure B-1 *Cont.*

2. After you are certain the code has been copied correctly, SAVE the program. Then, type RUN. You will be prompted to type the equation which must be entered in the form, Y = f(T). For example:

$$Y = 1 + 1.05 * 2.718^{\wedge}(-30 * T) * SIN(105 * T - 100)$$

3. After typing your equation, hit ⟨ENTER⟩.

4. You will then be asked to enter the maximum time (TMAX) scale you think you will need to completely plot your equation along the horizontal time axis so that it has time to settle out.

5. Next, you will be asked to make two entries on the same line separated by a comma. You must enter the minimum vertical scale needed to begin the curve (YMIN). Typically, this would be 0. You will also be asked to enter the maximum scale (YMAX) needed to contain the curve's highest point. If,

for example, you think 4 is enough for your YMAX, then you must make
your complete YMIN, YMAX entry as: 0,4. For the preceding example
curve, a TMAX of 0.25 and a YMIN,YMAX of 0,2 works well.

6. After hitting ⟨RTN⟩ (or ⟨ENTER⟩, whichever the case may be), the curve
   will become plotted in high-resolution mode. The horizontal line at the top
   of the graph represents your maximum vertical scale value, YMAX, while
   the lowest horizontal line represents your minimum vertical scale value,
   YMIN. The far right-hand vertical line on your graph is located at your
   TMAX value.

7. Following the plot, you must enter SCREEN 0 to go back to low res if you
   wish to run another plot. Also, your soft key line will be OFF at this time.
   Therefore, also enter KEY ON to return that line, if desired.

8. To run another plot, you must restore line 1000 and line 2000 in your program back to its original state. Do this by typing 1000 REM ⟨RTN⟩ and 2000
   REM ⟨RTN⟩ to restore the program. Check your entries by LISTing the
   program.

If you had entered the example equation given in step 2, you would obtain a
plot similar to the one shown in Figure B-2.

This is obviously a "bare-bones" program and not too "friendly." You may
want to modify it to have it perform in the manner you wish.

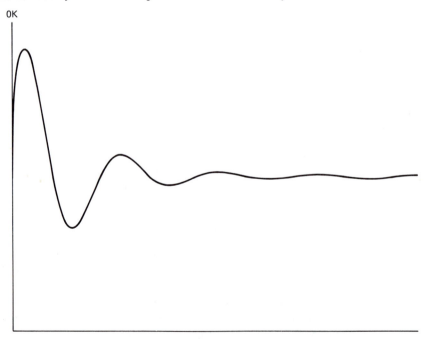

Figure B-2   Typical results from using PLOTFUN.

APPENDIX C

Computer Programs
for Finding Cube Roots

The two BASIC programs listed following have a rather interesting application. The program listed in Figure C-1 uses an iterative process in determining the roots for cubic equations. This is a process developed by Newton and is surprisingly accurate and efficient when done on a computer. The BASIC program is called FINDROOT.BAS. It can be used to find all three roots if desired. It does take a few moments to use, and a certain amount of patience is required to apply it. The program will not find roots less than −1000. The equation must be entered in one of the program lines along with the equation's first derivative. The process of finding the first derivative of a polynomial is rather easy. Here is how it is done. *The exponent associated with the variable is multiplied by the variable's coefficient. The exponent's value is then reduced by one.*
Here are some examples:

Find the first derivative of $5y^6$.
Solution: First derivative = $6 \cdot 5y^{6-1} = 30y^5$.

Find the first derivative of $123x$.
Solution: First derivative = $1 \cdot 123x^{1-1} = 123$.

Find the first derivative of $14p^{-4}$.
Solution: First derivative = $-4 \cdot 14p^{-4-1} = -56p^{-5}$.

Find the first derivative of -56 (which can be written as $-56x^0$).
Solution: First derivative = $0 \cdot -56x^{0-1} = 0$.
This tells us that *the derivative of any constant will be 0.*

Find the first derivative of the function $y = 3x^2 + 7x - 2$.
Solution: First derivative $= y_1 = 6x + 7$.

Now that we know how to find the first derivative of a polynomial expression, what does it do for us? The first derivative is nothing more than the equation that allows us to calculate the slope of a line passing through the points at a point of tangency to the original curve. Placing those points into the derivative, $y1$, (or y' as it is sometimes called) and solving, gives us the slope of that line of tangency. FINDROOT.BAS lets you blindly guess at the first root of an equation. It then calculates the slope of the line based on your first guess and also determines the point of intersection of that slope line with the x-axis. The computer then plugs the value of that intersection point back into the original equation and recalculates $f(x)$ to see if the resultant value of $f(x)$ becomes smaller. If it does, the process is repeated until two successive values of $f(x)$ are separated by less than 1×10^{-6}. When the separation becomes that small, the value of x is declared a root for $f(x)$.

Obviously, there are three roots for a cubic equation. As a result, you may want to guess at other values of x to see if the computer will converge on the other remaining ones. By the way, if the program doesn't converge within 50 attempts, this, most likely, is an indication that either the program is trying to converge on an imaginary root, or your guess was so far off that it is going to take more than 50 tries to bring in an answer. In either case, you will receive an error message stating "---- slope is zero", or "---- no convergence in XXX tries". If you should determine one root and don't wish to spend the time guessing for the others, you can then go to the next program, ROOTS.BAS, (listed following) to solve for the others. That program is more direct in finding the remaining roots.

FINDROOT.BAS is listed in Figure C-1. The following are the instructions for using the program:

1. Enter the program into your computer.
2. While still in BASIC or BASICA, type line number 8400 and enter the equation that you wish to find the roots of. Be sure to begin entry with "F = ".
3. Next, type line number 8420 and enter the first derivative of the preceding equation, beginning the entry with "F1 = ".
4. Type RUN. You will be asked to guess at the first root. Pick any number no less than $-1,000$. If you do, you will receive an error message.
5. All iterations will be displayed showing the converging process on the one root. Once that root is finally displayed, you can either guess at another or go to the next program for calculating the remaining roots in the case of cubic solutions.
6. In order to get out of the program, you must depress the CTRL-BREAK keys simultaneously.

```
10 REM This program is called FINDROOT.BAS and is based on using the Newton-
11 REM Raphson method.  This program was originally obtained from BASIC
12 REM PROGRAMS FOR SCIENTISTS AND ENGINEERS, by A. R. Miller, Sybex, 1981.
13 REM      D6        DX         delta x
14 REM      E1%       ERMES%     error flag
15 REM      F         FX         function
16 REM      F1        DFX        derivative of function
17 REM      F0%       FALSE%     zero
18 REM      H2        SMALL      small number
19 REM      M5%       MAXL%      maximum loops
20 REM      T0%       TRUE%      not false
21 REM      T1        TOL        tolerance
22 REM end of identifiers
23 REM
30 T1 = .000001
40 H2 = 1E-15
50 M5% = 50
60 F0% = 0
70 T0% = NOT F0%
80 INPUT "First guess ";X
90 IF (X < -1000) THEN 9999
100 GOSUB 8000
110 PRINT
120 IF (E1% = F0%) THEN PRINT "The solution is ";X
130 GOTO 80
8000 REM start of Newton's method
8010 E1% = F0%
8020 FOR L% = 1 TO M5%
8030 X1 = X
8040 GOSUB 8400
8050 IF (ABS(F1) > H2) THEN 8090
8060 PRINT "ERROR ---- slope is zero"
8070 E1% = T0%
8080 GOTO 8160
8090 D6 = F/F1
8100 X = X1 - D6
8110 PRINT "X = ";X1;", FX = "; F;" DFX = ";F1
8120 IF (ABS(D6) < = ABS(T1 * X)) THEN 8160
8130 NEXT L%
8140 PRINT "ERROR ---- no convergence in "; M5%;" tries"
8150 E1% = T0%
8160 RETURN : REM from Newton's method
8400 REM Enter cubic equation here, beginning with "F = ".
8420 REM Enter derivative of above equation here, beginning with "F1 = ".
8430 RETURN
9000 END
9999 PRINT "ERROR ---- you have exceeded the negative input range"
10000 PRINT "for your guess.  Please RUN program again and guess again."
10010 END
```

Figure C-1 BASIC code for FINDROOT.BAS. (*Reprinted with modifications and permission from BASIC PROGRAMS FOR SCIENTISTS AND ENGINEERS, A. R. Miller, Sybex Inc., copyright 1981, Alameda, Calif.*)

As an example, let's assume that you have to find the roots of the following cubic equation:

$$F = 3x^3 + 7x^2 - 4x - 3$$

We will need the first derivative of this equation in order to run the program, so let's determine that also.

$$F_1 = 9x^2 + 14x - 4$$

Placing these two expressions into lines 8400 and 8420, respectively, and remembering to use BASIC notation while we are doing it, let's use "500" as our first guess for the first root determination. Figure C-2 shows the results of this guess.

```
First guess ? 500
X =   500 , FX =   3.76748E+08  DFX =   2256996
X =   333.0755 , FX =   1.116287E+08  DFX =   1003112
X =   221.7931 , FX =   3.307493E+07  DFX =   445830.8
X =   147.606 , FX =   9799821  DFX =   198150.1
X =   98.14941 , FX =   2903548  DFX =   88069.86
X =   65.18073 , FX =   860242.1  DFX =   39145.27
X =   43.2051 , FX =   254841.2  DFX =   17400.99
X =   28.55988 , FX =   75478.49  DFX =   7736.841
X =   18.80416 , FX =   22344.21  DFX =   3441.626
X =   12.31182 , FX =   6607.528  DFX =   1532.593
X =   8.000481 , FX =   1949.329  DFX =   684.076
X =   5.150901 , FX =   572.1066  DFX =   306.8986
X =   3.286746 , FX =   165.9891  DFX =   139.2387
X =   2.094627 , FX =   46.90404  DFX =   64.81195
X =   1.370933 , FX =   12.4023  DFX =   32.10818
X =   .9846669 , FX =   2.712423  DFX =   18.51146
X =   .8381403 , FX =   .3311205  DFX =   14.05628
X =   .8145835 , FX =   8.031607E-03  DFX =   13.37609
X =   .813983 , FX =   4.529953E-06  DFX =   13.35888

The solution is   .8139827
First guess ?
1LIST    2RUN  3LOAD"  4SAVE"  5CONT 6,"LPT1 7TRON 8TROFF9KEY 0SCN
```

Figure C-2 Results from running FINDROOT.

We see that the first root is 0.8139827. It's interesting to study the displayed iterations used in finding this root. This will help you in understanding how the program works. To find the remaining roots, we could continue our guessing or we could go on to the next program. Let's continue guessing just to see what our results will be. This time, let's start with a very negative number such as "−500". Using this as our guess, we come up with a second root value of −2.69074. If we continued guessing, we would find our third root to be −0.45676. This was found using a guess of "−2".

As in the case of the previous program in Appendix B, FINDROOT.BAS lacks conveniency for entering data. You can revise the code to make it more suitable for you if you like. This program can also be used for determining the roots of many other kinds of equations, as long as you also enter the proper derivatives for those equations.

Another program for finding roots is called ROOTS.BAS. It allows you to find the remaining roots of any cubic equation regardless of whether those roots are real or imaginary. The program assumes that you know, within a value of

±0.5, the value of the first root. If your first root is outside that range value, the program will tell you in an error message. The listing for ROOTS.BAS is given in Figure C-3. The program is run by doing the following, assuming that you have it entered properly into your computer:

1. Type RUN.

2. You will first be prompted to enter what you believe is the first root. (This first root value could come from FINDROOT.BAS.) Hit ⟨ENTER⟩.

```
1 REM This program is called ROOTS.BAS
2 REM This program calculates the other two roots of a cubic equation when
3 REM given the first root, X1
5 INPUT "X1 ROOT = ";X1
10 INPUT "X-CUBED COEFFICIENT = ";S3
20 INPUT "X-SQUARED COEFFICIENT = ";S2
30 INPUT "X COEFFICIENT = ";S1
40 INPUT "EQUATION'S K-VALUE = ";K
50 PRINT S3
60 PRINT S2
70 PRINT S1
80 PRINT K
90 REM begin synthetic division
100 SYN1 = X1 * S3
110 SYN2 = S2 + SYN1
120 SYN3 = X1 * SYN2
130 SYN4 = S1 + SYN3
140 SYN5 = X1 * SYN4
150 SYN6 = K + SYN5
155 IF (ABS(SYN6) > .05) THEN 500
160 REM S3 now a-term for quadratic formula
170 REM SYN2 now b-term for quadratic formula
180 REM SYN4 now c-term for quadratic formula
190 BSQ = SYN2^2
200 AC4 = SYN4 * S3 * 4
210 HLD = BSQ - AC4
220 IF HLD >= 0 THEN 240
230 IF HLD < 0 THEN 300
240 RT = HLD^.5
250 X2 = -SYN2/(2*S3) + RT/(2*S3)
260 X3 = -SYN2/(2*S3) - RT/(2*S3)
270 PRINT "X2 = ";X2
280 PRINT "X3 = ";X3
290 END
300 HLD = AC4 - BSQ
310 RT = HLD^.5
340 PRINT "X2 = ";-SYN2/(2*S3)
350 PRINT "+J";RT/(2*S3)
360 PRINT "X3 = ";-SYN2/(2*S3)
370 PRINT "-J";RT/(2*S3)
380 END
500 PRINT "ERROR ---- first root, X1, not accurate enough.  Retry program"
510 PRINT "by entering RUN again and incrementing 'first guess' to X1"
520 PRINT "by a value of plus or minus .05."
540 END
```

Figure C-3   BASIC code for ROOTS.BAS.

3. Next, you will be asked to enter the coefficients associated with each variable in the cubic equation you are trying to solve. If a variable is missing, simply enter "0". Each entry is followed by ⟨ENTER⟩.

4. Your display will then show a listing of your coefficients that you have entered followed by the calculated remaining two root values.

5. To rerun the program, simply enter RUN.

To test the program, let's follow through with the example that we had begun in the first program, FINDROOT.BAS. Let's enter the first solved root of 0.8139827, along with the coefficient information of our cubic equation. The display should look like Figure C-4. If you wish, you can also enter the second and third roots found from the first program. You should find that these results compare with the displayed results in Figure C-3.

```
RUN
X1 ROOT = ? .8139827
X-CUBED COEFFICIENT = ? 3
X-SQUARED COEFFICIENT = ? 7
X COEFFICIENT = ? -4
EQUATION'S K-VALUE = ? -3
 3
 7
-4
-3
X2 = -.456576
X3 = -2.69074
Ok
```

```
1LIST    2RUN    3LOAD"  4SAVE   5CONT 6,"LPT1 7TRON 8TROFF9KEY 0SCN
```

Figure C-4   Results from running ROOTS using results from FINDROOT.

The ROOTS.BAS program will also return complex or imaginary values of roots. For example, find the roots for the cubic equation, $y = 0.125x^3 - 0.25x^2 + 0.5x + 3$ given that one of its roots was already found to be -2. ROOTS.BAS will return the display seen in Figure C-5. We see that the other two roots are $2 + j2.828427$ and $2 - j2.828427$.

```
RUN
X1 ROOT = ? -2
X-CUBED COEFFICIENT = ? .125
X-SQUARED COEFFICIENT = ? -.25
X COEFFICIENT = ? .5
EQUATION'S K-VALUE = ? 3
 .125
-.25
 .5
 3
X2 =  2
+J 2.828427
X3 =  2
-J 2.828427
Ok
```

```
1LIST    2RUN   3LOAD"   4SAVE"   5CONT  6,"LPT1 7TRON 8TROFF9KEY 0SCN
```

Figure C-5   Results of a complex root solution using ROOTS.

APPENDIX D

Euler's Equation

The purpose of this appendix is to present a proof for Eq. (8-22) in Section 8-6.1, which is called Euler's equation. This equation states that $e^{j\theta} = \cos\theta + j\sin\theta$. This is a very handy identity to know in coping with many of the mathematical expressions that have a mixture of trigonometry expressions and e^x expressions. This mixture occurs quite often in automatic control theory presentations.

To begin the proof, we first look at one of the math tools we need for this job. We rely on what is called the *Maclaurin's Series*. This is an expression that allows you convert a trigonometric expression into an algebraic expression. The Maclaurin's Series states:

$$f(x) = f(0) + f'(0)x + \frac{f''(0)x^2}{2!} + \frac{f'''(0)x^3}{3!} + \ldots$$
$$\frac{f^{(n)}(0)x^n}{n!}$$

As you can see, in order to use the preceding expression, you must be able to take multiple derivatives of terms. If you are unsure as to how this is done, you should obtain a good elementary text on differential calculus and review these procedures. You will find these procedures to be very straightforward, especially for sine and cosine expressions. For example, let's convert $\sin x$ into an algebraic expression.

$f(x) = \sin x \qquad$ when $x = 0 \qquad\qquad f(0) = \quad 0$
$f'(x) = \cos x$ (first deriv.) $\qquad\qquad\qquad\qquad f'(0) = \quad 1$

$f''(x) = -\sin x$ (second deriv.) $f''(0) = \quad 0$
$f'''(x) = -\cos x$ (third deriv.) $f'''(0) = -1$
$f^4(x) = \sin x$ (fourth deriv.) $f^4(0) = \quad 0$
$f^5(x) = \cos x$ (fifth deriv.) $f^5(0) = \quad 1$

Substituting the values from our table into the Maclaurin's Series, we get:

$$\sin x = 0 + (1)x + \frac{(0)x^2}{2!} + \frac{(-1)x^3}{3!} + \frac{(0)x^4}{4!} + \frac{(1)x^5}{5!}$$

$$= x - \frac{x^3}{6} + \frac{x^5}{120}$$

Using the same approach for $\cos x$, we would get:

$$\cos x = 1 - \frac{x^2}{2!} + \frac{x^4}{3!} - \frac{x^6}{4!} + \dots$$

And, if we did the same for e^x we would get:

$$e^x = 1 + x + \frac{x^2}{2!} + \frac{x^3}{3!} + \dots$$

Now, let $x = jx$ in the preceding e^x expression, remembering that

$$j = j$$
$$j^2 = -1$$
$$j^3 = -j$$
$$j^4 = \quad 1$$
$$j^5 = \quad j$$

Then,

$$e^{jx} = 1 + jx - \frac{x^2}{2!} - \frac{jx^3}{3!} + \frac{x^4}{4!} + \frac{jx^5}{5!}$$

Rearranging the terms, we get:

$$= \left| 1 - \frac{x^2}{2!} + \frac{x^4}{4!} \right| + j \left| x - \frac{x^3}{3!} + \frac{x^5}{5!} \right|$$

and $\cos x \longleftarrow \rfloor$ $\lfloor \longrightarrow \sin x$

Therefore, $e^{jx} = \cos x + j\sin x$.

Similarly, we can prove that $e^{-jx} = \cos x - j\sin x$.

AUTOMATIC
CONTROL
SYSTEMS
AND COMPONENTS

Index